ETHICS IN ENGINEERING

ETHICS IN ENGINEERING

SECOND EDITION

Mike W. Martin

Department of Philosophy
Chapman College

Roland Schinzinger

School of Engineering
University of California, Irvine

McGRAW-HILL BOOK COMPANY

New York St. Louis San Francisco Auckland Bogotá
Caracas Colorado Springs Hamburg Lisbon London Madrid Mexico
Milan Montreal New Delhi Oklahoma City Panama Paris San Juan
São Paulo Singapore Sydney Tokyo Toronto

This book was set in Palatino by the College Composition Unit
in cooperation with Ruttle Shaw & Wetherill, Inc.
The editors were Anne T. Brown, Lyn Beamesderfer, and Bernadette Boylan;
the cover designer was Carla Bauer;
the production supervisor was Friederich W. Schulte.
New drawings were done by Accurate Art, Inc.
Arcata Graphics/Halliday was printer and binder.

ETHICS IN ENGINEERING

3 4 5 6 7 8 9 0 H A L H A L 8 9 3 2 1 0

ISBN 0-07-040719-3

Library of Congress Cataloging-in-Publication Data

Martin, Mike W., (date).
Ethics in engineering.
Bibliography: p.
Includes index.
1. Engineering ethics. I. Schinzinger, Roland.
II. Title.
TA157.M33 1989 174'.962 88–13073
ISBN 0-07-040719-3

ABOUT THE AUTHORS

Roland Schinzinger and Mike W. Martin participated as an engineer-philosopher team in the National Project on Philosophy and Engineering Ethics during 1978–1980. Since then they have coauthored articles, team-taught courses, and together given presentations to audiences of philosophers and engineers.

Mike W. Martin received his B.S. (1969, Phi Beta Kappa and Phi Kappa Phi) and M.A. (1972) from the University of Utah, and his Ph.D. (1977) from the University of California, Irvine. Currently he is a professor of philosophy and honors at Chapman College, where he has served as chair of the philosophy department and chair of the Chapman Faculty (in charge of the faculty governance system). He serves on the Bioethics committee of his community hospital.

Dr. Martin received a Matchette Foundation Award for teaching (as a graduate student) and an Arnold L. and Lois S. Graves Award for Teachers in the Humanities (1983). He has held two fellowships from the National Endowment for the Humanities and two grants from the Association of American Colleges. His articles on business and professional ethics, social philosophy, aesthetics, and literary criticism have appeared in many books and journals. He is editor of *Self-Deception and Self-Understanding* (1985) and author of *Self-Deception and Morality* (1986) and *Everyday Morality* (1989).

Roland Schinzinger was born and raised in Japan where his parents were teachers of German. He came to the United States in 1949 at the age of twenty-two and attended the University of California at Berkeley, where he received his B.S. (1953, Tau Beta Pi, Eta Kappa Nu), M.S. (1954), and Ph.D. (1966, National Science Fellow). He worked for the Westinghouse Electric Corporation in East Pittsburgh, Pennsylvania (1954–1958), taught engineering in Turkey (1958–1963), and in 1965 came to the University of California at Irvine, where he is Professor of Electrical Engineering and former Associate Dean of Engineering.

Dr. Schinzinger has authored numerous papers on electric energy systems, conformal mapping, optimization, and engineering ethics. Currently he is completing a coauthored book on conformal mapping. He is a recipient of the IEEE Centennial Medal. Outside the University he does some safety-related consulting, is active in professional societies, and has served local government as planning commissioner and as a member of citizens advisory committees on energy and transportation.

For Shannon, for Ruth L. Martin,
and in memory of Theodore R. Martin.
Mike W. Martin

For Jane, for Robert Schinzinger,
and in memory of Annelise Schinzinger.
Roland Schinzinger

CONTENTS

PREFACE

Technology has a pervasive and profound effect on the contemporary world, and engineering plays a central role in all aspects of the development of technology. Because of this it is vital that there be an understanding of the ethical implications of engineers' work. Engineers must be aware of their social responsibilities and equip themselves to reflect critically on the moral dilemmas they will confront. Managers must be responsive to the rights of engineers to exercise their consciences responsibly. The public must acquire an understanding of the extent and limits of the responsibilities of engineers; it must be prepared to shoulder its own responsibilities where those of the engineers end.

This book is written for students preparing to function within the engineering profession and for their colleagues in philosophy and the humanities who seek to apply ethical theory to the pressing problems of everyday life in a technological society. Its intended audience is also practicing engineers, social scientists, philosophers, and all those engaged in the enterprise of technology, including the general public.

Purpose

Ethics in Engineering provides an introduction to the basic issues in engineering ethics, with emphasis given to the moral problems engineers face in the corporate setting. It places those issues within a wider philosophical framework than has been customary in the past, and it seeks to exhibit both their social importance and their intellectually challenging nature. The primary goal is to stimulate critical reflection on the moral issues surrounding engineering practice and to provide the conceptual tools necessary for pursuing those issues. The book is intended to be a teaching instrument while also serving to advance the field of engineering ethics.

In large measure we proceed by clarifying key concepts, sketching alternative views, and providing relevant case study material. Yet in places we argue for particular positions which in a subject like ethics can only be controversial. We do so because it better serves our goal of encouraging critical judgment than would a mere digest of others' views. Accordingly, our aim is not to force conviction, but to provoke either reasoned acceptance or reasoned rejection of what we say. We are confident that such reasoning is possible in ethics, and that, through lively and tolerant dialogue, progress can be made in dealing with what at first might seem irresolvable difficulties.

Courses

Sufficient material is provided for full courses devoted to the topic. The book can also be used for a several-week module on engineering ethics within a variety of other courses in engineering, philosophy, business, and the social sciences—courses which typically include such topics as general professional ethics, business ethics, applied philosophical ethics, engineering law, business management, values and technology, engineering design, technology assessment, and safety. Despite the intensity of the engineering curriculum, engineering ethics should enter in several contexts to ensure that students perceive it as a genuine concern of the faculty.

Outline

Ethics in Engineering is divided into four main parts, which emphasize, respectively, applied ethical theory, safety and risk, the corporate setting, and global awareness and career choice.

Part I provides an introduction to basic concepts and theories of ethics. Chapter 1 introduces and defines the field of engineering ethics as it relates to applied philosophical ethics. The roles of describing, evaluating, and clarifying in ethics are explained. The goal of fostering moral autonomy in studying ethics is emphasized in discussing the psychology of moral development. Chapter 2 introduces further key concepts: moral dilemmas, responsibility, and relativism. Three fundamental types of theories about right action—emphasizing duties, rights, and good consequences—are introduced. There is also a discussion of character, or virtue, ethics: the ethics of good and bad traits of character.

Part II deals with the moral issues surrounding safety assessments within the inherently risky activity of engineering. Chapter 3 develops a perspective on engineering as an experiment on a societal scale involving human subjects. This model provides a framework for discussing various aspects of responsible engineering practice: imaginative foreseeing of possible side effects, careful monitoring of projects, and respecting the rights of clients and the public to make informed decisions about the products which affect them. The model also provides a context for discussing the importance and limitations of laws and codes of ethics in engineering. Chapter 4 explores some of the moral complexities of safety and risk decisions.

Part III examines some of the special moral issues arising for the nearly 90 percent of engineers who are employed by corporations. Chapter 5 approaches those issues by clarifying the ideas of professionalism, loyalty to employers, and employer authority. It also contains discussions of employee obligations to employers related to conflicts of interest, confidentiality, unionism, and white-collar crime. Chapter 6 focuses on the various rights of engineers: professional, employee, and human rights. Special topics include whistle-blowing, freedom of conscience, due process from employers, and discrimination or preferential treatment.

Part IV sketches further issues relating to global awareness and career choice. Chapter 7 explores connections between engineering ethics and international corporations, environmental ethics, computer ethics, and nuclear deterrence and weapons development. Chapter 8 examines the role of morality in making career choices in engineering. It also outlines a few additional issues concerning the responsibilities of the engineering profession.

The Appendix contains sample codes of ethics of several major engineering professional societies. Study questions are provided at the end of major sections of each chapter.

Second Edition

Many recent case studies have been added for this edition. They include the space shuttle *Challenger,* Chernobyl, the Bhopal chemical disaster, acid rain, and industrial espionage. Several topics are new or given expanded treatment, such as white-collar crime, technology transfer, environmental ethics, character (or virtue) ethics, and the psychology of moral development. Many sections have been reorganized and rewritten to improve teachability and more effectively integrate theory and practical applications. This is especially true of Parts I and IV.

ACKNOWLEDGMENTS

This book was written after but inspired by our participation as a philosopher-engineer team in The National Project on Philosophy and Engineering Ethics from 1978 to 1980. We gratefully acknowledge the sponsor of that project, the National Endowment for the Humanities, and the project director, Robert J. Baum.

Mike Martin's work on the book was largely made possible by a fellowship from the National Endowment for the Humanities from 1981 to 1982. He is indebted to Nelson Pike and the Department of Philosophy at the University of California, Irvine, for granting him the status of Visiting Scholar during that time. In working on the second edition, he also was helped by a reduced teaching load from Chapman College.

We greatly benefited from the criticism and suggestions of participants in the classes we have taught together during the past several years. The formal course work ranged from segments of design courses given at the U.C.I. School of Engineering to full courses on engineering ethics taught in the U.C.I. Department of Philosophy and for the Masters Degree in Engineering Program of California State Polytechnic University, Pomona, taught at Fluor Corporation. An additional seminar brought together practicing engineers under the auspices of the Institute of Electrical and Electronics Engineers, Orange County Section. We have also been helped by the comments of audience members at lectures we gave at meetings of the American Philosophical Association, the Institute of Management Science, the American Society of Engineering Education and several other engineering societies at the Second and Third National Conferences on Engineering Ethics, and at the School of Engineering at the University of California, Santa Barbara.

Many colleagues and friends have influenced our thinking about the issues in engineering ethics. Our special thanks extend to Frank Alderman, Jim Andersen, Robert Anderson, Jean-Louis Armand, Paul D. Arthur, Robert J. Baum, Taft Broome, Robert Bruder, Richard T. De George, Paul Durbin, John Fielder, Albert Flores, Leslie Francis, Gilbert Geis, Jerry Gravander, Milton Gross, Jack Hagman, Michael Hodges, Lawrence Hollander, Jacqueline A. Hynes, Elizabeth Jacks, David James, Andrew Jameton, Edwin Jones, Kenneth Kipnis, Rob Kling, Robert Kunde, Robert Ladenson, Bruce Landesman, Edwin Layton, John Leckie, Barry Lichter, William Litle, Thomas Long, Ruth Macklin, Bill McIlvaine, Frank Manley, A. I. Melden, Robert Miles, John Mingle, Martha Montgomery, Carl Nelson, Richard Nesbit, Andrew Oldenquist, Sterling Olmsted, James Otten, David Palmer, Victor Paschkis, Susan Peterson, Michael Pritchard, Charles Reagan, Robert Redmon, Tom Riley, Tom Rogers, Jean Runzo, Joseph Runzo, Robert Saunders, Fay Sawyier, Ken Schneider, Robert Schultz, Sharon Schwarze, William Schwarze, Edward Slowter, Carol Ann Smith, Robert Smith, Sheri Smith, James Thorpe, Paul Torda, Stephen Unger, Vivian Weil, Carolyn Whitbeck, Donald Wilson, and Janet Zimmerman.

McGraw-Hill and the author would like to thank Thompson M. Faller, University of Portland; Robert L. Johnson, Lehigh University; Stothe Kezios, Georgia Institute of Technology; Maurice K. Kurtz, Jr., Florida Institute of Technology; Heinz C. Luegenbiehl, Rose–Hulman Institute of Technology; John D. Miles, University of Notre Dame, and Carl Mitcham, Polytechnic University for reviewing this text during the course of development.

Finally, we wish to thank the many authors and publishers who granted us permission to use copyrighted material as acknowledged in the bibliographical entries. We also thank the professional societies that allowed us to print their codes of ethics in the Appendix.

Mike W. Martin

Roland Schinzinger

ETHICS IN ENGINEERING

PART I

THE SCOPE OF ENGINEERING ETHICS

Whether or not it draws on new scientific research, technology is a branch of moral philosophy, not of science. It aims at prudent goods for the commonweal and to provide efficient means for these goods.... As a moral philosopher, a technician should be able to criticize the programs given him [or her] to implement.

Paul Goodman

Philosophy, though unable to tell us with certainty what is the true answer to the doubts which it raises, is able to suggest many possibilities which enlarge our thoughts and free them from the tyranny of custom.

Bertrand Russell

Morality...provides one possibility of settling conflict, a way of encompassing conflict which allows the continuance of personal relationships against the hard and apparently inevitable fact of misunderstanding, mutually incompatible wishes, commitments, loyalties, interests and needs.... We do not have to agree with one another in order to live in the same moral world, but we do have to know and respect one another's differences.

Stanley Cavell

CHAPTER 1

INTRODUCTION

Engineers create products and processes to satisfy basic needs for food and shelter—and in addition enhance the convenience, power, and beauty of our everyday lives. They even make possible spectacular human triumphs once only dreamed of in myth and science fiction. A century ago in *From the Earth to the Moon,* Jules Verne imagined American space travelers being launched from Florida, circling the moon, and returning to splash down in the Pacific Ocean. In December of 1968, three astronauts aboard an Apollo spacecraft did exactly that. Seven months later, on July 20, 1969, Neil Armstrong took the first human steps on the moon. This extraordinary event was shared with millions of earthbound people who watched the live broadcast on television. Engineering had transformed our sense of connection with the cosmos and even fostered dreams of routine space travel for ordinary citizens.

Those dreams were widely shared on the morning of January 28, 1986, when schoolteacher Christa McAuliffe joined 6 astronauts for a voyage aboard space shuttle *Challenger.* But two small events that took place during the launch doomed *Challenger* and its crew. A few milliseconds after ignition, a simple seal joining two segments of a booster rocket failed to contain hot gases from the burning fuel. Soon after, vibration from the launch jarred a backup seal from its proper position, allowing flames to spew out near the enormous fuel tank. Less than a minute and a half into its flight, *Challenger* burst in a fiery explosion watched by horrified children who were following the telecast in classrooms and auditoriums.

Public shock and grief over the tragedy were quickly compounded by anger. It was learned that the night before the launch fourteen engineers at

Morton Thiokol, the manufacturer of the booster rocket, had unanimously and vigorously voiced opposition to the launch. They warned that temperatures at the launch site were well below the tested safety range. Low temperature could lessen the pliability of the rubber seals, causing them to fail. Moreover, the engineers were well aware of a history of concern over the gaskets, which had shown alarming erosion in previous launches. They were already redesigning the seals between segments of the booster rockets. Yet on the eve of the launch, their concerns were overridden by top managers at Thiokol who together with executives at NASA failed to convey the engineers' concerns to the NASA administrators responsible for making the decision to launch.

Later we will enter into the details about *Challenger*. We will also explore other tragedies in which it should have been known in advance that safety was being compromised beyond the level of acceptable risk: the accidents at the nuclear plants at Chernobyl and Three Mile Island and the chemical plant at Bhopal, the chemical dumping at Love Canal, exploding Pinto gas tanks, deadly all-terrain vehicles, and indiscriminate uses of asbestos in manufacturing and construction, to name just a few examples. But it is already clear that the work of engineers has moral dimensions which should be of interest to us all. These implications should also be the highest priority for engineers and other professionals involved in technology.

It is equally clear, however, that the moral aspects of engineering are complex. Most engineering takes place within profit-making corporations which in turn are embedded in an intricate structure of society and government regulation. This will have to be taken into account in understanding what can and cannot be morally required of engineers.

Hence we should not expect a quick and simple answer to the question of who was responsible for the *Challenger* disaster or to the question of how similar events can be prevented. Before we rush to blame the fourteen engineers for not trying to prevent the disaster by blowing the whistle, we need to appreciate our own possible failure as citizens to see to it that whistle-blowing engineers are not routinely fired and persecuted by their employers. We need to ask how corporations can be better structured to allow responsible engineers to act on their moral convictions and professional judgments. And we need an enriched understanding of what engineers can and cannot do to improve their own working conditions. We need, in short, to engage in the study of engineering ethics.

WHAT IS ENGINEERING ETHICS?

Engineering ethics is (1) the study of the moral isssues and decisions confronting individuals and organizations involved in engineering and (2) the study of related questions about moral conduct, character, ideals, and relationships of people and organizations involved in technological development. Perhaps it is inevitable that moral problems, especially the perplexing moral dilemmas, will preoccupy us. After all, it is usually the response to specific problems that prods us to make the world better. Yet we should bear in mind that character, general ideals, and moral relationships are equally important foci in approaching engineering ethics.

While the emphasis of this book will be upon engineers, the ethics of engineering is wider in scope than the ethics of engineers. It applies also to the decisions made by others engaged in the technological enterprise, including scientists, managers, production workers and their supervisors, technicians, technical writers, government officials, lawyers, and the general public.

Historical Note

Ethical concern among engineers began as early as the profession of engineering. In the late nineteenth century, newly emerging professional societies for engineers formally expressed this concern by writing codes of ethics. Nevertheless, engineering ethics has traditionally been too narrowly conceived. Too often it has been regarded as encompassing little more than the drafting and promulgating by professional societies of official prescriptions in the forms of codes, guidelines, and opinions. These activities are vitally important. But they are only one aspect of engineering ethics, not its full substance.

As a discipline or area of extensive inquiry, engineering ethics is still young, certainly younger than medical ethics and legal ethics. Only since the late 1970s has systematic attention been devoted to it by engineers and members of several other scholarly disciplines. Earlier books by Harding and Canfield (1936), Mantell (1964), and Alger et al. (1965), as well as journals such as the *Professional Engineer*, covered the traditional aspects of engineering ethics very well, but they did not examine its wider implications. In the middle years of the 1970s changes could be noticed in several engineering periodicals, such as *Issues in Engineering*, published by the American Society of Civil Engineers, and the *Newsletter of the Committee on Social Implications of Technology* (now *Technology and Society Magazine*), published by the Institute of Electrical and Electronics Engineers. Today it is increasingly common to find articles on ethics in journals published by these and other professional societies, such as the American Society of Chemical Engineers and the American Society of Mechanical Engineers.

The conception of engineering ethics as an "interdisciplinary discipline" involving philosophy, social science, law, and business theory, in addition to engineering theory, became more clearly defined with the National Project on Philosophy and Engineering Ethics carried out under the direction of Robert Baum from 1978 to 1980. The first interdisciplinary conference took place in 1980 at Rensselaer Polytechnic Institute, and there have been several others since then. The first scholarly bibliography on engineering ethics was also published in 1980 at the Illinois Institute of Technology (Ladenson, 1980). The first interdisciplinary journal to emphasize articles on engineering ethics, the *Business and Professional Ethics Journal,* was created in 1981.

This late development of the discipline is ironic. Engineering is the largest profession numerically and affects all of us in most areas of our lives. The skill of a surgeon's hand affects one patient at a time; the judgment of a design engineer can influence hundreds of lives at once. Medicine manifests itself in the yearly checkup, the miraculous cure, and the contents of the household medicine chest; but the products of engineering confront us virtually everywhere we turn our eyes and every time we do something.

Why has the general significance of engineering ethics, with its focus on personal decision making and responsibility, only recently become appreciated? In spite of the dramatic impact of engineering on our safety and well-being, we have tended to stereotype it as a tool of vast impersonal organizations. Individuals involved in it have frequently been viewed as cogs in machines rather than as responsible decision makers. Emphasis has usually been upon products and their effects on society rather than upon the human drama behind their production. Yet engineering products derive from personal creative activity in which responsible conduct can make the difference between large-scale benefit or large-scale harm, up to and including life and death. Engineering ethics is the discipline which examines the moral import of that creative activity. It explores the moral dimensions of technology "from the inside."

Variety of Moral Issues

There are two contrasting approaches to engineering ethics, one of which emphasizes small everyday problems and the other larger social problems. Those of us who tend to view the world from the microcosm of our immediate surroundings may become preoccupied with the frequently petty, but nevertheless persistent and nagging, moral problems of everyday life and work. What we must do is to reach beyond those problems to seek an understanding of their root causes.

Others among us are more inclined toward a macroscopic view involving reflections on the moral condition of society. What we need is the discipline to reconcile the broad view with specific circumstances as they present themselves in different everyday settings.

Which approach is better? Neither by itself. What is required is ongoing interaction between the two. In this vein we will approach our topic, taking as examples some cases involving a wide social impact and others of a narrower or more routine nature.

An engineered product or project goes through various stages of design, manufacture or construction, testing, sales, and service. Engineers carry out or supervise the appropriate activities at whatever stage of this process a convenient division of labor has assigned them. The nature of the activity or project will generally dictate whether the engineers involved are by training civil, electrical, mechanical, or chemical engineers, to name only the major fields, but every field involves moral problems.

For example, as engineers carry out their tasks, there will be times when their activities will ultimately lead to a product which is less than useful or safe. This may happen intentionally, or under pressure, or in ignorance. A product may be intentionally designed for early obsolescence; an inferior material may be substituted under pressure of time or lack of money; or a product's eventual harmful effects may not be foreseen. Then too, because of the size of a project, or because of the large numbers of a product sold on the mass market, many people may be affected. And these problems arise quite apart from the temptations of bribes and other forms of outright corruption.

The following four specific examples (some of which led to regulatory changes) hint at a few of the areas covered by engineering ethics:

1 An inspector discovered faulty construction equipment and applied a violation tag preventing its continued use. The inspector's superior, viewing as minor the infraction of a relatively insignificant regulation, ordered the tag removed so the project would not be delayed. The inspector objected and was threatened with disciplinary action.

2 An electric utility company applied for a permit to operate a nuclear power plant. The licensing agency was interested in knowing what emergency measures had been established for human safety in case of reactor malfunction. The utility engineers described the alarm system and arrangements with local hospitals for treatment. They did not emphasize that these measures applied to plant personnel only and that they had no plans for the surrounding population. "That is someone else's responsibility," they claimed upon being questioned about this omission.

3 A chemical plant dumped wastes in a landfill. Hazardous substances found their way into the underground water table. The plant's engineers were aware of the situation but did not change the disposal method because their competitors did it the same cheap way, no law explicitly forbade the practice, and local government was not alert to the danger.

4 Electronics company ABC geared up for production of its own version of a popular new item. The product was not yet ready for sale, but even so, pictures and impressive specifications appeared in advertisements. Prospective customers were led to believe that it was available off the shelf and were drawn away from competing lines.

These examples show how ethical problems arise most often when there are differences of judgment or expectations as to what constitutes the true state of affairs or a proper course of action. The engineer may be faced with contrary opinions from within the firm, from the client, from other firms within the industry, or from government. And not to be left out, as indicated in Fig. 1-1, are still other possible positions that can be taken by the profession of engineering itself, usually embodied by or represented in the form of

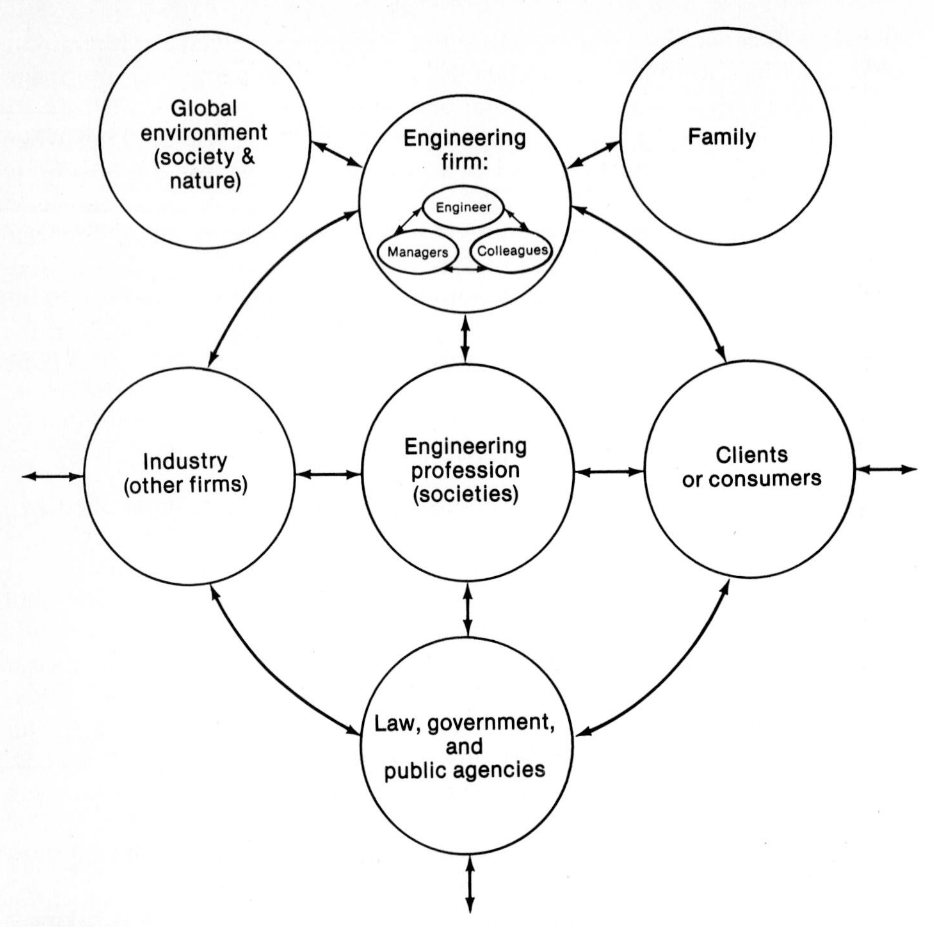

FIGURE 1-1
Contexts of potential professional disagreements engineers may encounter.

a professional society. We will have more to say about the relationships between engineers and their colleagues, their managers, and their clients later, particularly in connection with professional freedoms, rights, and obligations.

The four cases have already raised a number of pertinent moral questions. To what extent should an employer's or supervisor's directives be the authoritative guide to an engineer's conduct? What does one do when there are differences of judgment? Is it fair to be expected to put one's job on the line? Should one always follow the law to the letter? Is an engineer to do no more than what the specifications say, even if there are problems more serious than those initially anticipated? How far does an engineer's responsibility extend into the realm of anticipating and influencing the social impact of the projects he or she participates in?

The case of the inspector told to ignore rules has numerous variations: in the testing of prototypes or finished products in a factory, in the handling of financial audits, or in the response to unsolicited reports of dangerous situations, to name but a few.

The nuclear power plant situation demonstrates several ways in which a product's ultimate use and possible failures may be overlooked by its designers. Lack of coordination in planning for its effects on everyone connected with it can lead to catastrophic results; limiting one's attention only to the narrow specifications can result in the delivery of a product that satisfies the contract but does not serve the needs of the customer or the public in the long run.

The questionable conduct of the chemical plant and, to a greater degree, the ABC company falls into the category of "ethics cases" which traditionally have occupied the ethics review boards of professional societies. Prescriptive codes of conduct can be established for such infractions of common decency, and for that reason those instances constitute the majority of cases in which action has been taken. This has encouraged the identification of engineering ethics with a finite set of specific maxims and regulations designed to assure moral conduct. Yet as we shall see, the moral problems arising in this field are not always manageable in this straightforward manner.

Normative, Conceptual, and Descriptive Inquiries

The study of the moral questions in engineering can be viewed as involving three distinct kinds of inquiry: *normative, conceptual,* and *descriptive.*

First, and most centrally, engineering ethics involves normative inquiries. "Normative" refers to the moral norms or standards which are desirable for actions, attitudes, policies, organizational structures, and individual character traits. The primary aim of engineering ethics is to identify and justify the moral obligations, rights, and ideals of individuals and organizations engaged in engineering. It seeks to establish which prescriptions of basic duty and higher ideal ought to be endorsed on moral grounds, and attempts to apply those results to specific situations in a manner that yields practical guidance.

For example, it asks the following: How far does the obligation of engineers to protect public safety extend in given situations? What are the bases of engineers' obligations to their employers, their clients, and the general public? When, if ever, should engineers be expected to blow the whistle on dangerous practices of the employers for whom they work? Is whistleblowing a minimal moral duty or more a matter of heroism that goes beyond the requirements of an engineer's basic duty? Whose values ought to be primary in making judgments about acceptable risks in a design for a public transport system: those of management, senior engineers, government, voters, or some combination of these? Which particular laws and organizational procedures affecting engineering practice are morally warranted? What moral rights should engineers be recognized as having in order to help them fulfill their professional obligations?

Dealing effectively with normative issues, however, requires two further types of inquiry. One of these is *conceptual,* directed toward clarifying the basic concepts or ideas, principles, problems, and types of arguments used in discussing moral issues in engineering. One example of a conceptual inquiry is seeking to define "safety" and its relationship to such ideas as "risk." A related conceptual inquiry attempts to sharpen understanding of entries in codes of ethics such as the following: Engineers should "protect the safety, health, and welfare of the public," while they should also "act as faithful agents or trustees for their employers or clients."

The third type of inquiry seeks to uncover factual information bearing upon conceptual and normative issues. Since it is concerned with specifying and gathering relevant facts, it is called a *descriptive inquiry*. Where possible, researchers attempt to conduct descriptive inquiries using proven scientific techniques. Topics of special interest in the context of our discussion are the business realities of contemporary engineering practice, the history of the engineering profession, the effectiveness of professional societies in fostering moral conduct, the procedures used in making risk assessments, and psychological profiles of engineers. Determining the facts in these areas provides an understanding of the background conditions which generate moral problems. It also enables us to deal realistically with alternative ways of resolving those problems.

Interrelatedness of the Three Inquiries

To summarize, engineering ethics involves three distinct types of inquiry:

1 Normative, whose practical aim is to provide reasoned evaluations of the conduct and character of individuals, the functioning of organizations, and the alternative responses available to solve concrete problems. Interwoven with this is the more theoretical aim of justifying the major moral principles which ought to be affirmed by individuals and organizations involved in engineering.

2 Conceptual, concerned with clarifying basic ideas, principles, issues, and types of argument concerning the moral problems in engineering.

3 Descriptive, which seeks to provide factual information needed for understanding and dealing with both conceptual and normative issues.

These three types of study are complementary and usually closely interrelated, as the following example illustrates. Consider a young engineer who becomes convinced that the level of pollutants her company is pouring into a stream is dangerously high, given that children are using the river downstream for swimming. She expresses her view to her immediate supervisor, who says her fears are unfounded because the pollution has caused no complaints in the past. Is she required to do more?

Obviously this question poses a practical normative issue. But we will also want to know additional facts about the case, such as the nature of the pollutants in this instance. Then we should want to know about the costs of controlling them, and whether the engineer's conviction that they are dangerous is well founded. Hence a descriptive inquiry is called for.

Perhaps enough has already been said, however, to suggest that the engineer definitely should take further action, since her professional judgment has been overruled without a compelling reason. But should she merely request of the supervisor a more detailed response? Should she go above the supervisor to higher management? Should she write to the local mayor, or would going outside the normal organizational channels constitute disloyalty to the company? And what is the basis for holding that she even has the right to a cogently reasoned response from her supervisor?

These are further normative questions since they call for making moral evaluations of her responsibilities and rights. Answering them will require an understanding of the general moral obligations of engineers. And if this understanding is to be more than an undefended opinion or intuition, it will have to be grounded upon reasons as to why various specific moral principles ought to be affirmed. Providing these reasons is the task of a more theoretical inquiry, primarily normative in nature. But this inquiry will also involve clarification of ideas about safety, loyalty to companies, and professional freedom and autonomy. Hence it will involve conceptual inquiries.

Descriptive inquiries, moreover, enter the case yet again when we attempt to discover the realistic options open to the engineer, as well as to provide some estimate of their likely consequences. Those inquiries will help us understand the business, social, and political realities in which the company operates. Only with this knowledge can we make a sound recommendation about what the engineer should do.

Senses of "Engineering Ethics"

The word "ethics" (like the word "morals") has several distinct although related meanings. Corresponding to them are various senses of the expression "engineering ethics."

First, in the main sense used so far and in what follows, ethics is a discipline or area of study dealing with moral problems. Engineering ethics, accordingly, is the discipline or study of the moral issues arising in and surrounding engineering. As we have just seen, it involves normative (evaluative) inquiries, conceptual (meaning) inquiries, and descriptive (factual) inquiries into these moral issues. As understood here, the normative inquiries are central, with the conceptual and factual inquiries entering where relevant to support the normative inquiries.

Second, when we speak of ethical problems, issues, and controversies, we mean to distinguish them from nonethical or nonmoral problems. Here the word "ethical" marks a contrast between moral questions and questions of a political, legal, and artistic nature. Engineering ethics in this sense would refer to the set of specifically moral problems and issues related to engineering.

Third, sometimes the word "ethics" is used to refer to the particular set of beliefs, attitudes, and habits which a person or group displays in the area of morality. Thus we say that Mussolini's ethics were different from those of Dwight Eisenhower, and both differed from those of Karl Marx. We also speak of the Romans' ethics, the Victorian ethic, the Protestant work ethic, and socialist ethics. In doing so we are referring to people's actual outlooks on moral issues. This sense is linked directly to the original sense of the Greek word *ethos,* which meant customs (as did *mores,* the Latin root of "morals"). As such, engineering ethics would be the currently accepted standards endorsed by various groups of engineers and engineering societies. The discipline of engineering ethics has the task of examining these accepted conventions to see if they are clear, justified, and sufficiently comprehensive.

This third sense of "engineering ethics" is purely descriptive because it concerns merely the facts about what engineers and others believe as regards moral problems in engineering. It is a usage in which social scientists especially might be interested, since they seek to describe and explain beliefs and actions related to morality. We will avoid it, however, because our main interest is in normative questions about correct beliefs.

Fourth, the word "ethics" and its grammatical variants can be used as synonyms for "morally correct." For example, people's actions and principles of conduct can be spoken of as either ethical (right, good, or permissible) or unethical (immoral), and individuals can be evaluated as ethical (decent, having moral integrity) or unethical (unscrupulous). In this usage, engineering ethics would amount to the set of justified moral principles of obligation, rights, and ideals which ought to be endorsed by those engaged in engineering. Discovering such principles and applying them to concrete situations is the central goal of the discipline of engineering ethics.

In order to simplify matters and avoid confusion, we shall restrict the expression "engineering ethics" to the sense in which it names a discipline. Nevertheless, we shall occasionally use the word "ethical" in its other senses when we speak of ethical problems, an individual's or group's ethics, and ethical conduct. The context will make it clear which of these is meant.

Engineering Ethics and Philosophy

Much of engineering ethics can be viewed as part of applied philosophical ethics. Like medical, legal, and business ethics,* it is focused upon practical moral problems but seeks where possible to apply methods and theories derived from more general philosophical principles.

Thus, applied ethics necessarily makes contact in many places with ethical

*We might note here that the closest cousin of engineering ethics is business ethics, which is also now undergoing rapid development. Since most engineers are salaried employees, and since engineering decisions are usually tied to business decisions, it is not surprising that moral problems in engineering and business often overlap. The question, for example, of what is wrong with offering and accepting more than nominal gifts when negotiating contracts for engineering services is at once an issue in engineering ethics and business ethics.

theory. This is because applied ethics is concerned with uncovering cogent moral reasons for beliefs and actions, as opposed to accepting uncritically whatever beliefs or actions might happen to strike one's fancy as being correct at a given moment. And the general moral principles to which such reasons make reference either explicitly or implicitly are directly linked to ethical theory.

We say, for example, that a given action is wrong because it amounts to accepting a bribe in the context of negotiating a contract, and bribes are wrong because they unfairly influence judgment and decisions. This claim is based on a general principle that contract negotiations ought to be impartially centered on the merits of the contract alone. And that general principle, in turn, will be justified by appealing to a higher-order principle to the effect that we ought to be fair and impartial in certain situations. Such higher-order principles, when developed and integrated within theories of justice, constitute the broad philosophical perspectives on conduct that ethical theory can provide.

An analogy might help clarify the relationship between applied and general philosophical ethics. Engineering is itself an applied science which draws upon principles from general physics and mathematics. It is "applied" in the sense that it is aimed at finding practical solutions to concrete problems by applying either theoretical knowledge or the more intuitive understanding of a field gained after years of working within it. "Pure" science and mathematics, by contrast, are directed toward obtaining new knowledge, usually of a general sort. Practical engineering solutions are often obtainable without having to await final proofs of theoretical principles. On occasion particular engineering solutions will have a wide importance, however, and will provide insights that can be fed back into the more theoretical levels of science. Thus there arises a dynamic interplay between applied and so-called "pure" science.

Much the same can be said of the relationship between applied and general ethics, remembering that here the concern is with moral instead of scientific issues. General ethics tends to emphasize theoretical knowledge and examines practical cases only in order to illustrate and test theories. Applied ethics, by contrast, focuses upon concrete problems for their own sake, and invokes general theory where helpful in dealing with those problems. It is not the task of applied ethics per se to resolve long-standing philosophical disputes over the validity of general moral perspectives.

On occasion, however, applied ethics may well yield insights having a more theoretical importance. Future developments in engineering ethics seem likely to provide fruitful directions for thinking about enduring philosophical questions concerning the nature of collective responsibility, legitimate authority, justice within economic systems, and moral rights within complex institutions.

Philosophy for its part plays a major role in supplying the basic concepts and theories needed to clarify the nature of ethical problems and issues in engineering. Sometimes it can yield a heightened perception of those problems by placing them in perspective as aspects or instances of broader problems. For example, we shall see later how some appeals to codes of engineering ethics when trying to answer moral questions are problematic because the appeals presuppose a narrow kind of conventionalism in ethics. At other times philosophy can clarify general concepts which arise when thinking about any moral issue. An example of this to be dealt with in Chap. 2 is the concept of ethical dilemmas.

Study Questions

1 Cite examples of ethical problems which can arise in the production of a new type of motor bicycle as it passes through the stages of design, manufacture, testing, sales, and service. Use your imagination in developing the details of at least one problem at each stage.

2 The following three illustrations are based upon actual cases. With respect to each, first state what you see as the normative, or evaluative, issues involved. Then identify any descriptive inquiries which you think might be needed in making a reliable judgment about the case. Finally, are there any ideas or concepts involved in dealing with the moral issues that it would be useful to clarify?

   a A County Engineer in Virginia demanded a 25 percent kickback in secret payments for highway work contracts he issued. In 1967 he made such an offer to Allan Kammerer, a 32-year-old civil engineer who was vice president of a young and struggling consulting firm greatly in need of the work. Kammerer discussed the offer with others in the firm, who told him it was his decision to make. Finally Kammerer agreed to the deal, citing as a main reason his concern for getting sufficient work to retain his current employees (Fairweather, 54–55).

   b In 1970 Carl Houston was assigned as a welding superintendent to a nuclear power plant under construction. According to Houston, even his preliminary observations revealed numerous poor welds resulting from the use of poor procedures and improper materials. He considered this to be the result of improper training given to the welders and told his manager that if the situation was not corrected he would write to the main headquarters. Houston then described the situation to some of the subcontractors of the project, an act which led to his being fired for insubordination (Houston, 1975, 25–30).

   c The principal project finance officer for a large firm learned that an engineer had been charging his personal long-distance phone calls to a client for several months. The officer informed the engineer's supervisor about this. The charges were around $50 a month. It was extremely unlikely that the client would learn of them since the contract involved several million dollars, and if he did it would be easy to explain the situation as a bookkeeping error. The engineer was one of the firm's most valued employees, someone the supervisor regarded as the key to much of the hoped-for success of the firm during the next 2 years. The supervisor decided to ignore the situation. (From an unpublished case study written up by William Litle.)

3 In response to a request by the *Monthly Newsletter* of The National Project on Philosophy and Engineering Ethics for brief definitions of engineering ethics, the following three suggestions were submitted and printed in the May–June 1979 issue. Which, if any, of the four senses distinguished above does each author seem to have in mind?

a "Moral principles accepted by the profession relative to the practice of engineering." (Don Wilson, Engineer at Michael Baker, Jr., Inc.)
b "The rights and responsibilities of those persons who practice the profession of engineering." (Albert Flores, Department of Philosophy, California State University, Fullerton.)
c "If the task of *defining* engineering ethics is supposed to result in the articulation of a set of ethical principles which are individually necessary and jointly sufficient to distinguish engineering ethics from, say, business ethics, then there seems little hope that such a result is possible. For there will be an *engineering* ethics in this sense, that is, an ethics with principles which are engineering-specific, if and only if there are ethical aspects of engineering practice which are *unique* to the profession.... The ethical issues which bother the reflective engineer today revolve around questions of truth-telling, confidentiality, and obligations to employer/client vs. obligations to the public. But these ethical issues have been addressed by physicians for hundreds of years...." (Thomas A. Long, Department of Philosophy, University of Cincinnati, used with permission.)

4 A full definition of engineering ethics will be linked with a definition of engineering. One broad definition of engineering says it is the application of science in the use of natural resources for the benefit of society or humanity. Such a notion would be presumptuous if it implied that engineers are the only ones who participate in engineering as so defined (Ramo, 12). But if this implication is avoided, the definition has the advantage of emphasizing how all of us—citizens, engineers, management, scientists, technicians, government regulators—are participants in engineering and as such should be concerned with the field of engineering ethics.
Consult a dictionary, an encyclopedia, and an introductory engineering textbook to see how they define engineering. Then carefully state the definition you prefer in connection with thinking about engineering ethics.

5 The terms "moral," "virtue," and "character" have acquired meanings in colloquial use which make them awkward to use at times. Find their etymological roots and describe how today's use often departs from their original or former meanings.

AIMS IN STUDYING ENGINEERING ETHICS

What is the point in studying engineering ethics? And what can be gained from taking a course or a segment of a course devoted to it?

As suggested above, studying engineering ethics should increase the ability of engineers, managers, citizens, and others to confront the urgent moral questions raised by technological activity. Yet more needs to be said about how this is to be achieved through college courses, continuing education, or individual study.

Should the study of engineering ethics aim at inculcating particular moral beliefs? We do not think so. Instead, the aim should be to empower individuals to reason more effectively concerning moral questions. The aim should be to strengthen moral autonomy.

Moral Autonomy

Of course there are many possible topics, and each of us will have special interests in probing different issues. Yet we believe that a shared practical goal in studying ethics should be to think clearly and critically about moral issues. To invoke a term widely used in ethics, the unifying goal should be to increase one's *moral autonomy* in developing, expressing, and acting on reasoned moral views.

"Autonomy" literally means "self-ruling" or "independent." But not just any kind of independent reflection about ethics amounts to moral autonomy. Moral autonomy can be viewed as the ability to think, and the habit of thinking, rationally about ethical issues on the basis of moral concern. This foundation of moral concern, or general responsiveness to moral values, derives primarily from the training we receive as children in sensitivity to and consideration of the needs and rights of others. Where such training is absent, as it often is with abused or neglected children, the tragic result can be an adult sociopath capable of murdering without compunction. Sociopaths, who by definition lack a sense of moral concern and guilt, are never morally autonomous—no matter how "independent" their intellectual reasoning about ethics may be.

Adult moral concern, of course, can be evoked (or numbed) on specific occasions by any number of influences: friends, ministers, social events, novels, movies, and inspiring teachers of whatever subjects. It would be surprising if the topics dealt with in ethics courses did not call forth moral concern. Nevertheless, the main point of taking a course on applied ethics should be to improve the ability to reflect critically on moral issues.

This can be accomplished by improving various practical skills that will help produce effective independent thought about moral issues. As related to engineering ethics these skills include the following:

1 Proficiency in recognizing moral problems and issues in engineering. This involves being able to distinguish them from, as well as relate them to, problems in law, economics, religious doctrine, or the descriptions of physical systems.

2 Skill in comprehending, clarifying, and critically assessing arguments on opposing sides of moral issues.

3 The ability to form consistent and comprehensive viewpoints based upon consideration of relevant facts.

4 Imaginative awareness of alternative responses to the issues and creative solutions for practical difficulties.

5 Sensitivity to genuine difficulties and subtleties. This includes a willingness to undergo and tolerate some uncertainty in making troublesome moral judgments or decisions.

6 Increased precision in the use of a common ethical language, which is necessary in order to be able to express and defend one's moral views adequately to others.

7 Enriched appreciation of both the possibilities of using rational dialogue in resolving moral conflicts and of the need for tolerance of differences of perspective among morally reasonable people.

8 An awakened sense of the importance of integrating one's professional life and personal convictions—that is, the importance of maintaining one's moral integrity.

Most of us value moral autonomy for its own sake. Its exercise is central to what we think of as the possession of a mature moral outlook, that is, one which is something more than secondhand and passively adopted. Yet we also value it because we believe that it tends to lead to morally responsible conduct or is an integral part of being a responsible person. Certainly some faith in the ability of people to reflect on moral issues as morally autonomous individuals is presupposed in any attempt to engage in serious dialogue about ethics. And an affirmation of the individual's right to develop and exercise reasoned moral perspectives presupposes that most people have a large capacity for acting responsibly out of humane values. It is in this spirit that we, as authors, wish to enter into the discussion of engineering ethics.

Kohlberg's Theory of Moral Development

These comments on moral autonomy can be related to recent work in the psychology of moral development. In particular, they pertain to the psychological theories of moral development set forth by Lawrence Kohlberg and Carol Gilligan.

Building on the pioneering work of Jean Piaget, Lawrence Kohlberg suggested there are three main levels of moral development (Kohlberg, 1971). They are distinguished by the degree of moral cognitive development, that is, by the kinds of reasoning and motivation an individual adopts in response to moral questions.

Most primitive is the *Preconventional Level,* in which right conduct is regarded as whatever directly benefits oneself. Individuals are motivated primarily by the desire to avoid punishment, by unquestioning deference to power, or by a desire to satisfy their own needs. This is the level of development of all young children and a few adults who never manage to go beyond it.

Next is the *Conventional Level,* in which the norms of one's family, group, or society are accepted as the final standard of morality. These norms or conventions are adopted uncritically as being correct because they represent authority. Individuals at this level are motivated by the desire to please others and to meet the expectations of the social unit, regardless of immediate effects on their self-interest. Loyalty and close identification with others have overriding importance. Kohlberg's studies reveal that most adults never mature much beyond this stage.

Finally, the *Postconventional Level* is attained when an individual comes to regard the standard of right and wrong as a set of principles having to do with rights and the general good that are not reducible to self-interest or to social conventions. Kohlberg calls these individuals *autonomous* because they think for themselves and do not assume that customs are always right. They also seek to reason and live by general principles, such as the Golden Rule ("Do unto others as you would have them do unto you"), which apply universally to all people in all cultures. Their motivation is to do what is morally reasonable for its own sake (rather than solely from ulterior motives), together with a desire to maintain their moral integrity, self-respect, and the respect of other autonomous individuals.

Kohlberg's schema of moral development has an obvious connection with what we said about the goals of studying ethics in college. To be morally responsible, one must be able and willing to exercise moral reasoning, and this in turn requires overcoming passive acceptance of the dominant conventions in one's society or "in group." Yet moral responsibility emerges from a foundation of early moral training by one's parents and culture. This early training, which involves submitting to the power of one's parents, makes possible later growth beyond complete self-centeredness (at the Preconventional Level) and uncritical acceptance of customs (at the Conventional Level) toward respect for the rights of other people (at the Postconventional Level).

But how does Kohlberg know that these are the correct stages specifying moral development or growth? How does he know that his "higher" levels represent more advanced stages of moral maturity? He contends that advanced stages constitute a "better cognitive organization" that embodies more distinctions and represents a more universalized perspective. But why does that amount to moral progress, and not merely to increased intellectual sophistication?

The answer to these questions is not found by a mere appeal to facts about the numbers of people who reach each level. As we noted, Kohlberg thinks that relatively few people reach the Postconventional Level, and hence his schema does not record the path of moral development that the majority of people follow. Instead, Kohlberg seems to base his schema on the fundamental assumption that movement toward autonomy is morally desirable. This is consistent with the view expressed earlier that moral autonomy is an inherently valuable general character trait essential for exercising moral responsibility. But it deserves emphasis that this is a normative claim, not a purely descriptive or psychological claim.

Could it be that Kohlberg bases his work on other moral assumptions that are somewhat more controversial? In particular, what should be said about his emphasis on abstract universal rules and rights?

Gilligan's Theory of Moral Development

One of Kohlberg's former students and colleagues has challenged his work on precisely this point and done so in a manner relevant to contemporary debates over male and female approaches to morality. In her book, *In a Different Voice,* Carol Gilligan charges that Kohlberg's studies are distorted by a male bias. Not only did he conduct his studies primarily with male subjects, but according to Gilligan he approached his studies with a typically male preoccupation with general rules and rights.

Gilligan's own studies suggest that there is some tendency for men to be more interested in trying to solve moral problems by applying abstract moral principles. Males tend to resolve moral dilemmas by determining which moral rule is most important and should override other moral rules relevant to the dilemma. Women, by contrast, try harder to preserve personal relationships with all people involved in a situation. In order to do so, they focus greater attention on the details of the context in which the dilemma arises, rather than invoking and trying to rank general rules. Gilligan refers to this context-oriented emphasis on maintaining personal relationships as the *ethics of care,* and contrasts it with an *ethics of rules and rights.*

Both males and females sometimes use both kinds of ethics. Gilligan wishes to draw attention only to a difference in emphasis, not a strict difference based on gender. Moreover, she does not attempt to answer the question of whether this difference in emphasis is due to biology (genetic determination) or to social conditioning.

In order to make Gilligan's criticism of Kohlberg clearer, consider the most famous example that Kohlberg used in his questionnaires and interviews. This example, called "Heinz's Dilemma," involves a woman living in Europe who will die from cancer unless she obtains an expensive drug which the doctors think will help her. Her husband, Heinz, cannot afford to purchase the drug. The local pharmacist is charging 10 times the cost of making the drug. He also invented the drug and remains the sole source for obtaining it. The husband goes to everyone he knows seeking to borrow money, but he manages to raise only half the money needed to purchase the drug. When he asks the pharmacist to sell the drug at a cheaper price or to let him pay for it later, the pharmacist refuses. In desperation, Heinz breaks into the pharmacy and steals the drug. Was the theft morally right or wrong?

Applying his schema of moral development, Kohlberg ranked experimental subjects according to the kinds of reasoning they used about the dilemma (and not depending on their specific answers or conclusions). For example, the subjects who said that Heinz did wrong because he broke the law are reasoning at the Conventional Level, in which right conduct is regarded as simply obeying the law. Also at this level are subjects who said the husband did right because according to their religious beliefs God commanded that human life is sacred and God should be obeyed. By contrast, subjects who said that the right to life of the wife is inherently more important than the property right of the pharmacist are reasoning at the Postconventional Level.

Women, interestingly enough, tended to cluster more frequently than men at Kohlberg's Conventional Level. This was because they showed greater hesitancy about stealing the drug and searched for alternative solutions in terms of the context. For example, they recommended further attempts to reason with the pharmacist and to find creative ways to raise the necessary money. Kohlberg inferred that the women were overly preoccupied with conventional rules against stealing and that they were also wishy-washy in applying general principles about the right to live.

Gilligan, however, drew a very different conclusion from this data. She contended that it reveals a greater sensitivity to people and personal relationships, including the relationship with the pharmacist and the wife (who would not be helped if the husband ended up in jail for stealing the drug). She also saw value in the context-oriented reasoning used by women who did not locate the solution of the dilemma in abstract general rules ranked in order of importance.

Drawing on such reinterpretations of Kohlberg's experimental data, and combining them with her own studies of women, Gilligan offered a strikingly different schema of moral development. She recast Kohlberg's three levels of moral development as stages of growth toward an ethic of caring. Gilligan's recasting looked something like this:

The Preconventional Level This is roughly the same as Kohlberg's first level in that the person is preoccupied with self-centered reasoning. Right conduct is viewed in a selfish manner as solely what is good for oneself.

The Conventional Level Here there is the opposite preoccupation with not hurting others and with a willingness to sacrifice one's own interests in order to help or nurture others. Women are especially prone to fall prey to the cultural stereotypes that pressure them always to be willing to give up their personal interests in order to serve the needs of others.

The Postconventional Level The individual becomes able to strike a reasoned balance between caring about other people and pursuing one's own self-interest while exercising one's rights. The aim is to balance one's own needs with the needs of others, while maintaining relationships based on mutual caring. This is achieved through context-oriented reasoning, rather than by applying abstract rules ranked in a hierarchy of importance.

How does Gilligan's theory of moral development relate to our emphasis on moral autonomy as a goal of studying ethics at the college level? Like Kohlberg's theory, it is entirely compatible. Note that we did not define autonomy as separateness from other people. On the contrary, we said that autonomy requires independent reasoning on the basis of moral concern. That concern is often best understood in terms of caring for others and trying to maintain personal relationships with them. Moral autonomy may well have as much to do with caring for other people within a community based on personal relationships (as Gilligan says) as it does with being sensitive to general principles and human rights (as Kohlberg says). And it surely has to do with sensitivity to the subtleties of special situations (consistent with Gilligan's emphasis on context-oriented reasoning), just as it does with appreciation of general moral principles and rights (consistent with Kohlberg).

Consensus and Controversy

When individuals exercise moral autonomy, there is no assurance that they will arrive at either the truth or the same verdicts as other people exercising *their* moral autonomy. Indeed, there seem to be some basic moral differences which are not reducible to disagreements over facts or errors in logical inference. Perhaps this is inevitable with a subject like morality which is not as precise and clear-cut as arithmetic (Aristotle, 936). Tolerance requires us to allow room for disagreement among autonomous, reasonable, and responsible persons.

This suggests that the aim of teaching engineering ethics should not be to produce a unanimous conformity of outlook, even if such conformity could be achieved by resorting to indoctrination, authoritarian and dogmatic teaching, hypnotism, or other autonomy-destroying techniques. Indeed, just as one major goal of investigators in the field of engineering ethics should be to uncover ways of promoting tolerance in the exercise of moral autonomy by engineers, the same goal should be sought in courses on engineering ethics.

This similarity between the goals of courses on engineering ethics and the goals of responsible engineering can be extended. In both the classroom and the workplace there is a need for authority: teachers having authority over students, and managers having authority over engineers. Both situations presuppose the need for some consensus concerning the role of authority. Is such a consensus undermined by stressing the moral autonomy of individuals to develop and express their own moral views?

Part III of this book responds to this question in detail. Indeed, the question is tacitly considered throughout the book as we discuss specific issues and examples. Here we wish to note two general points about the relationship between autonomy and authority, illustrating them by reference to the classroom.

The first point is that moral autonomy and respect for authority are not inherently incompatible. Moral autonomy, by definition, is exercised on the basis of moral concern for other people and recognition of good moral reasons. In addition, valuing moral autonomy presupposes faith in most people's capacity for moral reasonableness. Now, there is a very good reason for accepting authority in the classroom. Authority provides the framework in which learning can take place. This reason, and not sheer coercion, underlies the acceptance of authority by most students and professors. Without this rough consensus among autonomous members of the academic community, classes could not be conducted in orderly ways, cheating would be encouraged, and trust and respect between faculty and students would be eroded. Considered in this way, the constraints inherent in respecting authority of professors are not much different from those imposed by a conductor on the musicians in an orchestra.

Nevertheless, and this is the second point, sometimes a tension does arise between the need for autonomy of individuals and the need for consensus about authority. For one thing, authority and the rules it generates are not always clear-cut. There may arise good-faith differences among students and faculty as to what is consistent with rules of a given class, differences that need to be discussed openly whenever possible.

Cheating, for example, is clearly forbidden. But is it always clear what cheating is? It is easy to give a general definition: *cheating* is dishonesty in trying to gain something underserved or in falsely representing one's work. But is it dishonest to work in groups on an assignment? Is it dishonest to look at a previous exam that is widely circulated among students? Does it matter that only a few students may have access to a previous exam if it is known that the professor rarely repeats exams?

Most of us know from our own experience that there are gray areas concerning course requirements and that autonomous reflection does not always result in everyone seeing eye to eye about them. Additional conflicts between autonomy and authority arise when authority is abused, or seems to be abused. In small classes it is usually assumed that students should be allowed to express their own views, and that authority is abused when discussion is discouraged by a professor's intimidating approach. Yet there may be reasonable differences concerning how much time should be allowed for discussion and also concerning what is an intimidating approach.

Study Questions

1 Most of us agree that the dogmatic teaching of ethics can threaten the exercise of moral autonomy. Does this mean that college teachers should withhold expression of their own views on moral issues?

2 Describe a clear-cut example of a situation in which one person tries to indoctrinate another into holding a particular belief or value. Does the attempt center more on what is said (for example, false views, one-sided views, etc.) or on how the view is presented (for example, with intimidation, intolerance, the suppression of criticism, etc.)?

3 Respond to the following argument: "The primary goal of a course on engineering ethics ought to be to have students master the standards of professional conduct specified in the major engineering codes of ethics. This is because the codes are formulated by the engineering societies which officially speak for the engineering professional on moral issues. Encouraging engineers to think autonomously threatens to produce chaos within organizations."

4 Present and defend your view about Heinz's Dilemma: Should Heinz have stolen the drug in order to help his wife? Also, using the dilemma as a focus, explain whether you find Kohlberg's or Gilligan's theory more illuminating as an account of morally mature reasoning about the dilemma.

5 In 1959, C. P. Snow, an English scientist turned novelist, delivered a famous lecture entitled "The Two Cultures" (Snow). In it he warned of an increasing gap in communication and mutual appreciation between people educated primarily in science and those educated in the humanities. If you have experienced such a gap, what harm do you see it causing? How can education contribute to bridging the gap, and in particular how might the study of engineering ethics serve as one bridge between the "two cultures"? Do you see a different gap between the practice-oriented

disciplines, which include engineering and applied ethics, and the more theory-oriented disciplines, such as pure mathematics and more esoteric branches of philosophy?

SUMMARY

Various things may be meant by the expression "engineering ethics." In this book *engineering ethics* will mean the examination of the moral issues in engineering and the field of study which results from that examination. It centers on a *normative,* or evaluative, inquiry into how people and organizations involved in engineering ought to be and to act, and what laws, codes of ethics, and institutional norms morally ought to be in effect. But it also involves *conceptual* inquiries aimed at clarifying the meanings of key ideas and issues and *descriptive* inquiries designed to provide relevant factual information.

The issues in engineering ethics are wider in scope than the moral problems confronted specifically by engineers, for they include the moral problems bearing on engineering faced by many other people, including consumers, managers, scientists, technical writers, lawyers, and government officials. Even though the emphasis in this book will be on those problems as they relate to engineers, it will become clear in the course of our discussion that the responsibilities of engineers must be understood in conjunction with the rights and obligations of these other people, especially consumers and managers.

Engineering ethics is intimately tied to philosophical ethics, and it can be viewed as a branch of applied philosophical ethics. This is especially true wherever general ethical theories and distinctions are applied in order to (1) enrich our understanding of the nature of the moral problems and issues in engineering, (2) provide reasoned responses to those problems, and (3) clarify the meanings of key concepts and distinctions.

The practical aim in studying and teaching engineering ethics is to help foster moral autonomy. *Moral autonomy* is the ability to arrive at reasoned moral views based on the responsiveness to humane values most of us were taught as children. This does not require that a person always reach the "correct" moral view. But it involves displaying competencies and sensitivities such as the following: the abilities to discern moral problems and to clarify them, to work out reasoned and sometimes creative responses to them while taking account of opposing viewpoints, and to exercise the verbal and communicative skills relevant to discussing one's views with others.

Moral autonomy is an achievement made possible in part by a foundation of early moral training that helps instill moral concern. Psychologists Lawrence Kohlberg and Carol Gilligan trace the development of autonomy in three major stages: the Preconventional Level of self-centeredness, the Conventional Level of respect for conventional rules and authority, and the Postconventional Level of autonomy. Kohlberg gives greater emphasis to recognizing rights and abstract universal rules, whereas Gilligan stresses the importance of maintaining personal relationships based on mutual caring.

CHAPTER 2

MORAL REASONING

On October 10, 1973, Spiro T. Agnew resigned as Vice President of the United States amidst charges of bribery and tax evasion related to his previous service as County Executive of Baltimore County. A civil engineer and lawyer, he had risen to influential positions in local government. As County Executive during 1962 to 1966 he had the authority to award contracts for public works projects to engineering firms. In exercising that authority he functioned at the top of a lucrative kickback scheme (Cohen 1974).

Lester Matz and John Childs were two of the many engineers who participated in that scheme. Their consulting firm was given special consideration in receiving contracts for public-works projects so long as they made secret payments to Agnew of 5 percent of fees from clients. Even though their firm was doing reasonably well, they entered into the arrangement in order to expand their business. They felt that in the past they had been denied contracts from the county because of their lack of political connections.

Moral Problems and Dilemmas

If we say that Matz and Childs confronted a moral problem, we might mean several things. We could be calling attention to the fact that they were in a situation calling for a decision involving moral considerations about fair practices, honesty, and avoiding deception. Or we could mean they were tempted to ignore these moral considerations and violate moral rules like "Do not cheat." But we would probably not intend to imply there was any serious doubt or difficulty in determining what they were morally required to do—that was clear-cut. The only uncertainty was whether they would decide to do what was morally proper.

Usually, however, when we speak of moral problems we have in mind situations where what ought to be done is not so straightforward and obvious. Those situations may involve two sorts of murkiness or complexity which set them in sharp contrast to the case of Agnew, Matz, and Childs.

First, it may be unclear to the individuals involved which, if any, moral considerations or principles apply to their situation. An engineer starting a new job, for example, may have doubts about whether it is morally permissible to accept an expensive desk set as a gift from a salesperson with whom her company does business. Would this be accepting a bribe? Would it create a conflict of interest? Perhaps a conversation with a colleague will answer these questions. But there will always be troublesome cases where there is considerable vagueness about whether the "gift" is an innocent amenity or an unacceptable bribe.

Second, it may be perfectly clear which moral principles apply to one's situation. The difficulty instead might be that two different moral principles, both of which apply to one's situation, come into conflict or that one principle seems to point simultaneously in two different directions. These kinds of moral problems are called *moral dilemmas.*

Stated more fully, *moral dilemmas* are situations in which two or more moral obligations, duties, rights, goods, or ideals come into conflict with one another, and at least on the surface it appears that not all of them can be fulfilled or respected. It is also possible for one moral principle to have two or more incompatible applications in a given situation.

Moral dilemmas occur frequently, although usually there is only moderate difficulty in seeing what should be done. We make a promise to a friend, thereby creating an obligation to do what we have promised. Our parents become ill and staying home to help them prevents us from keeping the promise. The dilemma, which consists of a conflict between the duty to keep promises and an obligation to one's parents, can usually be resolved by an apologetic phone call to the friend. Or again, we make one promise to our employer and another to a colleague, and it turns out that we cannot keep both. Here the general duty to keep promises has two incompatible applications. Once again, an apology to the offended party will often settle the matter.

Yet dilemmas are not always so easily dealt with. Resolving some of them can require searching, even agonizing, reflection. Contemporary engineering practice makes it virtually inevitable that nearly all engineers will be confronted with some moral dilemmas during their careers. Indeed, this is true of all professionals, including physicians, lawyers, and teachers.

In the next section we will introduce four ethical theories that can be useful in confronting moral dilemmas. But first it will be helpful to sharpen our understanding of what morality is.

WHAT IS MORALITY?

We defined engineering ethics as the study of moral problems in engineering and the study of related questions about character, moral ideals, and moral relationships. But what is the meaning of the word "moral" in this definition?

One suggestion frequently given in dictionaries is that *morality* concerns what *ought* or *ought not* to be done in a given situation, what is *right* or *wrong* about the handling of it, or what is *good* or *bad* about the actions of the people involved in it. But this definition is inadequate for two reasons. On the one hand, morality concerns not just actions, but also good and bad character (what people ought to *be,* not just *do*), relationships (that ought to be sustained), and ideals (to which we ought to aspire).

On the other hand, mere reference to words like "ought," "right," and "good" does not suffice to define even the dimension of morality concerned with conduct. For there are many nonmoral usages of these words and ideas. Thus, in order to start a car a person *ought* to put the key in the ignition—that is the *right* thing to do. People *ought* to brush their teeth before leaving for school or work, and it is *good* to avoid drinking so much coffee that one becomes jittery. To increase profits a company *ought* to try to cut unnecessary costs while increasing efficiency. None of these judgments would typically be counted as moral ones. Moral judgments are about what *morally* ought or ought not to be done, what is *morally* right or wrong, and what is *morally* good or bad. Since the word "morally" is but the adverbial form of the adjective we are trying to define, we are caught in a circle if we cannot move beyond these general evaluative notions.

Some progress can be made if we characterize moral judgments in terms of the particular sorts of reasons used to justify them (Frankena, 1973, 110). This is because such reasons will typically differ significantly from the grounds we give in justifying other types of value judgments. If we ask why persons ought to brush their teeth, the answer will be in terms of health and social etiquette. If we ask why a certain painting is judged to be a good one, the answer will be in terms of its striking lines, color, unity, symbolism, and so on. Giving these reasons in support of judgments makes it explicit that the judgments are nonmoral ones. Thus, too, if an engineering design is said to be a good one simply because it is simple, elegant, or cost-effective, we know that technical and business values are at issue rather than specifically moral reasons.

What, then, are *moral* reasons and ideals? If this question calls for a comprehensive characterization, there is no easy answer. Morality is complex and not easily encapsulated. The theories about right action presented in the next section can be viewed as attempts to offer precise characterizations of morality, but as we shall see, even they are controversial.

If the question, by contrast, calls for examples of moral reasons and ideals, those are easy enough to provide. We are all familiar with a variety of such examples. They concern, for instance, respecting persons by being fair and just with them, respecting their rights, keeping our promises, avoiding unnecessary offense and pain to them, avoiding cheating and dishonesty (of the sort Matz, Childs, and Agnew engaged in). They also concern caring for others by being sometimes willing to help them (especially when they are in distress), to show gratitude for favors, to be compassionate in response to their suffering. There are also high ideals that inspire us to go beyond the minimal requirements of moral duty, such as the ideals of selfless devotion to causes like ending world hunger, promoting world peace, and helping to advance disadvantaged groups. Such ideals may involve a highly personal dimension of morality that is desirable, though optional, for people to pursue. But there are also professional ideals, as we shall see, that should be fostered among practicing engineers.

Moral reasons and ideals, we have said, form a distinct category of value, different from other categories of values. Let us make this clearer by relating and contrasting moral values to three other types of values: self-interest (one's personal good), law, and religion. In each case, we will consider attempts to reduce morality to these other types of value. Following this discussion we turn to theories of right action.

Self-Interest and Ethical Egoism

It seemed easy to recognize that Matz and Childs should not have participated in the kickback scheme supervised by Spiro Agnew. That was because they were pursuing their own private good in a situation where they should have had regard to the requirements of fairness and honesty in business.

But suppose this judgment were challenged as follows: If Matz and Childs did wrong, it was because they adopted an overly narrow view of their own self-interest. Self-interest is what is good for oneself *in the long run.* Matz and Childs took foolish risks, which in the long run resulted in their being caught and probably harmed more than if they had not paid the kickback money. In general, people should always and only pursue their self-interest, but in doing so they should be careful to assess that interest rationally in light of the facts.

This view is called *Ethical Egoism:* "ethical" because it is a theory about morality, and "egoism" because it says that the sole duty of each of us is to maximize our own good. According to its proponents, moral values are reduced to concern for oneself (prudence), but always a "rational" concern requiring consideration of one's long-term interests (Rand, 1964).

Defenders of Ethical Egoism draw a distinction between narrower and wider forms of self-interest. To be selfishly preoccupied with one's own private good to the point of indifference and disregard for the good of others will generally cut one off from rewarding friendships and love. Thus the "paradox of happiness": To seek happiness by blinding oneself to other people's happiness leads to one's own unhappiness. Personal well-being generally requires taking some wider interest in others, although the rational egoist insists that the only reason for showing an interest in others is for the sake of oneself.

There is a problem with this last claim, however. Friendship and love seem to require—by definition—caring for other people at least in part for themselves, not from ulterior motives of self-concern. They require valuing other people for their sakes, and not solely because they serve one's personal ends.

Ethical egoists also try to defend their position by contending that an ironic consequence of everyone rationally pursuing their self-interest is that everyone benefits. For example, the classical economist Adam Smith and his contemporary defender Milton Friedman believe that society benefits most when (1) individuals pursue their private good, and (2) corporations (as expressions of many individual wills) pursue maximum profits in a competitive free market (Smith; Friedman, 1962). The idea is that this will make the economy prosper, thereby benefiting everyone, because each individual and corporation is in the best position to know what is good for them and how best to pursue that good.

We will return to the views of Smith and Friedman in Chap. 6. Here we express our strong doubt that private pursuit of self-interest, whether at the individual or the corporate level, always works out to everyone's advantage. To be sure, often it does, and often the requirements of morality and prudence point in the same direction. For example, the prudent employee and the morally conscientious engineer for the most part look alike in their conduct—but only for the most part. Morality requires a willingness on the part of both individuals and corporations to place some restraints on the pursuit of private interests. Acceptance of some such constraints is presupposed in what is meant by moral concern. Engineering ethics has as one task uncovering the moral limits on the pursuit of self-interest in the profession of engineering.

Of course, these remarks do not constitute a refutation of Ethical Egoism; and we shall not attempt a further refutation beyond making the following suggestion. At the very least, morality requires that we value and are concerned for the good of other people. (It also requires not being cruel to animals.) This means that Ethical Egoism is not really a plausible theory about what morality is, but instead a skeptical rejection of morality. It amounts to claiming that what are ordinarily viewed as moral reasons (for example, respecting other people's rights or caring about their well-being for their sake) should be disregarded except where they happen to coincide with looking out for one's own neck. "Number 1" is all that counts. Such a view denies the validity of moral reasons.

Laws and Ethical Conventionalism

A different challenge to the distinctiveness of moral values is the idea that morality reduces to law or to the customs and conventions of a society. According to this view, which is called *Ethical Conventionalism,* an act is morally right when it is approved by law or convention; it is wrong when it violates laws or customs.

Why would anyone believe this view? One reason is that laws seem so tangible and clear-cut. They provide a public way of cutting through seemingly endless disputes about right and wrong that at times seem little more than assertions of prejudice. Laws seem to be an "objective" way to approach values.

Ironically, a second rationale for this legalistic approach to morality points in the opposite direction. Ethical Conventionalism seems attractive to some people because it treats values as subjective at the cultural level. They insist that moral standards vary dramatically from culture to culture. The only kind of objectivity possible is limited to and "relative to" a given set of laws in a given society. Acknowledging this relativity of morality, they think, encourages tolerance of differences among societies.

Both of these arguments for Ethical Conventionalism are flawed in a manner that allows them to be turned on their heads. The first argument underestimates the extent to which moral reasons are objective so as to make them transcend individual prejudice and bias. Moral reasons, in fact, allow objective criticisms of given laws as immoral or morally inadequate. For example, moral reasons are used in criticizing the apartheid laws of South Africa, which flagrantly violate the human rights of the majority of black citizens. These human rights are not given legal protection, but they ought to be—*morally* ought to be. Apartheid laws represent precisely the harmful kind of subjectivity that morality helps us to overcome by calling for respect for all people, independent of race.

The second argument embodies two confusions. On the one hand it suggests that because laws, customs, and beliefs about morality differ from society to society it follows in effect that all of them are right and none of them are wrong. But there is nothing self-certifying about laws and beliefs. To use an extreme illustration, Ethical Conventionalism would allow that Hitler and his followers acted correctly when they murdered 6 million Jews, for their laws, customs, and beliefs were based on antisemitism.

This same illustration shows why Ethical Conventionalism is anything but the tolerant doctrine it claims to be. There is nothing tolerant—in any admirable sense—in refusing to criticize Nazi beliefs about morality. Sanctioning intolerant antisemitic beliefs is not an act of tolerance. It is true, as we shall emphasize in Chap. 8, that judgments about other cultures have to be based on understanding of and sensitivity to special cultural circumstances. But that is because those circumstances are objectively relevant to morality, and not because whatever a culture adopts as its laws or customs is automatically justified.

Defenders of Ethical Conventionalism generally add that an action is right "for cultures" that believe it is right—it is right "for them," though not "for us." Or moral beliefs are "true for" those cultures who hold them, though not true "for us." But these are needless shuffles. If the expressions "right for them" and "true for them" mean merely that those cultures believed something was right and that this belief played a key role in their lives (such as when we say of the ancients that the earth was "flat for them"), then this shaky view reduces to the truism that they believed what they believed. If, by contrast, it means their beliefs justified their actions, then we are left with the false claim that believing something to be right makes it right. Beliefs, however customary or widely shared, are not self-certifying. Nor, we might add, are attitudes concerning morality.

Religion and Divine Command Ethics

Moral reasons are not reducible to matters of self-interest, nor to law and custom. But how about religion? *Divine Command Ethics* is the view that to say an act is right means it is commanded by God, and to say it is wrong means it is forbidden by God. Accordingly, if there were no God to issue commands, then there would be no morality.

One difficulty raised by this view, of course, is how to know precisely what God's commands are. Another difficulty is knowing whether God exists. In fact, there are religions which do not emphasize belief in God, unlike the theistic (God-centered) religions like Judaism, Christianity, and Islam. Buddhism, Taoism, and Confucianism, for example, call for faith in a *right path* from which is derived a code of ethics. In Buddhism, for instance, the right path incorporates eight steps: right understanding, right intention, right speech, right action, right livelihood, right effort, right mindfulness, and right concentration. Part of the right path for many nontheistic religions is the need for contemplation and a feeling of oneness with nature in place of communion with a deity.

Questions about belief in God, however, are not the main difficulty with Divine Command Ethics. In fact, most theologians (such as St. Thomas Aquinas) reject the view for a different reason. This reason pertains to a question asked long ago by Socrates (Plato). Socrates asked, in effect: *Why* does God make certain commands and not others? Are the commands made on the basis of whim? Surely not, for God is supposed to be morally good and hence would neither approve of nor command such acts as wanton killing, rape, torture—and other immoralities.

Stated in another way, suppose that a man claimed that God commanded him to kill people randomly. (The "Son-of-Sam" murderer who randomly shot people on the streets of New York claimed this.) Without having to make any kind of religious inquiry, we would know the man was mistaken. Wanton killing is a clear-cut example of immorality and we know that a morally good deity would not command that kind of act (by definition of a "morally good deity").

It follows that Divine Command Ethics has things backwards. A morally good deity commands on the basis of moral reasons that determine the wrongness of certain kinds of actions and the rightness of others. Instead of the commands creating moral reasons, moral reasons are presupposed as the foundation for making certain commands rather than others.

Nothing in this argument against Divine Command Ethics should be viewed as a threat to religion. On the contrary, religious belief can be regarded as supporting morality by providing further motivation for being moral. We are not referring to self-interested motives like the fear of damnation or other types of punishment. Religious faith, or at least religious hope, implies trust: trust that we can receive insight into what should govern right action and that we can be sustained in that action. Hence it brings an added inspiration to be moral, even though many people are moral without having religious beliefs.

Consider in this connection the definition of religion given by C. J. Ducasse:

> A religion...is any set of articles of faith—together with the observances, attitudes, obligations, and feelings tied up therewith—which, in so far as it is influential in a person, tends to perform two functions, one social and the other personal. The social function is to provide motivation for the individual to conduct himself altruistically on occasions when his individual interest conflicts with that of society and when neither his spontaneous altruistic impulses, nor the sanctions of the laws or of public opinion, are potent enough by themselves or together to motivate such conduct. The personal function, on the other hand, is to give the individual in some measure the serene assurance out of which flows courage on occasions of fear, endurance in adversity, strength in moments of weakness, dignity in defeat, humility in success, conscientiousness and moderation in the exercise of power (Ducasse, 1953, 115).

Ducasse points out that the main social function of religion is to motivate right action, which involves the notion of ethics per se. Likewise the personal function of religion is very important. It has sustained many people in trying to follow their convictions, and it can promote tolerance and moral concern for others when those motivated by it are confronted with the wide variety of beliefs and individual needs to be found in the world. Many engineers are certainly among those so motivated, which is why these paragraphs on religion are an appropriate part of our larger topic, engineering ethics.

Minimal Conception of Morality

The discussion in this section has moved us toward what might be called a *minimal conception of morality* (Rachels, 1986). The conception is minimal in that it does not presuppose any one of the theories about right action to be discussed in the next section. Instead, it is the starting point for developing an ethical theory. That is, any sound ethical theory should presuppose it and in part be about it.

According to the minimal conception as it pertains to actions, morality concerns *reasons* for the desirability of certain kinds of actions and the undesirability of others. To say that an act is right is not to express a mere feeling or bias, but instead to assert that the best moral reasons support doing it.

What is a moral reason? It is a reason which requires us to respect other people, to care for their good as well as our own. In addition, moral reasons are such that they set limits to the legitimate pursuit of self-interest. They can be used to evaluate laws, to praise some and criticize others. They are not reducible to religious matters, although religious belief may provide an additional motivation for responding to them.

Study Questions

1 What is the moral dilemma involved in each of the following situations? Identify the moral obligations which come into conflict. How do you think the dilemma should be resolved? Would an ethical egoist and an ethical conventionalist agree with you about how to resolve the dilemma?

a "Bill, a process engineer, learns from a former classmate who is now an OSHA regional compliance officer that there will be an unannounced inspection of Bill's plant. Bill believes that unsafe practices are often tolerated in the plant, especially in the handling of toxic chemicals. Although there have been small spills, no serious accidents have occurred in the plant during the past few years. What should Bill do?" (Matley, Greene, McCauley, 1987, 115)

b "On a midnight shift, a botched solution of sodium cyanide, a reactant in an organic synthesis, is temporarily stored in drums for reprocessing. Two weeks later, the day shift foreperson cannot find the drums. Roy, the plant manager, finds out that the batch has been illegally dumped into the sanitary sewer. He severely disciplines the night shift foreperson. Upon making discreet inquires, he finds out that no apparent harm has resulted from the dumping" (Matley, Greene, McCauley, 1987, 117). Should Roy inform government authorities, as is required by law in this kind of situation?

2 Interview someone involved in engineering, preferably an engineer. Ask them to describe one or more moral problems they have confronted. Is the problem they describe a moral dilemma? If so, explain which competing moral obligations, ideals, or principles are involved and how you would have resolved it.

3 Locate and read a complete work of fiction in which the subject matter is related to engineering and in which the plot involves a moral dilemma. Then write an essay in which you (*a*) discuss the origin and nature of the dilemma, and (*b*) explain whether or not you agree with how the person involved resolved the dilemma, giving reasons for your opinion.

You might consider the following works: Pierre Boulle, *The Bridge over the River Kwai;* Eugene Burdick and Harvey Wheeler, *Fail-Safe;* William Golding, *The Spire;* Henrik Ibsen, *The Master Builder;* John D. MacDonald, *Condominium;* Louis V. McIntyre and Marion B. McIntyre, *Scientists and Engineers: The Professionals Who Are Not;* Nevil Shute, *No Highway;* Kurt Vonnegut, Jr., *Player Piano;* Burton Wohl, *The China Syndrome.* Additional possibilities are given in Samuel C. Florman's *The Existential Pleasures of Engineering.*

THEORIES ABOUT MORALITY

Moral conduct is based on concern for other people; it is not reducible to self-interest, law, or religion. But more precisely, what makes some actions morally right and others wrong? What are the most basic reasons why we morally ought or ought not to do certain things?

Four Types of Moral Theories

More than two millenia of philosophical reflection since Socrates have not led to a consensus about how to answer these questions. Nevertheless, there is widespread agreement that there are four main types of theories about morality. These theories differ according to what they treat as the most fundamental moral concept: good consequences for all, duties, human rights, or virtue.

As an illustration to help introduce these theories, return to the kickback scheme described at the beginning of this chapter. We assumed it was clear that the actions of the participants were unethical. But what reasons can be given to support this assumption? *Why* was it wrong for the engineers to make secret payments to Spiro Agnew in return for being given preference in the awarding of contracts for public projects?

One answer is that more bad than good resulted. Other engineering firms were harmed by not having a chance to obtain the contracts they may have been best qualified to receive. The system also removed the potential benefits of healthy competition among a wider range of firms, benefits such as lower costs and better products for the public. Equally significant, discovery of the scheme led to a loss of trust in public officials, a trust important for the well-functioning of government. And the perpetrators themselves eventually suffered greatly.

Let us define *utility* as the overall balance of good over bad consequences. *High utility* will usually mean much good and little bad (although it can also mean the lesser of two evils). *Utilitarianism* holds that we ought always to produce the most utility, taking into equal account everyone affected by our actions. Good and bad consequences are the only relevant moral considerations, and hence all moral principles reduce to one: "We ought to maximize utility."

A different answer to what was wrong with engaging in the kickback scheme would have us focus directly on the actions involved, rather than their consequences. The actions were intended to keep outsiders deceived about what was going on. They were also inherently unfair to other people who were denied equality of opportunity to bid for the contracts. Hence the actions, irrespective of their actual or probable consequences, violated at least two basic principles of duty: "Avoid deceiving others" and "Be fair." *Duty ethics* asserts there are duties like these which ought to be performed even though doing so may not always produce the most good.

Yet another answer to why it was wrong to participate in the kickback scheme is that it violated the rights of other people. A shared understanding exists that there will be equality of opportunity in seeking public contracts and that elected officials will grant contracts based on merit, not bribes. Against this background, qualified persons or firms acquire a right to unbiased consideration of their contract proposals, and these rights were violated by the kickback scheme. It might also be argued that the public's rights to the benefits of fair competition were violated as well.

Rights ethics views actions as wrong when they violate moral rights. Like duty ethics, it denies that good consequences are the only moral consideration. But rights ethics says we have duties to other people *because* people have rights that ought to be respected, whereas duty ethics says rights are created by duties.

A very different answer to why it was wrong to enter into the kickback scheme makes reference to virtues and vices, that is, to good and bad traits of character. Agnew displayed unfairness, dishonesty, and greed—that is the kind of person he showed himself to be. Matz and Childs displayed moral weakness, deceptiveness, dishonesty, and perhaps cowardice in the face of temptation. Morally better people would have manifested virtues such as courage, honesty, fairness, and conscientiousness.

Virtue ethics regards actions as wrong insofar as they manifest bad character traits (vices) and right insofar as they display or support good character traits (virtues). Here the fundamental concept is a morally good person, rather than right actions. Virtue ethicists are primarily interested in what kind of people we ought to be, to emulate, and to inspire others to become. Right actions are simply those which express, build, or reinforce virtues.

The following table lists the four main types of theories about morality. In the remainder of this section we will discuss one classical and one contemporary defender of each of the first three types introduced: utilitarianism, duty ethics, and rights ethics. In a later section virtue ethics will be discussed in more detail.

Theory about morality	Basic concepts
Utilitarianism	Most good for the most people
Duty ethics	Duties
Rights ethics	Human rights
Virtue ethics	Virtues and vices

Mill: Act-Utilitarianism and Happiness

Utilitarianism is the view that we ought to produce the most good for the most people, giving equal consideration to everyone affected. The standard of right conduct is maximization of goodness. At first glance, this seems simple enough. But what is the goodness that is to be maximized? And how is the "production" of goodness related to everyday moral rules? Depending on how these questions are answered, utilitarianism can be developed in different directions.

Act-utilitarianism says we should focus on individual actions, rather than general rules. An act is right if it is likely to produce the most good for the most people involved in the particular situation. Everyday maxims like "Keep your promises," "Don't deceive," and "Don't bribe" are only rough guidelines. According to John Stuart Mill (1806–1873), these maxims are useful rules of thumb that summarize past human experience about the types of actions which usually maximize utility (Mill). But the rules should be broken whenever doing so will produce the most good in a specific situation.

If the standard of right action is maximizing goodness, what is goodness? Mill believed that happiness is the only *intrinsic good,* that is, something good in-and-of-itself or desirable for its own sake. All other good things are *instrumental goods* in that they provide means ("instruments") for happiness. A trip to the dentist, for example, is an instrumental good that promotes happiness by avoiding or removing the pain of toothache.

In Mill's view, a happy life is comprised of many pleasures in great variety, mixed with some inevitable brief pains. The happiest life is also rich in *higher pleasures.* Higher pleasures are preferable in quality or *in kind* to other pleasures. For example, Mill contended that the pleasures derived through intellectual inquiry, creative accomplishment, appreciation of beauty, and friendship and love are inherently better than the bodily pleasures derived from eating, sex, and exercise.

How did Mill know these are the "higher" pleasures? He offered the following test: One kind of pleasure is preferable to another if the majority of people who have experienced both kinds favor it. Using this test, it is no surprise that he and his contemporaries living in the Victorian Age should rank pleasures of the mind and of personal relationships over those of the body. Contemporary utilitarians often reject this aspect of Mill's thought as biased.

Brandt: Rule-Utilitarianism and Rational Desires

Rule-utilitarianism, which is the second main version of utilitarianism, regards moral rules as primary. According to it, we ought always to act on those rules which if generally followed would produce the most good for the most people. Individual actions are right when they conform to such rules. Thus, we ought to keep promises and avoid bribes, even when those acts do not have the best consequences in a particular situation, because the general practices of promising and not bribing produce the most overall good (compared to other practices).

Richard Brandt is an influential contemporary rule-utilitarian. Brandt believes that rules should be considered in sets which he calls *moral codes* (Brandt). A moral code is justified when it is the *optimal code* which (if adopted and followed) would maximize the public good more than alternative codes would. The codes may be society-wide standards or special codes for a profession like engineering.

There are debates over precisely how much act- and rule-utilitarianism differ from each other. Yet they do seem to lead to different conclusions in some situations. Rule-utilitarianism, for example, gives a more straightforward condemnation of participation in kickback schemes. Matz and Childs acted on a rule something like "Engage in secret payoffs when necessary for profitable business ventures." If this rule were generally followed, it would cause a breakdown of trust between business people and their clients. Again, general adherence to Agnew's principle of action, which was something like "Break the law when you can personally profit from doing so," would produce a mentality that would have devastating consequences.

Act-utilitarianism, by contrast, leaves it open whether participation in some kickback schemes may produce overall good. It all depends on the particular context: who is hurt and how much, and what are the chances of being caught? Because act-utilitarianism seems to open "loopholes" licensing unfair exceptions, many utilitarians have abandoned it in favor of rule-utilitarianism.

Many contemporary utilitarians also disagree with Mill's view that happiness is the only intrinsically good thing. They often regard friendship, love, understanding, and appreciation of beauty as intrinsically good, even when they do not lead to happiness.

Brandt, however, believes that such things are good because they satisfy rational desires. *Rational desires* are those we would have and approve of if we scrutinized our desires in light of all relevant information about the world and our own psychology. Some self-destructive desires, such as the desire to use dangerous drugs, are not rational since if we saw their full implications we would not approve of them.

Still other utilitarians, especially economists, are concerned with difficulties about how to identify and measure desires and the pleasures they yield. They seek an objective way to determine the good. Economists base their cost-benefit analyses on the preferences that people express through their buying habits. In this version, utilitarianism becomes that view that right actions are those producing the greatest satisfaction of the preferences of people affected.

Kant: Duties and Respect for Persons

Immanuel Kant (1724–1804) is the most famous of the ethicists who regard duties, rather than good consequences, as fundamental. In his view, right actions are those required by a list of duties such as: be honest, keep your promises, don't inflict suffering on other people, be fair, make reparation when you have been unfair, show gratitude for kindness extended by others.

There are also duties to ourselves: seek to improve one's own intelligence and character, develop one's talents, don't commit suicide.

Why are these our duties? According to Kant, it is because they meet three conditions: each expresses respect for persons, each expresses an unqualified command for autonomous moral agents, and each is a universal principle. We will now examine Kant's three conditions in more detail.

First, in contrast to Mill, who said happiness is the only intrinsic good, Kant valued the *good will:* the intention to do one's duty. Thus, Kant greatly valued the honest and conscientious effort to fulfill duties. Moreover, people also have inherent worth as rational beings insofar as they have the capacity for a good will. This capacity makes people worthy of respect. To respect people is to seek to fulfill our duties to them, and to respect oneself is to seek to fulfill our duties to ourselves.

This sounds rather abstract until we look at examples of what happens when duties are disregarded. Consider the deceiving, bribing, and coercing involved in the Agnew case. These activities are various forms of manipulation. They constitute treating people as means to one's own ends, rather than as rational beings who have purposes of their own (or who are "ends-in-themselves," to use Kant's expression). Violent acts such as murder, rape, and torture are even more flagrant ways of treating people as mere objects serving our own purposes.

Second, duties prescribe certain actions categorically, without qualifications or conditions attached. Duties are *categorical imperatives.* These commands are best understood by contrasting them with nonmoral commands which Kant called *hypothetical imperatives.* Hypothetical imperatives command on the basis of some condition or "hypothesis." For example, "If you desire to become healthier, then stop overeating" and "If you want to be happy, you ought to enrich your life by developing friendships." Another example is the mugger with a gun who commands, "Your money or your life." Here there is an implicit condition: "If you want to avoid being killed, then hand over your money."

Moral imperatives are different in that they have no such conditions attached. They require us to do certain things whether we want to or not. Thus, we ought to avoid cheating and other forms of dishonesty simply because we ought to—period! It is our duty, independently of whether it will make us happy. Stated another way, duty should be followed because of our autonomous commitment to morality itself, rather than because of ulterior motives.

Third, categorical imperatives are binding on us only if they are also applicable to everyone. That is, moral reasons and principles are those which we are willing to have everyone act upon and which we can conceive of all people heeding. In this sense they must be *universalizable.*

Most everyday moral rules pass this test. For example, we can imagine and favor having everyone obey the command "Keep promises." By contrast, we cannot imagine all people obeying the command "Keep promises except when you don't feel like it." If everyone did that, promises would no longer be possible. Whenever someone attempted to give their solemn word to us by uttering the words "I promise," we would merely laugh. Serious promises are understood as not being subject to the whims of those who seek to gain advantages for themselves. Thus we are not *willing* to encourage people to break promises whenever they find it convenient, and we become caught in a *contradiction* when we try to imagine a situation in which everyone acted that way.

The kickback scheme provides a second illustration. Consider the principle "Engage in secret kickback schemes whenever you can profit and get away with it." If everyone followed this principle, people would no longer be able to make legal business contracts at all. Contracts are possible because of an underlying basis of trust among at least most participants. Whereas utilitarians objected to kickbacks because of their actual or probable consequences, Kant says they are wrong because they cannot be willed to be universal principles applying to all rational beings.

Prima Facie Duties

One difficulty with Kant's view is that he thought principles of duty were *absolute* in the sense of never having justifiable exceptions. He failed to be sensitive to how principles of duty can conflict with each other, thereby creating moral dilemmas. Contemporary duty ethicists recognize that some moral dilemmas are resolvable only by making exceptions to simple principles of duty. Thus, "Do not deceive" is a duty, but it has exceptions when it conflicts with the moral principle "Protect innocent life." One ought to deceive a kidnapper if that is the only way to keep a hostage alive until the police can intervene. Principles of duty that have exceptions are called *prima facie duties* (W. D. Ross, 1946). Most duties are in fact prima facie ones.

How do we tell which duties should override others when they come into conflict? Some recent duty ethicists emphasize the importance of careful reflection on each situation, weighing all relevant duties in light of all the facts, and trying to arrive at a sound judgment or intuition. They also stress that some principles, such as "Do not kill" and "Protect innocent life," clearly involve more pressing kinds of respect for persons than other principles, such as "Keep promises." Other duty ethicists, like John Rawls, have tried to formulate general principles that can be ranked in order of importance without having to rely on intuitive judgments.

Duty Ethics: Rawls's Two Principles

John Rawls is a leading contemporary ethicist who has developed Kant's ideas in fresh directions. According to Rawls, valid principles of duty are those which would be voluntarily agreed upon by all rational persons in an imaginary "contracting" situation.

The persons in this hypothetical situation are characterized by several features:

1 They lack all specific knowledge about themselves—for example, about their particular desires, intelligence, and achievements. This ensures they will not be biased by self-interest in their deliberations.
2 They do have general knowledge about human psychology, the economics and politics of society, and science.
3 They have a rational concern for promoting their long-term interests.
4 They seek to agree with each other about the principles they will voluntarily and autonomously follow as a group. That is, they form a moral agreement or contract to abide by principles to which they all subscribe.

Rawls believes that placing ourselves (in imagination) in this hypothetical contracting situation helps us to reason more easily and honestly about moral principles. It enables us to check our intuitions and to set aside our biases. His view is Kantian in that it emphasizes the autonomy each person exercises in forming hypothetical agreements with other rational people.

All rational people, Rawls argues, will agree in this hypothetical situation to abide by two basic moral principles applicable to societies and social institutions like professions: (1) Each person is entitled to the most extensive amount of liberty compatible with an equal amount for others. (2) Differences in social power and economic benefits are justified only when they are likely to benefit everyone, including members of the most disadvantaged groups (Rawls, 1971, 60).

The first principle is most important and should be satisfied first. Without basic liberties no other economic or social benefits can be sustained in the long run. The second principle is also very important, however. It insists that allowing some people great wealth and power is justified only when all other groups benefit. Thus, it might be argued that allowing differences of this sort within the free enterprise system is permissible insofar as it provides the capital needed for businesses to prosper, thereby providing job opportunities and taxes to fund a welfare system to help the poor.

Locke: Liberty Rights

The third type of ethical theory, *human rights ethics,* is familiar and can be introduced more briefly. Human rights ethicists assert that duties arise because people have rights, not vice versa. For example, individuals do not have rights to life because others have duties not to kill them. Instead, possessing the right to life is the reason why others ought not to kill them.

John Locke (1632–1704) argued that to be a person entails having rights—human rights—to life, liberty, and the property generated by one's labor. His views had an enormous impact at the time of the French and American revolutions. The words in the Declaration of Independence are not far from his own: "We hold these truths to be self-evident; that all men are created equal; that they are endowed by their creator with inherent and inalienable rights; that among these are life, liberty, and the pursuit of happiness."

Locke's own version of a human rights ethics was highly individualistic. He viewed rights primarily as entitlements that prevent other people from meddling in one's life. These are referred to as *liberty rights* or *negative rights* that place duties on other people *not* to interfere with one's life.

This aspect of Locke's thought is reflected on the contemporary political scene in the libertarian ideology, with its emphasis on protection of private property and the condemnation of welfare systems (Nozick, 1974). Libertarians take a harsh view of taxes and government involvement beyond the bare minimum necessary for national defense and preservation of free enterprise. This perspective contrasts sharply with Rawls's concern for the disadvantaged members of society. It also contrasts with a second version of human rights ethics.

Melden: Welfare Rights

This second version of rights ethics conceives of human rights as intimately related to communities of people. A contemporary philosopher, A. I. Melden, has argued that having moral rights presupposes the capacity to show concern for others and to be accountable within a moral community (Melden, 1977). The extent of rights, in his view, always has to be determined in terms of interrelationships among persons. Melden's account allows for more "positive" *welfare rights,* which he defined as rights to community benefits needed for living a minimally decent human life. Thus it lays the groundwork for recognizing a social welfare system such as the United States currently has.

Not all moral rights are human rights. Some arise from special relationships and roles which people might have. A promise, for example, gives rise to the special right to have the promise kept. But rights ethicists seek to justify special rights by reference to human rights. Thus, according to Melden, promises create special rights because people have human rights to liberty, and because breaking a promise is a way of interfering with the liberty of the person to whom one has committed one's help by making a promise.

Many of the rights we will examine later arise within institutions and professions, such as the right of engineers to warn the public about unsafe technological products. And there are rights of all participants in competitive situations to be treated fairly, rights which were violated in the Agnew kickback scheme. Later we shall see how basic human rights can be used as a basis for some of these special rights.

Testing Ethical Theories

Our intent is not to evaluate which of these theories is best. In fact, we believe that each of them has insights to offer, and we are more impressed by how they complement each other than by how they differ. For example, Kant and Locke disagree sharply over whether duties or rights are most fundamental. But we are more interested in how for every duty there is a corresponding right, and vice versa. Thus, if you have a right to life, then I have a duty not to kill you; and if I have a duty to respect your freedom, then you have a right not to be interfered with. It follows that for practical purposes it matters little whether we adopt duties or rights as the starting point for moral reflection.

Again, rights ethics, duty ethics, and rule-utilitarianism for the most part all agree about the general principles we ought to follow (even though they give different justifications of those principles). As authors, we have reservations about act-utilitarianism as a sound moral theory. But we are confident that rule-utilitarianism, duty ethics, rights ethics, and virtue ethics all capture essential elements of sound moral reflection, and that for the purposes of engineering ethics all of them converge toward similar conclusions.

We have already seen this illustrated as we looked at the the Agnew kickback scheme. With the possible exception of act-utilitarianism (which left some loopholes for engaging in some such schemes), all the main ethical theories gave compelling and interrelated reasons for not participating in the scheme.

Perhaps someday an even more comprehensive moral theory will be developed that will reveal how all the theories are connected and have elements of truth. Yet even if there are ultimate moral disagreements that make such a unified theory impossible, there remain enough broad similarities between the existing theories to warrant invoking all of them as aids to practical moral reflection.

In what follows, therefore, we will draw freely on the language of duties, rights, utility, and virtue, wherever it aids practical reflections on moral dilemmas in engineering. Yet it is worth mentioning five widely used tests for evaluating ethical theories.

First, the theory must be clear and formulated with concepts that are coherent and applicable.

Second, it must be internally consistent in that none of its tenets contradicts any other.

Third, neither the theory nor its defense can rely upon false information.

Fourth, it must be sufficiently comprehensive to provide guidance in specific situations of interest to us.

Fifth, and perhaps most important, it must be compatible with our most carefully considered moral convictions (judgments, intuitions) about concrete situations. If an abstract ethical theory said it was all right to torture mentally handicapped children to make other people happy, that would be enough to show the theory was false.

Good theories, of course, may lead us upon reflection to modify some of our previously held views—one of their main uses is to correct mistaken judgments. In this way theories and concrete intuitions mutually interact, each serving as a test for the other. Ethical theories are developed to illuminate, unify, and correct commonsense judgments; and refined commonsense judgments about specific situations are used to test ethical theories (Rawls, 1971, 46–53).

Study Questions

1 Apply utilitarianism, duty ethics, and rights ethics in resolving the following moral problems. Be sure to consider alternative versions of each theory, such as act-utilitarianism and rule-utilitarianism. Do the theories lead to the same or different answers to the problems?

a A train is approaching a switch, and it is traveling too fast to stop before a tragedy occurs. Tied to one fork of the track are the leaders of three important nations (who are vital to current efforts to achieve world peace and prosperity). Tied to the other fork are four people who are your closest friends and relatives, but who have no international or even national social importance. If you were in control of the switch, which fork in the track ought you to select?

b A doctor can save the lives of three important national leaders by making transplants of the kidneys and heart of a local convicted mass murderer who is serving a life sentence. The operations would be done in secret and would involve the full cooperation of the local police officials, who would claim the murderer was killed while trying to escape from prison. Is it morally permissible (i.e., all right) to make the transplants?

c George had a bad reaction to an illegal drug he accepted from friends at a party. He calls in sick the day after, and when he returns to work the following day he looks ill. His supervisor asks him why he is not feeling well. Is it morally permissible for George to lie by telling his supervisor that he had a bad reaction to some medicine his doctor prescribed for him?

d Jillian was aware of a recent company memo reminding employees that office supplies were for use at work only. Yet she knew that most of the other engineers in her division thought nothing about occasionally taking home notepads, pens, computer discs, and other office "incidentals." Her 8-year-old daughter had asked her for a company-inscribed ledger like the one she saw her carrying. The ledger costs less than $20, and Jillian recalls that she has probably used that much from her personal stationery supplies during the past year for work purposes. Is it all right for her to take home a ledger for her daughter without asking her supervisor for permission?

e Robert is a third-year engineering student who has been placed on probation for a low grade-point average, even though he knows he is doing the best work he can. A friend offers to help him by sitting next to him and "sharing" his answers during the next exam. Robert has never cheated on an exam before, but this time he is desperate. Should he accept his friend's offer?

f Because he had been mugged before, Bernard Goetz (who happened to be an engineer) illegally carried a concealed revolver when he rode the New York subway. When several young men confronted him in a threatening way, asking for money, he drew the revolver and fired several shots that resulted in permanent injuries. Did his right to life and his right to defend himself justify his acts of (i) carrying the revolver and (ii) using it as he did?

2 Find, in a current newspaper or magazine, an article which raises a moral issue in engineering. State the issue in your own words, making clear why you view it as a moral one. Explain how the problem might be approached and resolved by drawing on utilitarianism, duty ethics, and rights ethics.

3 Consult the writings of a major ethicist and summarize the main ideas of the theories involved. For suggested sources see the bibliographical entries on Aristotle, Brandt, Kant, Melden, Mill, Oldenquist (1979), Rawls, and MacIntyre. For helpful secondary sources, see Frankena (1973), Rachels, and Taylor.

ETHICAL THEORY AND SAFETY OBLIGATIONS

Ethical theories have two main applications to engineering ethics. First, they help us to deal with practical moral problems, especially moral dilemmas. Second, they can be used to justify the general obligations of engineers and others involved in technological development. Using safety-related problems and obligations as examples, we will illustrate the applications of utilitarianism, rights ethics, and duty ethics, postponing virtue ethics to the next section of this chapter (see under "Responsibility and Virtue Ethics").

The DC-10 Case

In 1974 the first crash of a fully loaded DC-10 jumbo jet occurred over the suburbs of Paris; 346 people were killed, a record for a single-plane crash. It was known in advance that the crash was bound to occur because of the jet's defective design (Eddy, 1976; Godson, 1975).

The fuselage of the plane was developed by Convair, a subcontractor for McDonnell-Douglas. Two years earlier Convair's senior engineer directing the project, Dan Applegate, had written a memo to the vice president of the company itemizing the dangers that could result from the design. He accurately detailed several ways the cargo doors could burst open during flight, depressurize the cargo space, and thereby collapse the floor of the passenger cabin above. Since control lines ran along the cabin floor, this would mean a loss of control of the plane. Applegate recommended redesigning the doors and strengthening the cabin floor. Without such changes, he stated, it was inevitable that some DC-10 cargo doors would open in midair, resulting in crashes.

In responding to this memo, top management at Convair disputed neither the technical facts cited by Applegate nor his predictions. Company officers maintained, however, that the possible financial liabilities Convair might incur prohibited them from passing on this information to McDonnell-

Douglas. These liabilities could be severe since the cost of grounding the planes to make safety improvements would be very high and come at a time when McDonnell-Douglas would be placed at a competitive disadvantage (Newhouse, 1982).

It might be argued that as a loyal employee Applegate had an obligation to follow company directives, at least reasonable ones. Perhaps he also had family obligations which made it important for him not to jeopardize his job. Yet as an engineer he was obligated to protect the safety of those who would use or be affected by the products he designed. Thus the dilemma he confronted involved a clash between at least two general professional obligations—one to his employer and one to the public—and possibly a clash between professional and personal obligations as well.

While this is an extreme case, it nevertheless illustrates a common class of moral dilemmas in engineering. Given that the vast majority of engineers are salaried employees, it is very likely that duties to employers will on occasion conflict with duties to the public. For this reason we will take all of Chap. 5 to examine the relations between these kinds of obligations. Later we shall also comment on some of the managerial and regulatory voids revealed after the crash over Paris.

Steps in Confronting Moral Dilemmas

In approaching dilemmas like the one Dan Applegate had to face, several steps are important. The steps are distinct, even though they are interrelated and can often be taken in tandem.

1 Identify the relevant moral factors and reasons. What are the clashing duties, competing rights, alternative goods and bads, and virtues and vices involved?

2 Gather all available facts that are pertinent to the moral factors involved.

3 If possible, rank the moral considerations in order of importance as they apply to the situation.

4 Consider alternative courses of action as ways of resolving the dilemma, tracing the full implications of each.

5 Talk with colleagues (or friends or other students), seeking their suggestions and alternative perspectives on the dilemma.

6 Arrive at a carefully reasoned judgment by weighing all the relevant moral factors and reasons in light of the facts.

Ethical theories cannot be expected to provide simple resolutions of complex dilemmas. They are not moral algorithms that can be mechanically applied to remove perplexity. But they can help by providing frameworks for understanding and reflecting upon dilemmas. In fact, they can be useful at each of the above steps.

1 Ethical theories aid in identifying the moral considerations or reasons which constitute the dilemma. Thus, utilitarianism construes the Applegate dilemma in terms of competing goods: the safety of the public versus the economic benefits to Convair and the personal benefits to Applegate. Duty ethics indicates that he had competing duties to protect the public affected by his work and to respect his employer's legitimate authority to make management-level decisions about expenditures. And rights ethics emphasizes the rights of the public to be protected (or at least warned of dangers) and the rights of management to have their decisions respected.

2 Ethical theories provide a more precise sense of what kinds of information are relevant. All the theories, for example, agree that facts about the potential harm to the public are directly and urgently relevant. It would be improper to consider only the benefits to Convair and to Applegate in reaching a decision about the dilemma.

3 Sometimes the theories offer ways to rank the relevant moral considerations in order of importance. We shall argue in a moment that the theories suggest a priority of the obligation to protect the public, given (i) the special importance of rights to life and to informed consent concerning risks to one's life, (ii) the importance of duties to protect the vulnerable public, and (iii) the degree of badness involved in death and risk of death compared to economic benefits to corporations.

4 The theories help us identify the full moral ramifications of alternative courses of action, urging a wide perspective on the moral implications of the options, and providing a systematic framework for comparing the alternatives.

5 The theories augment the precision with which we use moral terms, and they provide frameworks for moral reasoning when discussing moral issues with colleagues.

6 By providing frameworks for development of moral arguments, the theories strengthen our ability to reach balanced and insightful judgments.

Foundations of Professional Obligations: Safety

The second use of ethical theories is in justifying the general obligations of engineers and others involved in technological development. We will illustrate this by asking, "Why do engineers have obligations to protect the safety of the public affected by their products and projects?" What reasons or justification can be given for our earlier claims that engineers have these obligations?

This question has wide relevance to engineering ethics, beyond its connection with the DC-10 example. In one way or another, safety is involved in most of the thorny issues in engineering ethics. Certainly it is the most pressing consideration in most situations involving whistle-blowing, confidentiality, and the exercise of professional autonomy. In fact, it is perhaps only a slight exaggeration to say that engineering ethics takes as its primary focus the promotion of safety while bringing useful technological products to the public, whereas medical ethics centers on the professional's role in promoting health within the bounds of patient autonomy, and legal ethics centers on the advocacy of clients' rights within the bounds set by law.

An architectural metaphor may help orient the reader to the idea of justifying the safety obligations of engineers. In the tower shown in Fig. 2-1, each of the four main stages or girders represents a type of moral claim. Girder 4 at the top represents claims about particular actions being right or obligatory.

FIGURE 2-1
Justifying moral claims about safety in engineering.

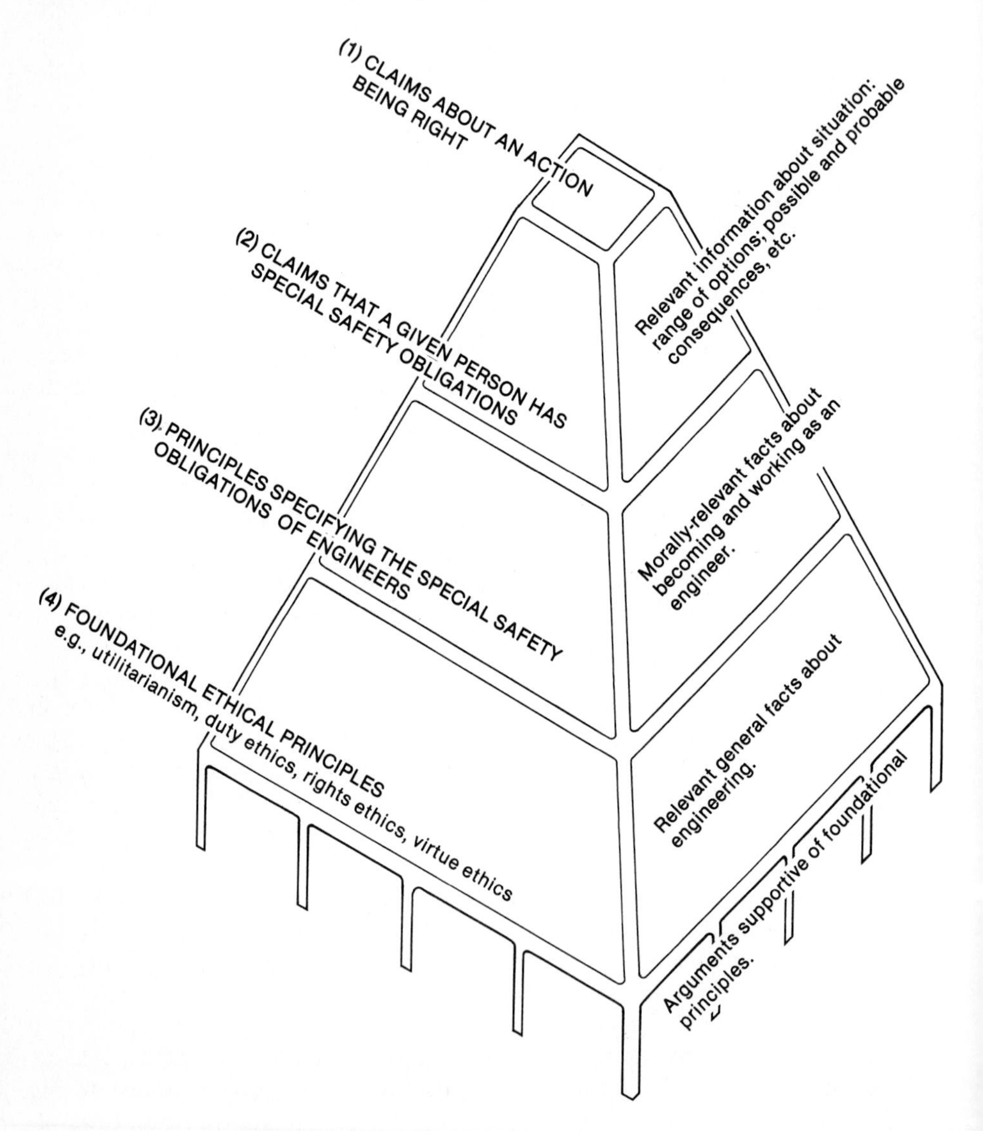

The next beam down symbolizes claims that specific engineers have special moral obligations concerning safety. Girder 2 represents the special safety principles themselves. Candidates for inclusion here would be items appearing in engineering codes of ethics, such as "Engineers shall hold paramount the safety, health and welfare of the public in the performance of their duties" and "Should the Engineers' professional judgment be overruled under circumstances where the safety, health, and welfare of the public are endangered, the Engineers shall inform their clients or employers of the possible consequences and notify other proper authority of the situation, as may be appropriate." (These are only "candidates," since they must be established as justified in order to be included.) The lowest stage, girder 1, is concerned with the most general and basic foundational or philosophical principles.

The columns connecting the girders represents the morally relevant information needed to move from one type of normative claim to another. For example, consider the move from level 3 to level 4. Suppose we agree that an engineer has an obligation to inform the proper authority of serious dangers to the public. In order to know how this obligation should be carried out correctly, then, we need to know who in the particular situation constitutes the proper authority and how that authority should be notified—by an anonymous phone call or by a formal memo delivered via certified mail?

The move from level 1 to level 2 consists in deriving the special obligations of engineers concerning safety from general ethical theory and the relevant facts about engineers' work.

Under *act-utilitarianism,* the special safety obligations of engineers would ultimately reduce to one general obligation: to act in each situation so as to maximize the good consequences for everyone affected by engineering projects and products. *Rule-utilitarianism,* on the other hand, would have engineers act according to those rules which if widely followed would produce the best consequences for everyone affected. *Duty ethics* would ground the obligations of engineers in one or more basic principles of duty. And *rights ethics* would emphasize how engineers' safety obligations are based on the requirement that professionals respect the moral rights of those affected by their work.

Rather than elaborating on each of these approaches, we will select the rights theory for further comment because of its special relevance to the perspective on engineers' responsibilities developed in Part II of this text.

A rights theory begins with the assumption that every person has an inherent right as a human being to pursue his or her legitimate interests, i.e., interests not harming others (Melden, 1977).

Does this imply an unqualified right not to be harmed by technological products? No. If people purchase hanggliders and then kill themselves by flying them carelessly or under unfortunate weather conditions, their rights have not been violated—so long as advertisements about the joys of hanggliding did not contain misleading information. But the basic right does imply a right not to be poisoned, maimed, or killed by technological products whose dangers are not obvious or are deliberately hidden. This in turn implies a right to informed consent when purchasing or using products or services that might be dangerous (e.g., buying an airplane ticket). We might think of this as a right to make an "informed purchase."

These rights place on those creating products, or engaging in projects, correlative obligations to acquire and transfer relevant safety information to consumers and others affected by the products or projects. The nature of such obligations, in general outline, will be shaped by the rights. Thus there is a direct link between basic human rights and the safety obligations of engineers, both in regard to what those obligations are and how they are acquired. And laws, professional codes, and employment agreements are important to that linkage insofar as they help ensure that the safety obligations are fulfilled. These issues will arise again for discussion in later chapters.

Professional Ethics and Ordinary Morality

As the above illustration suggests, the special obligations concerning safety that engineers acquire as a consequence of their work are intimately connected with ordinary or everyday morality. The same ethical theories that are useful in expressing everyday moral experience are also useful in justifying the obligations of professionals.

To underscore this fact, consider four other views concerning the origin and justification of the safety (and other) obligations of engineers.

1 The first view is that engineers acquire moral obligations concerning safety by being subject to laws or enforced codes that require them to be so obligated. This constitutes a legalistic approach to morality, that is, an attempt to model morality on the law or reduce it to legal and paralegal considerations. When generalized to all morality, it is a version of what we referred to earlier as Ethical Conventionalism: the doctrine that morality is nothing more than the dominant conventions, mores, or laws operating within a given society. And we reject this view, just as we rejected Ethical Conventionalism.

2 The second view is that engineers acquire special obligations by joining a professional society and thereby agreeing to live by that society's code of ethics.

This view differs greatly from the first by emphasizing the voluntary act of agreeing to abide by a code rather than passively being subject to enforced laws and codes. Like the next two views to be discussed, it places the origin of engineers' obligations to safety in a personal commitment to act according to principles implying ethical obligation. Thus it ties directly into our ordinary understanding of how promises and other self-committing acts generate obligations. Yet it is also doubly insufficient. What of the many engineers who choose not to join professional societies? Surely they are not released from the responsibility to meet obligations to ensure safety. But if the failure to join a society does not remove such obligations, then the act of joining cannot be the sole or main way they are acquired. Moreover, there is always the question of whether what is promised is ethical or not. Thus, if the code of a professional society happened to contain morally harmful entries—such as one restricting responsible criticism of other engineers' safety judgments—the promise to obey that code would either be nullified or overridden by other moral considerations.

3 The third view is that engineers acquire safety obligations through the contractual agreements by which they are hired by their companies or employers. Here we may agree immediately that *some* special safety obligations are acquired in this way. An engineer who is hired as a safety inspector surely does acquire special work responsibilities related to safety. And those responsibilities cannot be reduced merely to prudential concerns, i.e., reduced to the attitude that not meeting them would lead to the loss of one's job or a promotion. Rather they entail specific commitments to fulfill such job-related duties, and this means *moral* obligations have been generated.

Yet even explicit obligations to employers cannot be the sole basis for the safety obligations of engineers. For no engineer is obligated by her or his employee status to sacrifice safety by following an unscrupulous employer's directives to lie, cheat, forge, or directly risk innocent lives by producing or approving shoddy designs and constructions. It is not empty rhetoric, or at least it should not be, to insist that general safety obligations to the public can override obligations to employers. Rather it reflects the point made earlier that specific promises and voluntary commitments sometimes must give way to wider obligations.

4 A fourth view is that engineers, upon entering their careers, made a broad, tacit promise to the public to protect and safeguard it in the course of performing their tasks. In return the public has largely underwritten their education through financial support for schools and implicitly granted the profession as a whole certain privileges. For example, it allows professional societies to accredit schools of engineering, and to participate in setting standards for the title of "professional engineer," as well as establishing technical standards. In principle it could also grant individual engineers the right to zealously pursue public safety, backed by the support of professional societies. An analogy to this would be how the public has granted doctors and lawyers the special privilege of keeping patient and client information confidential so as to increase the trust needed for successful medical therapy and legal defense (Freedman, 1978, 13). Because of these privileges as well as the public's expectations, it is claimed it would be unfair of professionals not to reciprocate by committing themselves to promoting those aspects of the public good that fall within their sphere of activity—in the case of engineers, to promoting public safety.

This amounts to the appealing view that engineers have tacitly signed a kind of mutual contract with the public although the shared understanding needed to make sense of such a contract is currently rather limited. Nevertheless, if this idea of a contract is to be more than just a metaphor, it remains to be seen how such contracts are justified. That is, are such mutual agreements morally permissible or even obligatory? Answering this question takes us beyond the mere idea of agreements and reciprocal commitments to the issue of *justification,* i.e., to the question of whether those commitments ought to be made in the first place. That issue can only be resolved by reference to the kinds of general ethical theories that we have invoked. Hence each of these four views proves to be inadequate by itself, without reference to ethical theory.

In conclusion, we might distinguish between two different senses in which it is sometimes claimed that engineers have *special* safety obligations in regard to their work. If "special obligations" refers to obligations not grounded in the general human rights which play a central role in ordinary morality by placing obligations on *all* people, then the only special obligations of engineers are those arising out of special employment agreements or agreements with professional societies. But the main safety obligations of engineers do *not* arise from some special membership in a professional society, or from some special law, tradition, or employment condition inapplicable to nonengineers.

If, however, the word "special" is applied to obligations to give special care and attention to safety matters concerning the projects they engage in, then all engineers do have special safety obligations. They have them in virtue of how their particular expertise and functions are directly related to the rights of persons affected by their work. In this sense we can say that an examination of the special professional obligations of engineers in regard to safety meshes straightforwardly with an examination of human rights and other basic moral considerations, and this establishes a link between engineering and moral philosophy.

Study Questions

1 Sketch a rule-utilitarian justification of the special safety obligations of engineers listed in the National Society of Professional Engineers code (see Appendix). Then sketch a duty-based justification for those obligations. What would act-utilitarianism have to say about them?
2 According to Kenneth Kipnis, a professor of philosophy, the design engineers share the blame for the death of the passengers in the DC-10 crash described above. Kipnis contends that the engineers' overriding obligation was to obey the following principle: "Engineers shall not participate in projects that degrade ambient levels of public safety unless information concerning those degradations is made generally available" (Kipnis, 1981, 82). Do you agree or disagree with Kipnis, and why?
3 An engineer visits a construction site where a structure designed by him is being erected. He has not been hired to supervise the construction. Noticing some unsafe conditions (poor scaffolding and the like), he wonders whether or not to report

them. He remembers that on a previous job a colleague had early on reported some safety violations; then, when on later visits to the site she had not noticed additional safety violations which subsequently caused injuries to workers, she had been sued for carelessness. Her first reports had placed her in jeopardy! What should he do?

RESPONSIBILITY AND VIRTUE ETHICS

The ability to effectively confront problems and dilemmas related to right conduct is a vital part of professional ethics. Yet preoccupation with it should not lead us to neglect the heart and spirit of true professionalism. That has to do with the moral ideals to which a profession is dedicated and the moral character of its practitioners. Moral character, as defined by virtues and vices, has as much to do with motives, attitudes, aspirations, and ideals as it does with right and wrong conduct.

It will be useful, therefore, to turn from utilitarianism, duty ethics, and rights ethics to a fuller discussion of virtue ethics. After briefly discussing one classical and one contemporary virtue ethicist, we will consider in more detail the general virtue of being morally responsible and two specific virtues, trustworthiness and benevolence. This will lay a foundation for the conception of responsible engineering set forth in the next chapter.

Aristotle: Virtue and The Golden Mean

Aristotle (384–322 B.C.) defined virtues as acquired habits that enable us to engage effectively in rational activities—activities which define us as human beings. For example, foresight, efficiency, mental discipline, perseverance, and creativity are necessary for successful rational activities that range from engineering to philosophical inquiry. Aristotle called these particular qualities *intellectual virtues* to distinguish them from specifically *moral virtues*.

Moral virtues are tendencies, acquired through habit formation, to reach a proper balance between extremes in conduct, emotion, desire, and attitude. To use the phrase inspired by his theory, virtues are tendencies to find *The Golden Mean* between the extremes of too much (excess) and too little (deficiency).

For example, courage is the appropriate middle ground between foolhardiness (the excess of rashness) and cowardice (the deficiency of self-control and clear thought in the face of danger). Truthfulness is the mean between revealing just everything in violation of tact and confidentiality (excess) and being secretive or lacking in candor (deficiency). Generosity is the virtue lying between wasting one's resources (excess) and being miserly (deficiency). Friendliness is being agreeable and considerate without being annoyingly effusive (excess) or sulky and surly (deficiencies).

Moral virtues enable us to pursue a variety of social goods within a *com-*

munity—a concept that was especially important for citizens of ancient Greek city-states, since the city-state's survival depended on close cooperation of its citizens. Taken together, the moral virtues also enable us to fulfill ourselves as human beings. They enable us to attain *happiness,* by which Aristotle meant an active life in accordance with our reason (rather than a life of pleasure).

MacIntyre: Virtues and Practices

Virtue ethics has recently been revived and enriched by Alasdair MacIntyre, among others, and applied to professional ethics (MacIntyre, 1984). MacIntyre begins with the idea of *practices*—cooperative activities aimed toward achieving social goods that could not otherwise be achieved. These goods are *internal* to the practices in that they define what the practices are all about. Hence they differ from *external goods* like fame and prestige which can be achieved through many different kinds of activities and do not define any specific practice. For example, the primary internal goods of medicine are good health and respect for patients' autonomy; the primary internal good of law is social justice.

The primary internal good of engineering is the creation of useful and safe products while respecting the autonomy of clients and the public. The virtues and ideals especially for engineers are defined by reference to these end products. Only the conscientious, safety conscious, and imaginative engineer is likely to achieve the good outcomes expected of engineering. These virtues also make possible *integrity,* or moral unity, between the engineer's personal and professional life. Before developing these ideas in the next chapter, let us gain a richer understanding of moral responsibility.

Moral Responsibility

The notion of moral responsibility cuts across judgments about both right actions and people. In every case where moral responsibility is ascribed to someone, a moral judgment is being made; judgments may be of various types (Hart, 1973, 211–230). The interest may be in assessing (1) obligations to perform right actions, (2) general moral capacities of people, (3) the virtue of a person, or (4) liabilities and accountability for actions.

1 We speak of persons as *having* moral responsibilities. In this sense, responsibilities are simply obligations and duties to perform morally right acts. Some of those are shared by us all: for example, the responsibilities to be truthful, to be fair, and to promote justice. Others relate only to people performing within certain social roles or professions: For example parents have specific responsibilities to care for their children, a safety engineer might have responsibilities for making regular inspections at a building site, or an operations engineer might have special responsibilities for identifying potential benefits and risks of one system as compared to another.

2 Sometimes when we ascribe responsibility to a person viewed as a whole rather than in respect to a specific area of his or her conduct, we have in mind an active *capacity* for knowing how to act in morally appropriate ways. In this sense young children are not yet morally responsible. They gradually become so as they mature and learn how to be responsive to the needs and interests of others. Adult sociopaths who lack any sense of guilt for wrongdoing never become responsible in this sense.

3 At other times when we say someone is responsible, we mean to ascribe a general moral *virtue* to the person. We mean that he or she is regularly concerned to do the right thing, is conscientious and diligent in meeting obligations, and is someone who can be counted on to carry out duties or be considerate of others. In a moment we shall return to this sense.

4 Finally, "responsible" often means accountable, answerable, or liable for meeting obligations. In this sense, to say individuals are responsible for actions means they can be "held to account" for them: that is, they can be called upon to explain why they acted as they did; to provide excuses or justification if appropriate; and to be open to commendation or censure, praise or blame, or demands for compensation. We also hold ourselves accountable for our own actions, responding to them with emotions of self-esteem or shame, self-respect or guilt. This notion of responsibility will also be developed more fully in Chap. 3.

Accountability and Voluntary Action Usually when we hold a person accountable for an action we imply that the action was not completely involuntary. But it is not easy to know precisely what this requirement amounts to.

Aristotle suggested that involuntary acts are of two main kinds (Aristotle, 964–967). First, they include acts done in ignorance. If, unknowingly, we loan a car to a distraught friend who crashes it, we have not voluntarily contributed to the friend's death. The problem here is that we also hold people accountable when they *should have known* what they were doing and what the likely consequences of their action would be. Ignorance then, is not always an excuse. On the other hand, it is often difficult to judge fairly what people should have known, especially in the types of complicated situations that can arise in professions like engineering.

Second, Aristotle said acts are involuntary when performed under compulsion. He interpreted compulsion as an external force which determines our actions. Aristotle also noted that the mere existence of obstacles does not entirely negate voluntariness. Rather their presence limits the range of choices open to us. When those choices become sufficiently limited, we have to think in terms of degrees of voluntariness, and hence degrees of liability for harm done. An example would be the limitations placed on our decisions in response to a kidnapper who demands a ransom for not killing someone dear to us.

To make things even more complicated, many psychologists today would make a major addition to Aristotle's views: Acts may be involuntary when they are generated by uncontrollable inner compulsions, such as those motivating psychotics or pathological liars. Yet it is extremely difficult to make accurate assessments of other people at this level.

Causal and Legal Responsibility There are two other concepts of responsibility (Hart, 1973, 214–215). These should not be confused with moral responsibility in any of its four preceding senses. First, *causal responsibility* consists simply in being a cause of some event. In this sense we speak of lightning as being responsible for a house catching fire.

People can be causally responsible for an event without necessarily being morally responsible for it. For example, a 2-year-old child may cause a fire while playing with matches, but it is the parents who left the matches within the child's reach who are morally responsible for the fire.

Second, *legal responsibility* should also be distinguished from moral responsibility. An engineer or engineering firm can be held legally responsible for harm which was so unlikely and unforeseeable that little or no moral responsibility is involved.

One famous court case involved a farmer who lost an eye when a metal chip flew off the hammer he was using (Vaughn, 1977, 41–47). He had used the hammer without problems for 11 months before the accident. It was constructed from metals satisfying all the relevant safety regulations, and no specific defect was found in it. The manufacturer was held legally responsible and required to pay damages. The basis for the ruling was the doctrine of *strict legal liability,* which does not require proof of defect or negligence in design. Yet surely the manufacturer was not morally culpable or blameworthy for the harm done. If we say the manufacturer was morally responsible, we mean at most that the company has an obligation (based on the special relationship between it and the farmer created by the accident) to help remedy the problem caused by the defective hammer.

Conversely, it is also possible to be morally responsible for something one cannot be held legally responsible for. For example, because of the fine wording of a contract an engineer may be free from any legal liability for failing to report an observed danger at a construction site. Yet it may have been his or her professional and moral obligation to report that danger.

Motives and Professional Ethics

Let us now focus on responsibility as a virtue. Calling professionals responsible in this sense ascribes to them conscientious concern for the moral ideals and aims of their profession. A responsible physician is motivated (in part) by a concern for the health and autonomy of patients. A responsible engineer is motivated (in part) by respect for the safety and autonomy of the public and clients.

Of course, none of us is motivated by entirely simple motives. Very frequently we pursue a line of conduct from a combination of motives, some pertaining to morality and some not. A student's motives for attending college, for example, might include the desire to obtain a well-paying job, to gain social recognition, to please parents, and to prepare for a socially useful career. The last of these motives is often grounded in the morally admirable motive of altruism—the desire to contribute to the good of other people. Similarly, the desire to obtain a well-paying job so as not to be a burden on others or so as to be able to support a family is a morally admirable motive. Even the desire to find challenging work is related to the moral ideal of self-fulfillment.

Professionals are similarly motivated in their careers by a mixture of motives. This mixed motivation is not lamentable; instead it is desirable. Moral ideals are easier to achieve when moral motives are reinforced by self-interest. ("Self-interest" means concern for one's own good. It does not mean "selfishness"—that is, excessive concern for one's own good at the expense of other people.)

In addition to motives of moral concern and self-interest, there is another important category of professional motives: concern for achieving excellence in the technical aspects of one's work. The excitement of engineering, combined with a strong desire to see it done well, constitutes a potent stimulus for professional conduct.

In fact, the technical challenge of work is sometimes enough by itself to inspire right conduct throughout much of a career, even though some moral motivation seems essential to most careers. An interesting illustration of this is presented by Graham Greene in his novel *A Burnt-Out Case*.

Greene describes an architect who has reached the top of his profession without caring very much about the good of the public which has benefited from his work. The architect, world renowned, abandons a career in which he has made numerous brilliant contributions without any wrongdoing. He travels to Africa and meets a doctor who is practicing medicine on the basis of a concern for his patients. In one scene the architect explains to the doctor that his interest had always been in the "space and light and the proportion" of buildings, not in the people who might use them (Greene, 1977, 44). Jokingly, the doctor remarks that he would not have trusted the plumbing in the structure designed by the architect. But the architect presses his point. He confesses that of course he had to consider human needs, but only in the same way he had to consider the brick, glass, and other building materials. His sole motivation, however, was the creation of beautiful structures.

As this incident suggests, it is possible for a person to act on professional obligations from primarily nonmoral motives, such as a sheer pleasure in the beauty of the emerging product and excitement over the technical aspects of the work. The doctor's facetious remark about plumbing reminds us that architects and engineers must be concerned professionally to satisfy the needs of people they serve, but that concern may not be moral in origin. Accordingly, much of professional ethics focuses on the level of care the architect says he showed in his work—attention to the safety, well-being, and needs of those affected by the professional activities involved—no matter what the ultimate wellspring of that attention may be.

The story also illustrates, however, that long-term involvement in a career (avoiding early "burnout") may require moral concern. Moreover, things easily go amiss when preoccupation with the technical aspects of work leads to a disregard of moral obligations. An illustration of this point is found in William Golding's novel *The Spire*.

The Spire is a rich allegory about fidelity, creativity, and the way in which self-deception can warp concern for safety within engineering. Set in England during the Middle Ages, the plot revolves around the construction of a 400-foot spire atop an aging cathedral. Success in the project would mean developing technology well beyond its then current state, and this provides the motivation for the master builder commissioned to undertake the project.

Yet as the master builder assesses the weaknesses of the foundations supporting the church, he is led to suspect that the stone and glass spire cannot be supported properly. Thereafter his suspicions are repeatedly confirmed to the point where his best professional judgment—what he humbly calls his "guesses"—indicate the task is both futile and dangerous. Nevertheless, the priest who is his client desperately clings to a vision of the spire as a "prayer in stone" and urges the craftsman on.

The master builder gradually becomes biased as he allows his excitement over the project and the personal influence of the priest to lead him to disregard safety. As the story ends, the entire structure is slowly crumbling.

Personal Integrity and Virtues

There is a further reason why moral conduct is essential for professionals. This reason has to do with the maintenance of personal and moral integrity. Morality requires that our lives be unified where fundamental values are at stake, not compartmentalized. There must not be a cleavage between the working life and the public self of the sort Charles Reich described when he wrote, "It is this split that sometimes infuriates his children when they become of college age, for they see it as hypocrisy. The individual has two roles, two lives, two masks, two sets of values. . . . Neither the man at work nor the man at home is the whole man; it is impossible to know, talk to, or confront the whole man, for that wholeness is precisely what does not exist" (Reich, 1970, 78).

Virtues provide a bridge between private and professional life. Virtues are general patterns of action, emotion, and attitude that permeate all areas of life. They involve habits that constitute fundamental ways of relating to the world, not just to selected situations. Moral integrity (inner unity on the basis of moral commitments) is maintained when virtues are manifested across the line between personal and professional life.

This explains what is often wrong when an employee says, "Don't blame me; I was just doing my job." The implication is that the employee is a mere cog in the machinery of the workplace, or a mere tool to be used by an employer, rather than a responsible person whose life has moral coherence. Again, when people try to justify wrongdoing by saying, "If I don't do it, someone else will," they are failing to take responsibility for their actions (regardless of what other people do).

Trustworthiness and Benevolence

Taking responsibility for one's actions is a very general virtue. Trustworthiness and benevolence are two of the specific virtues it encompasses—virtues especially important in professions like engineering.

Trustworthiness is a fundamental virtue for those who engage in the relationships between engineers and their employers and clients. These relationships are based on trust—trust that engineers will effectively perform the services for which they are hired. Here is a list of even more specific virtues that trustworthiness involves (Bayles, 1981, 70–86).

Honesty in Acts: For example, not stealing, not padding expense sheets, not engaging in bribes and kickbacks

Honesty in Speech: Not deceiving; being candid by revealing all pertinent information

Competence: Being well prepared for the jobs one undertakes

Diligence: Zeal and careful attention to detail in performing tasks (by, for example, avoiding the defect of laziness and the excesses of the workaholic)

Loyalty: Acting faithfully on behalf of the interests of the employer or client (avoiding the defect of allowing self-interest to distort one's service and avoiding the excess of disregarding other important duties such as those to the public)

Discretion: Sensitivity to the legitimate areas of privacy of the employer or client, especially with regard to confidential information

Benevolence is also pertinent to the relationship between employers and clients, but it is especially important in thinking about obligations to third parties affected by one's work, in particular the public. *Benevolence* is the desire to promote the good of others based on an attitude of concern for their well-being. Hence much of the discussion of concern for others presented in this section applies to benevolence.

The following specific virtues are all aspects of benevolence.

Nonmaleficence: Not harming others

Beneficence: Doing good and preventing or removing harms to others

Generosity: Going beyond the minimal degrees of helping others

While each of these aspects of benevolence is important, the first is the most basic. It lies behind the oldest professional dictum, one embedded in the Hippocratic Oath taken by physicians: "Above all, do no harm." As we shall see, heeding that charge is a complicated task in engineering.

Study Questions

Kermit Vandivier had worked at B. F. Goodrich for 5 years, first in instrumentation and later as a data analyst and technical writer. In 1968 he was assigned to write a report on the performance of the Goodrich wheels and brakes commissioned by the Air Force for its new A7D light attack aircraft. According to his account, he became aware of the design's limitations and of serious irregularities in the qualification tests. The brake failed to meet Air Force specifications. Upon pointing out these problems, however, he was given a direct order to stop complaining and write a report which would show the brake qualified. He was led to believe that several layers of management were behind this demand and would accept whatever distortions might be needed because their engineering judgment assured them the brake was acceptable.

Vandivier then drafted a 200-page report with dozens of falsifications and misrepresentations. But he refused to sign it. Later he gave as excuses for his complicity the facts that he was 42 years old with a wife and six children. He had recently bought a home and felt financially unable to change jobs. He felt certain that he would have been fired if he had refused to participate in writing the report (Vandivier, 20–24).

1 Present and defend your view as to whether Vandivier was justified in writing the report or not. In doing so, draw upon one of the theories of right action discussed in the second section of this chapter.
2 Was Vandivier guilty or blameworthy? That is, even if his actions were wrong, is it appropriate to excuse him from blame because of circumstances beyond his control?
3 Is Vandivier responsible for what he did? In answering this question, distinguish between the various senses of "responsible" discussed in this section.
4 Which virtues did Vandivier not display, and what might those virtues have required of him in his situation?
5 Truthfulness and truth telling are key virtues for engineers as they interact with other participants in the technological enterprise (illustrated in Fig. 1-1, Chap. 1). Their meanings come into sharper focus when their antonyms are examined. These include lying, deception, and withholding information. (The latter two are often grouped as "disinformation" in government parlance.) Give examples from engineering, business, or other professions to illustrate these concepts.

SUMMARY

Moral problems, in the widest sense, are those that arise in any situation calling for decisions based upon moral reasons. Sometimes what ought to be done is a straightforward matter, and the only difficulty is in avoiding temptations to violate moral obligations. At other times it may be unclear whether a moral principle applies, as when deciding whether accepting some gifts from salespeople violates the rule "Do not accept bribes." *Moral dilemmas* are those moral problems in which two or more moral obligations, duties, rights, ideals, or applications of a single principle come into conflict in a situation in which not all of them can be respected or fulfilled. Duties which sometimes allow exceptions in such situations are called *prima facie duties*.

Moral values require that we be concerned about the good and the rights of other people. Hence morality is not reducible to matters of self-interest, law, or religion. For this reason we rejected *Ethical Egoism* (the view that right action consists in producing one's own good), *Ethical Conventionalism* (the view that right action is merely what the law and customs of one's society require), and *Divine Command Ethics* (the view that right action is defined by the commands of God, such that without a God there could be no moral values).

There are four main types of ethical theories which provide helpful frameworks for identifying the factors involved in moral dilemmas and for offering guidance.

Acts are morally right when:		
• They produce the most good for the most people.	Act-utilitarianism: Mill	Utilitarianism
• They fall under a rule which if widely followed would produce the most good for the most people.	Rule-utilitarianism: Brandt	
• They fall under principles of duty which respect the autonomy and rationality of persons, and which can be willed universally to apply to all people.	Kant	Duty Theories
• They fall under principles which would be agreed upon by all rational agents in a hypothetical contracting situation that assures impartiality.	Rawls	
• They are the best way to respect the human rights of everyone affected.	Locke and Melden	Rights Theories
• They most fully manifest or support relevant virtues, where virtues are traits of character making possible the achievement of social goods.	Aristotle and MacIntyre	Virtue Theories

These ethical theories give help in approaching moral dilemmas by providing frameworks for assessing the relevant moral factors involved and by offering guidance. They can also be applied to identify and justify the general obligations of engineers and other professionals.

Finally, if *actions* can be judged right or wrong, *people* can be judged as good or bad, virtuous or vicious, responsible or irresponsible. Underlying such assessments are moral virtues: good traits of character which involve habits and patterns of action, emotion, attitude, and desire. The most general professional virtue is moral responsibility, which differs from mere causal and legal responsibility. In addition to this "virtue" sense of responsibility, the concept of moral responsibility also refers sometimes to obligations, the general capacity to act in morally concerned ways, and accountability for actions. Two of the more specific virtues related to being responsible as a professional are trustworthiness (honesty in action and speech, competence, diligence, loyalty, and discretion) and benevolence (nonmaleficence, beneficence, and generosity).

PART II

THE EXPERIMENTAL NATURE OF ENGINEERING

To undertake a great work, and especially a work of a novel type, means carrying out an experiment. It means taking up a struggle with the forces of nature without the assurance of emerging as the victor after the first attack.

Louis Marie Henri Navier (1785–1836)
a founder of structural analysis

A ship in harbor is safe, but that is not what ships are built for.

John A. Shedd

Primum non nocere. (Above all, do no harm.)

Admonition to Physicians

CHAPTER 3

ENGINEERING AS SOCIAL EXPERIMENTATION

As it departed on its maiden voyage in April 1912 the *Titanic* was proclaimed the greatest engineering achievement ever. Not merely was it the largest ship the world had seen, having a length of two and a half football fields, it was also the most glamorous of ocean liners, complete with a tropical vinegarden restaurant and the first seagoing masseuse. It was supposed to be the first fully safe ship. Since the worst collision envisaged was at the juncture of two of its sixteen watertight compartments, and since it could float with any four compartments flooded, the *Titanic* was confidently believed to be virtually unsinkable.

Buoyed by such confidence, the captain allowed the ship to sail full speed at night in an area of reported icebergs, one of which tore a large gap in its side, directly or indirectly* flooding five compartments. Time remained to evacuate the ship, but there were not enough lifeboats to accommodate all the passengers and crew. British regulations then in effect did not foresee vessels of this size. Accordingly only 825 places were required in lifeboats, sufficient for a mere one-quarter of the *Titanic*'s capacity of 3547 passengers and crew. No extra precautions had seemed necessary for a practically unsinkable ship. The result: 1522 dead (drowned or frozen) out of the 2227 on board for the *Titanic*'s first trip (Lord, 1976; Wade, 1980; Davie, 1986).

In his poem written shortly after the event, "The Convergence of the Twain," Thomas Hardy portrayed the meeting of the ship and iceberg as de-

*Some investigators believe the *Titanic* left England with a coal fire on board, that this made the captain rush the ship to New York, and that water entering the coal bunkers through the gash caused an explosion and thereby greater damage to the compartments.

termined by unpredictable fate: "No mortal eye could see/The intimate welding of their later history." Yet greater imagination and prudence could have prevented the disaster.

Interestingly enough, novelists did not lack the imaginative foresight to describe scenarios that paralleled the later real events in shocking detail. In Morgan Robertson's 1898 novel *Futility,* a ship almost identical in size to the *Titanic* was wrecked by an iceberg on a cold April night. The ship in the book was named the *Titan;* it too had a less than sufficient number of lifeboats. Mayn Clew Garnett's story "The White Ghost of Disaster" was being readied for publication in *Popular Magazine* while the *Titanic* was on her maiden voyage. It is said that Garnett had dreamed the story while traveling on the *Titanic*'s sister ship, the *Olympic.* Again, circumstances similar to those surrounding the sinking of the *Titanic,* as well as an insufficient number of lifeboats to save all the passengers, were key elements in the narrative (Wade, 1980, 70–71).

The *Titanic* remains a haunting image of technological complacency. Perhaps all we can take for granted today is Murphy's law that if anything can go wrong, it will—sooner or later. All products of technology present some potential dangers, and thus engineering is an inherently risky activity. In order to underscore this fact and help in exploring its ethical implications, we suggest that engineering should be viewed as an experimental process. It is not, of course, an experiment conducted solely in a laboratory under controlled conditions. Rather, it is an experiment on a social scale involving human subjects.

ENGINEERING AS EXPERIMENTATION

Experimentation is commonly recognized to play an essential role in the design process. Preliminary tests or simulations are conducted from the time it is decided to convert a new engineering concept into its first rough design. Materials and processes are tried out, usually employing formal experimental techniques. Such tests serve as the basis for more detailed designs, which in turn are tested. At the production stage further tests are run, until a finished product evolves. The normal design process is thus iterative, carried out on trial designs with modifications being made on the basis of feedback information acquired from tests. Beyond those specific tests and experiments, however, each engineering project taken as a totality may itself be viewed as an experiment.

Similarities to Standard Experiments

Several features of virtually every kind of engineering practice combine to make it appropriate to view engineering projects as experiments. First, any project is carried out in partial ignorance. There are uncertainties in the abstract model used for the design calculations; there are uncertainties in the precise characteristics of the materials purchased; there are uncertainties about the nature of the stresses the finished product will encounter. Engineers do not have the luxury of waiting until all the relevant facts are in before commencing work. At some point theoretical exploration and laboratory testing must be bypassed for the sake of moving ahead on a project. Indeed, one talent crucial to an engineer's success lies precisely in the ability to accomplish tasks with only a partial knowledge of scientific laws about nature and society.

Second, the final outcomes of engineering projects, like those of experiments, are generally uncertain. Often in engineering it is not even known what the possible outcomes may be, and great risks may attend even seemingly benign projects. A reservoir may do damage to a region's social fabric or to its ecosystem. It may not even serve its intended purpose if the dam leaks or breaks. An aqueduct may bring about a population explosion in a region where it is the only source of water, creating dependency and vulnerability without adequate safeguards. An aircraft may become a status symbol that ultimately bankrupts its owners. A special-purpose fingerprint reader may find its main application in the identification and surveillance of dissidents by totalitarian regimes. A nuclear reactor, the scaled-up version of a successful smaller model, may exhibit unexpected problems that endanger the surrounding population, leading to its untimely shutdown at great cost to owner and consumers alike. A hair dryer may expose the unknowing or unwary user to lung damage from the asbestos insulation in its barrel.

Third, effective engineering relies upon knowledge gained about products both before and after they leave the factory—knowledge needed for improving current products and creating better ones. That is, ongoing success in engineering depends upon gaining new knowledge, just as does ongoing success in experimentation. Monitoring is thus as essential to engineering as it is to experimentation in general. *To monitor* is to make periodic observations and tests in order to check for both successful performance and unintended side effects. But since the ultimate test of a product's efficiency, safety, cost-effectiveness, environmental impact, and aesthetic value lies in how well that product functions within society, monitoring cannot be restricted to the development or testing phases of an engineering venture. It also extends to the stage of client use. Just as in experimentation, both the intermediate and final results of an engineering project deserve analysis if the correct lessons are to be learned from it.

Learning from the Past

It might be expected that engineers would learn not only from their own earlier design and operating results, but also from those of other engineers. Unfortunately that is frequently not the case. Lack of established channels of communication, misplaced pride in not asking for information, embarrassment at failure, and plain neglect often impede the flow of such information and lead to many repetitions of past mistakes. Here are a few examples:

1 The *Titanic* lacked a sufficient number of lifeboats decades after most of the passengers and crew on the steamship *Arctic* had perished because of the same problem (Wade, 1980, 417).

2 "Complete lack of protection against impact by shipping caused Sweden's worst ever bridge collapse on Friday as a result of which eight people were killed." Thus reported the *New Civil Engineer* on January 24, 1980. On May 15 of the same year it also reported the following: "Last Friday's disaster at Tampa Bay, Florida, was the largest and most tragic of a growing number of incidents of errant ships colliding with bridges over navigable waterways." While collisions of ships with bridges do occur—other well-known cases being those of the Maracaibo Bridge (Venezuela, 1964) and the Tasman Bridge (Australia, 1975)—Tampa's Sunshine Skyline Bridge was not designed with horizontal impact forces in mind because the code did not require it. Floating concrete bumpers which can deflect ships have been proposed by Laura and Nava (1981).

3 In June 1966 a section of the Milford Haven bridge in Wales collapsed during construction. A bridge of similar design was being erected by the same bridge builder (Freeman Fox and Partners) in Melbourne, Australia, when it too partially collapsed, killing thirty-three people and injuring nineteen. This happened in October of the same year, shortly after chief construction engineer Jack Hindshaw (also a casualty) had assured worried workers that the bridge was safe (Yarrow Bridge, 415).

4 Valves are notorious for being among the least reliable components of hydraulic systems. It was a pressure relief valve, and lack of positive information regarding its open or shut state, which helped lead to the nuclear reactor accident at Three Mile Island on March 28, 1979. Similar malfunctions had occurred with identical valves on nuclear reactors at other locations. The required reports had been filed with Babcock and Wilcox, the reactor's manufacturer, but no attention had been given to them (Sugarman, 1979, 72).

5 The Bureau of Reclamation, which built the ill-fated Teton Dam, allowed it to be filled rapidly, thus failing to provide sufficient time to monitor for the presence of leaks in a project constructed out of less than ideal soil. The Bureau did not heed the lesson of its Fontenelle Dam, where 10 years earlier massive leaks had also developed and caused a partial collapse (Shaw, 1977; Boffey, 1977).

These examples, and others to be given in later chapters, illustrate why it is not sufficient for engineers to rely on handbooks alone. Engineering, just like experimentation, demands practitioners who remain alert and well informed at every stage of a project's history.

Contrasts with Standard Experiments

To be sure, engineering differs in some respects from standard experimentation. Some of those very differences help to highlight the engineer's special responsibilities. And exploring the differences can also aid our thinking about the moral responsibilities of all those engaged in engineering.

Experimental Control One great difference has to do with experimental control. In a standard experiment this involves the selection, at random, of members for two different groups. The members of one group receive the special, experimental treatment. Members of the other group, called the *control group,* do not receive that special treatment although they are subjected to the same environment as the first group in every other respect.

In engineering this is not the usual practice, unless the project is confined to laboratory experimentation, because the experimental subjects are humans out of the range of the experimenter's control. Indeed, clients and consumers exercise most of the control because it is they who choose the product or item they wish to use. This makes it impossible to obtain a random selection of participants from various groups. Nor can parallel control groups be established based on random sampling. Thus no careful study of the effects of changing variables on two or more comparison groups is possible, and one must simply work with the available historical and retrospective data about various groups that use the product.

This suggests that the view of engineering as a social experiment involves a somewhat extended usage of the concept of experimentation. Nevertheless, "engineering as social experimentation" should not be dismissed as a merely metaphorical notion. There are other fields where it is not uncommon to speak of experiments whose original purpose was not experimental in nature and that involve no control groups.

For example, social scientists monitor and collect data on differences and similarities between existing educational systems that were not initially set up as systematic experiments. In doing so they regard the current diversity of systems as constituting what has been called a "natural experiment" (as opposed to a deliberately initiated one) (Rivlin, 1970, 70). Similarly, we think that engineering can be appropriately viewed as just such a "natural experiment" using human subjects, despite the fact that most engineers do not currently consider it in that light.

Informed Consent Viewing engineering as an experiment on a societal scale places the focus where it should be: on the human beings affected by technology. For the experiment is performed on persons, not on inanimate objects. In this respect, albeit on a much larger scale, engineering closely parallels medical testing of new drugs and techniques on human subjects.

Society has recently come to recognize the primacy of the subject's safety and freedom of choice as to whether to participate in medical experiments. Ever since the revelations of prison and concentration camp horrors in the name of medicine, an increasing number of moral and legal safeguards have arisen to ensure that subjects in experiments participate on the basis of informed consent.

But while current medical practice has increasingly tended to accept as fundamental the subject's moral and legal rights to give informed consent before participating in an experiment, contemporary engineering practice is only beginning to recognize those rights. We believe that the problem of informed consent, which is so vital to the concept of a properly conducted experiment involving human subjects, should be the keystone in the interaction between engineers and the public. We are talking about the *lay public*. When a manufacturer sells a new device to a knowledgeable firm which has its own engineering staff, there is usually an agreement regarding the shared risks and benefits of trying out the technological innovation.

Informed consent is understood as including two main elements: knowledge and voluntariness. First, subjects should be given not only the information they request, but all the information which is needed for making a reasonable decision. Second, subjects must enter into the experiment without being subjected to force, fraud, or deception. Respect for the fundamental rights of dissenting minorities and compensation for harmful effects are taken for granted here.

The mere purchase of a product does not constitute informed consent, any more than does the act of showing up on the occasion of a medical examination. The public and clients must be given information about the practical risks and benefits of the product in terms they can understand. Supplying complete information about the product is neither necessary nor in most cases possible. In both medicine and engineering there may be an enormous gap between the experimenter's and the subject's understanding of the complexities of an experiment. But while this gap most likely cannot be closed, it should be possible to convey all pertinent information needed for making a reasonable decision on whether to participate or not.

We do not propose a proliferation of lengthy environmental impact reports. We favor the kind of sound advice a responsible physician gives a patient when prescribing a course of drug treatment that has possible side effects. The physician must search beyond the typical sales brochures from drug manufacturers for adequate information; hospital management must allow the physician the freedom to undertake different treatments for different patients, as each case may constitute a different "experiment" involving different circumstances; finally, the patient must be readied to receive the information.

Likewise, an engineer cannot succeed in providing essential information about a project or product unless there is cooperation by management and a receptivity on the part of those who should have the information. Management is often understandably reluctant to provide more information than current laws require, fearing disclosure to potential competitors and exposure to potential lawsuits. Moreover, it is possible that, paralleling the experience in medicine, clients or the public may not be interested in all of the relevant information about an engineering project, at least not until a crisis looms. It is important nevertheless that all avenues for disseminating such information be kept open and ready.

We note that the matter of informed consent is surfacing indirectly in the continuing debate over acceptable forms of energy. Representatives of the nuclear industry can be heard expressing their impatience with critics who worry about reactor malfunction while engaging in statistically more hazardous activities such as driving automobiles and smoking cigarettes. But what is being overlooked by those representatives is the common enough human readiness to accept risks voluntarily undertaken (as in daring sports), even while objecting to involuntary risks resulting from activities in which the individual is neither a direct participant nor a decision maker. In other words, we all prefer to be the subjects of our own experiments rather than those of somebody else. When it comes to approving a nearby oil-drilling platform or a nuclear plant, affected parties expect their consent to be sought no less than it is when a doctor contemplates surgery.

Prior consultation of the kind suggested can be effective. When Northern States Power Company (Minnesota) was planning a new power plant, it got in touch with local citizens and environmental groups before it committed large sums of money to preliminary design studies. The company was able to present convincing evidence regarding the need for a new plant and then suggested several sites. Citizen groups responded with a site proposal of their own. The latter was found acceptable by the company. Thus informed consent was sought from and voluntarily given by those the project affected, and the acrimonious and protracted battle so common in other cases where a company has already invested heavily in decisions based on engineering studies alone was avoided (Borrelli, 36–39). Note that the utility company interacted with groups that could serve as proxy for various segments of the ratepaying public. Obviously it would have been difficult to involve the ratepayers individually.

We endorse a broad notion of informed consent, or what some would call *valid consent* (Culver and Gert), defined by the following conditions:

1 The consent was given voluntarily.

2 The consent was based on the information that a rational person would want, together with any other information requested, presented to them in understandable form.

3 The consenter was competent (not too young or mentally ill, for instance) to process the information and make rational decisions.

We suggest two requirements for situations in which the subject cannot be readily identified as an individual:

4 Information that a rational person would need, stated in understandable form, has been widely disseminated.

5 The subject's consent was offered in proxy by a group that collectively represents many subjects of like interests, concerns, and exposure to risk.

Knowledge Gained

Scientific experiments are conducted to gain new knowledge, while "engineering projects are experiments that are not necessarily designed to produce very much knowledge," according to a valuable interpretation of our paradigm by Broome (1987). When we carry out an engineering activity as if it were an experiment, we are primarily preparing ourselves for unexpected outcomes. The best outcome in this sense is one which tells us nothing new but merely affirms that we are right about something. Unexpected outcomes send us on a search for new knowledge—possibly involving an experiment of the first (scientific) type. For the purposes of our model the distinction is not vital because we are concerned about the manner in which the experiment is conducted, such as that valid consent of human subjects is sought, safety measures are taken, and means exist for terminating the experiment at any time and providing all participants a safe exit.

Study Questions

1 On June 5, 1976, Idaho's Teton Dam collapsed, killing eleven people and causing $400 million in damage. Drawing upon the concept of engineering as social experimentation, discuss the following facts uncovered by the General Accounting Office and reported in the press.
   a Because of the designers' confidence in the basic design of Teton Dam, it was believed that no significant water seepage would occur. Thus sufficient instrumentation to detect water erosion was not installed.
   b Significant information suggesting the possibility of water seepage was acquired at the dam site 6 weeks before the collapse. It was sent through routine channels from the project supervisors to the designers, and arrived at the designers the day after the collapse.
   c During the important stage of filling the reservoir, there was no around-the-clock observation of the dam. As a result the leak was detected only 5 hours before the collapse. Even then the main outlet could not be opened to prevent the collapse because a contractor was behind schedule in completing the outlet structure (Shaw, 1977, 3; Boffee, 1977, 270–272).
2 The University of California uses tax dollars to develop farm machinery such as tomato, lettuce, melon harvesters, and fruit tree shakers. Such machinery reduces the need for farm labor and raises farm productivity. It definitely benefits the growers. It is also said to benefit all of society. Farm workers, however, claim that replacing an adequate and willing work force with machines will generate social costs not offset by higher productivity. Among the costs they cite are the need to retrain farm workers for other jobs and the loss of small farms. Discuss if and how continuing farm mechanization may be viewed as an experiment.
3 Apply the social experimentation model to the DC-10 case described in Chap. 2. Specifically, in order to facilitate informed consent concerning dangers entailed by the plane's design, should Dan Applegate have been allowed to convey information to public representatives (in government or consumer groups) or directly (via newspapers) to the public who must decide whether to fly on DC-10 airplanes?
4 Models often influence thinking by effectively organizing and guiding reflection and crystallizing attitudes. Yet they usually have limitations and can themselves be misleading to some degree. Write a short essay in which you critically assess the strengths and weaknesses you see in the social experimentation model.

   One possible criticism you might consider is whether the model focuses too much on the creation of new products, whereas a great deal of engineering involves the routine application of results from past work and projects. Another point to consider is how informed consent is to be measured in situations where groups are involved, as in the construction of a nuclear power plant near a community of people having mixed views about the advisability of constructing the plant.

5 Debates over responsibility for safety in regard to technological products often turn on whether the consumer should be considered mainly responsible ("buyer beware") or the manufacturer ("seller beware"). How might an emphasis on the idea of informed consent influence thinking about this question?

6 In the following passage from *A Nation of Guinea Pigs,* Marshall Shapo applies the concept of experimentation to the marketing of drugs. Comment on parallels and dissimilarities you see between the moral aspects of social experimentation in engineering and in drug marketing.

> ...experimentation is a label which connotes an attempt to solve problems in a fresh and novel way, using the subjects of the attempt as means to gather information. The image that the term conveys in the context of hazards involving products and processes tends to be a laboratory image. But much experimentation goes beyond the laboratory. In the process of testing and marketing new drugs, after procedures first limited to testing for toxicity and pharmacological effects, it takes place with increasingly large groups of patients in clinical trials. And although we do not conventionally attach the label "experimental" to the general marketing of products, it is clear that widespread distribution in fact involves a continuous process of experimentation. Especially with goods that are scientifically complex, the information-collecting goal of the experimenter is never attained in the formally investigational stages of the process. Some hazards may become apparent only after the products are used by millions of people, and over extended periods of time (Shapo, 1979, 30).

7 Engineering and medical practice are intimately linked in medical engineering. Its products range from artificial limbs and organs to heart pacers and x-ray machines. Its engineers and medical experts are experimenters with excellent track records, but failures do occur. For example, the State University of New York at Albany admitted that its psychology department had conducted electroshock experiments on patients who were not given fair explanations of risks and whose consent had not been obtained. The machine itself was unsafe (R. J. Smith, 1977). Discuss the ethical implications of this case.

8 "On Being One's Own Rabbit" is the title of an essay by J. B. Haldin, who conducted many risky medical experiments on his own body (quoted in Mullan, 1987). Seek examples of engineers and inventors who served as their own subjects and discuss to what extent such practice is desirable or not. (Example: Wright Brothers)

9 *"Primum non nocere"* ("Above all, do not harm") is an admonition to medical students and practitioners. What should engineers do when hired to carry out tasks they feel might cause harm? Are clients not entitled to engineering services in the same way that we insist on legal services being available to everyone, including crooks? In certain restricted cases it might be morally justifiable for engineers to proceed with the requested task. Baum (1980) made the concept of informed consent central to thinking about engineering ethics in connection with such circumstances. Describe a real or hypothetical situation where engineer, client, and affected parties might disagree and another case where they might agree.

ENGINEERS AS RESPONSIBLE EXPERIMENTERS

What are the responsibilities of engineers to society? Viewing engineering as social experimentation does not by itself answer this question. For while engineers are the main technical enablers or facilitators, they are far from being the sole experimenters. Their responsibility is shared with management, the public, and others. Yet their expertise places them in a unique position to monitor projects, to identify risks, and to provide clients and the public with the information needed to make reasonable decisions.

The detailed content of engineers' responsibilities, in the sense of obligations, will be explored throughout the remainder of this book. At present we are interested in another of the senses of "responsibility" distinguished in Chap. 2. We want to know what is involved in displaying the virtue of being a responsible person while acting as an engineer. From the perspective of engineering as social experimentation, what are the general features of morally responsible engineers?

At least four elements are pertinent: a conscientious commitment to live by moral values, a comprehensive perspective, autonomy, and accountability (Haydon, 1978, 50–53). Or, stated in greater detail as applied to engineering projects conceived as social experiments:

1 A primary obligation to protect the safety of and respect the right of consent of human subjects

2 A constant awareness of the experimental nature of any project, imaginative forecasting of its possible side effects, and a reasonable effort to monitor them

3 Autonomous, personal involvement in all steps of a project

4 Accepting accountability for the results of a project

It is implied in the foregoing that engineers should also display technical competence and other attributes of professionalism. Inclusion of these four requirements as part of engineering practice would then earmark a definite "style" of engineering. In elaborating upon this style, we will note some of the contemporary threats to it.

Conscientiousness

People act responsibly to the extent that they conscientiously commit themselves to live according to moral values. But moving beyond this truism leads immediately to controversy over the precise nature of those values. In Chap. 1 we adopted the minimal thesis that moral values transcend a consuming preoccupation with a narrowly conceived self-interest. Accordingly, individuals who think solely of their own good to the exclusion of the good of others are not moral agents. By *conscientious* moral commitment is meant a sensitivity to the full range of moral values and responsibilities that are relevant to a given situation, and the willingness to develop the skill and expend the effort needed to reach the best balance possible among those considerations.

The contemporary working conditions of engineers tend to narrow moral vision solely to the obligations that accompany employee status. As stated earlier, some 90 percent of engineers are salaried employees, most of whom work within large bureaucracies under great pressure to function smoothly within the organization. There are obvious benefits in terms of prudential self-interest and concern for one's family that make it easy to emphasize as primary the obligations to one's employer. Gradually the minimal negative duties, such as not falsifying data, not violating patent rights, and not breaching confidentiality, may come to be viewed as the full extent of moral aspiration.

Conceiving engineering as social experimentation restores the vision of engineers as guardians of the public interest, whose professional duty it is to guard the welfare and safety of those affected by engineering projects. And this helps to ensure that such safety and welfare will not be disregarded in the quest for new knowledge, the rush for profits, a narrow adherence to rules, or a concern over benefits for the many that ignores harm to the few.

The role of social guardian should not suggest that engineers force, paternalistically, their own views of the social good upon society. For, as with medical experimentation on humans, the social experimentation involved in engineering should be restricted by the participant's consent—voluntary and informed consent.

Relevant Information

Conscientiousness is blind without relevant factual information. Hence showing moral concern involves a commitment to obtain and properly assess all available information pertinent to meeting one's moral obligations. This means, as a first step, fully grasping the *context* of one's work which makes it *count* as an activity having a moral import.

For example, there is nothing wrong in itself with being concerned to design a good heat exchanger. But if I ignore the fact that the heat exchanger will be used as part of a still involved in the manufacture of a potent, illegal hallucinogen, I am showing a lack of moral concern. It is this requirement that one be aware of the wider implications of one's work which makes participation in, say, a design project for a superweapon morally problematic—and which makes it sometimes convenient for engineers self-deceivingly to ignore the wider context of their activities, a context that may rest uneasily with an active conscience.

Another way of blurring the context of one's work results from the ever increasing specialization and division of labor which makes it easy to think of someone else in the organization as responsible for what otherwise might be a bothersome personal problem. For example, a company may produce items with obsolescence built into them, or the items might promote unnecessary energy usage. It is easy to place the burden on the sales department: "Let *them* inform the customers—if the customers ask." It may be natural to thus rationalize one's neglect of safety or cost considerations, but it shows no moral concern.

These ways of losing perspective on the nature of one's work also hinder acquiring a full perspective along a second dimension of factual information: the *consequences* of what one does. And so while regarding engineering as social experimentation points out the importance of context, it also urges the engineer to view his or her specialized activities in a project as part of a larger whole having a social impact—an impact that may involve a variety of unintended effects. Accordingly, it emphasizes the need for wide training in disciplines related to engineering and its results, as well as the need for a constant effort to imaginatively foresee dangers.

It might be said that the goal is to practice what Chauncey Starr once called "defensive engineering." Or perhaps more fundamental is the concept of "preventive technology" as described by Ruth Davis, who could have addressed the following lines equally well to engineers as she did to scientists and physicians:

> The solution to the problem is not in successive cures to successive science-caused problems; it is in their prevention. Unfortunately, cures for scientific ills are generally more interesting to scientists than is the prevention of those ills. We have the unhappy history of the medical community to show us the difficulties associated with trying to establish preventive medicine as a specialty.
>
> Scientists probably had more fun developing scientific defenses against nuclear weapons (that is, cures) than they would have had practicing preventive nuclear science during the development of the atomic bomb. Computer scientists find it more attractive to develop technological safeguards, after the fact, to prevent invasions of privacy associated with computer data banks than to develop good information practices along with the computer systems.
>
> However, it now seems quite clear that public patience with the cure always following after the ill has worn thin. The public wants to see some preventive measures taken. Indeed, individuals have taken what can be called preventive technology into their own hands. We have seen the public in action in this way in its handling of the supersonic transport issue and its reaction toward siting of nuclear power plants. This is the reactive mode of practicing preventive technology, and it hinges on public recognition that technology is fallible (Davis, 1975, 213).

No amount of disciplined and imaginative foresight, however, can serve to anticipate all dangers. Because engineering projects are inherently experimental in nature, it is crucial for them to be monitored on an ongoing basis from the time they are put into effect. While individual practitioners cannot privately conduct full-blown environmental and social impact studies, they can choose to make the extra effort needed to keep in touch with the course of a project after it has officially left their hands. This is a mark of *personal* identification with one's work, a notion that leads to the next aspect of moral responsibility.

Moral Autonomy

People are morally autonomous when their moral conduct and principles of action are their *own*, in a special sense deriving from Kant. That, is, moral beliefs and attitudes must be held on the basis of critical reflection rather than merely through passive adoption of the particular conventions of one's society, church, or profession. This is often what is meant by "authenticity" in one's commitment to moral values.

Those beliefs and attitudes, moreover, must be integrated into the core of an individual's personality in a manner that leads to committed action. They cannot be agreed to abstractly and formally and adhered to merely verbally. Thus, just as one's principles are not passively imbibed from others when one is morally autonomous, so too one's actions are not treated as something alien and apart from oneself.

It is a comfortable illusion to think that in working for an employer, and thereby performing acts directly serving a company's interests, one is no longer morally and personally identified with one's actions. Selling one's labor and skills may make it seem that one has thereby disowned and forfeited power over one's actions (Lachs, 1978, 201–213).

Viewing engineering as social experimentation can help one overcome this tendency and can help restore a sense of autonomous participation in one's work. As an experimenter, an engineer is exercising the sophisticated training that forms the core of his or her identity as a professional. Moreover, viewing an engineering project as an experiment that can result in unknown consequences should help inspire a critical and questioning attitude about the adequacy of current economic and safety standards. This also can lead to a greater sense of personal involvement with one's work.

The attitude of management plays a decisive role in how much moral autonomy engineers feel they have. It would be in the long-term interest of a high-technology firm to grant its engineers a great deal of latitude in exercising their professional judgment on moral issues relevant to their jobs (and, indeed, on, technical issues as well). But the yardsticks by which a manager's performance is judged on a quarterly or yearly basis most often militate against this. This is particularly true in our age of conglomerates, when near-term profitability is more important than consistent quality and long-term retention of satisfied customers.

In government-sponsored projects it is frequently a deadline which becomes the ruling factor, along with fears of interagency or foreign competition. Tight schedules contributed to the loss of the U.S. space shuttle *Challenger* as we shall see later.

Accordingly engineers are compelled to look to their professional societies and other outside organizations for moral support. Yet it is no exaggeration to claim that the blue-collar worker with union backing has greater leverage at present in exercising moral autonomy than do many employed professionals. A steel plant worker, for instance, who refused to dump oil into a river in an unauthorized manner was threatened with dismissal, but his union saw to it that the threat was never carried out (Nader, 1972, 189). Or take the case of the automobile plant inspector who repeatedly warned his supervisors about poorly welded panels which allowed carbon monoxide from the exhaust to leak into the cab. Receiving no satisfactory response from the company, he blew the whistle. The company wanted to fire him, but pressure from the union allowed him to keep his job. (The union, however, did not concern itself with the safety issue. It was probably as surprised as the company by the number of eventual fatalities traceable to the defect, the recall order those deaths necessitated, and the tremendous financial loss ultimately incurred by the company.) (Nader, 1972, 75–89)

Professional societies, originally organized as learned societies dedicated to the exchange of technical information, lack comparable power to protect their members, although most engineers have no other group to rely on for such protection. Only now is the need for moral and legal support of members in the exercise of their professional obligations being recognized by those societies. Unger (1987) describes how engineering societies can proceed, even in the face of difficulties such as litigation.

Accountability

Finally, responsible people accept moral responsibility for their actions. Too often "accountable" is understood in the overly narrow sense of being culpable and blameworthy for misdeeds. But the term more properly refers to the general disposition of being willing to submit one's actions to moral scrutiny and be open and responsive to the assessments of others. It involves a willingness to present morally cogent reasons for one's conduct when called upon to do so in appropriate circumstances.

Submission to an employer's authority, or any authority for that matter, creates in many people a narrowed sense of accountability for the consequences of their actions. This was documented by some famous experiments conducted by Stanley Milgram during the 1960s (Milgram, 1974). Subjects would come to a laboratory believing they were to participate in a memory and learning test. In one variation two other people were involved, the "experimenter" and the "learner." The experimenter was regarded by the subject as an authority figure, representing the scientific community. He or she would give the subject orders to administer electric shocks to the "learner" whenever the latter failed in the memory test. The subject was told the shocks were to be increased in magnitude with each memory failure. All this, however, was a deception—a "setup." There were no real shocks and the apparent "learner" and the "experimenter" were merely acting parts in a ruse designed to see how far the unknowing experimental subject was willing to go in following orders from an authority figure.

The results were astounding. When the subjects were placed in an adjoining room separated from the "learner" by a shaded glass window, over half were willing to follow orders to the full extent: giving the maximum electric jolt of 450 volts. This was in spite of seeing the "learner," who was strapped in a chair, writhing in (apparent) agony. The same results occurred when the subjects were allowed to hear the (apparently) pained screams and protests of the "learner," screams and protests which became intense from 130 volts on. There was a striking difference, however, when subjects were placed in the same room within touching distance of the "learner." Then the number of subjects willing to continue to the maximum shock dropped by one-half.

Milgram explained these results by citing a strong psychological tendency in people to be willing to abandon personal accountability when placed under authority. He saw his subjects ascribing all initiative, and thereby all accountability, to what they viewed as legitimate authority. And he noted that the closer the physical proximity, the more difficult it becomes to divest oneself of personal accountability.

The divorce between causal influence and moral accountability is common in business and the professions, and engineering is no exception. Such a psychological schism is encouraged by several prominent features of contemporary engineering practice.

First, large-scale engineering projects involve fragmentation of work. Each person makes only a small contribution to something much vaster. Moreover, the final product is often physically removed from one's immediate workplace, creating the kind of "distancing" that Milgram identified as encouraging a lessened sense of personal accountability.

Second, corresponding to the fragmentation of work is a vast diffusion of accountability within large institutions. The often massive bureaucracies within which most engineers work are designed to diffuse and delimit areas of personal accountability within hierarchies of authority.

Third, there is frequently pressure to move on to a new project before the current one has been operating long enough to be observed carefully. This promotes a sense of being accountable only for meeting schedules.

Fourth, the contagion of malpractice suits currently afflicting the medical profession is carrying over into engineering. With this comes a crippling preoccupation with legalities, a preoccupation which makes one wary of becoming morally involved in matters beyond one's strictly defined institutional role.

We do not mean to underestimate the very real difficulties these conditions pose for engineers who seek to act as morally accountable people on their jobs. Much less do we wish to say engineers are blameworthy for all the bad side effects of the projects they work on, even though they partially cause those effects simply by working on the projects. That would be to confuse accountability with blameworthiness, and also to confuse causal responsibility with moral responsibility. But we do claim that engineers who endorse the perspective of engineering as a social experiment will find it more difficult to divorce themselves psychologically from personal responsibility for their work. Such an attitude will deepen their awareness of how engineers daily cooperate in a risky enterprise in which they exercise their personal expertise toward goals they are especially qualified to attain, and for which they are also accountable.

Study Questions

1 A common excuse for carrying out a morally questionable project is, "If I don't do it somebody else will." This rationale may be tempting for engineers who typically

work in situations where someone else might be ready to replace them on a project. Do you view it as a legitimate excuse for engaging in projects which might be unethical? (In your answer, comment upon the concept of responsible conduct developed in this section.)

2 Another commonly used phrase, "I only work here," implies that one is not personally accountable for the company rules since one does not make them. It also suggests that one wishes to restrict one's area of responsibility within tight bounds as defined by those rules (Lachs, 1978, 201–213). In light of the discussion in this section, respond to the potential implications of this phrase and the attitude represented by it when exhibited by engineers.

3 You have been asked to design an electronic vote counter for a legislative body. You have no difficulty with the physical features of the machine, but you begin to ask yourself some questions. If heretofore all votes, except for secret ballots, were by a show of hands, should a display board be provided indicating each individual vote? Or would total tallies be sufficient, thereby assuring anonymity of voting on each occasion? What would be the implication of each option in terms of respecting the public's right to know? How much need you worry about unauthorized tampering with such a machine? Describe to what extent the model of social experimentation can be applied to the introduction of the vote counter. (For a short report on the West German parliament's reluctance to put a vote counter to use in 1971, see the *Los Angeles Times* article by Joe Alex Morris, Jr., cited in the Bibliography.)

4 Threats to a sense of personal responsibility are neither unique to nor more acute for engineers than they are for others involved with engineering and its results. The reason is that, in general, *public* accountability also tends to lessen as professional roles become narrowly differentiated. With this in mind, critique each of the remarks made in the following dialogue. Is the remark true, or partially true? What needs to be added to make it accurate?

Engineer: My responsibility is to receive directives and to create products within specifications set by others. The decision about what products to make and their general specifications are economic in nature and made by management.

Scientist: My responsibility is to gain knowledge. How the knowledge is applied is an economic decision made by management, or else a political decision made by elected representatives in government.

Manager: My responsibility is solely to make profits for stockholders.

Stockholder: I invest my money for the purpose of making a profit. It is up to managers to make decisions about the directions of technological development.

Consumer: My responsibility is to my family. Government should make sure corporations do not harm me with dangerous products, harmful side effects of technology, or dishonest claims.

Government regulator: By current reckoning, government has strangled the economy through overregulation of business. Accordingly at present on my job, especially given decreasing budget allotments, I must back off from the idea that business should be policed, and urge corporations to assume greater public responsibility.

5 Cancer therapy machines were discarded at dump sites in Juarez, Mexico (R.J. Smith, 1984) , and Goiânia, Brazil (L. Roberts, 1987). The radioactive isotopes, removed from their canisters, exposed many people. At least one child died. Discuss

the responsibility of the manufacturers' and hospitals' engineers for safe disposal of such apparatus. Is this part of the monitoring function set forth in the engineering as experiment paradigm?

6 Is this a true experiment, wishful thinking, or a scam? The Cryonics movement believes in keeping fresh corpses frozen, with blood replaced by glycol antifreeze, until advances in medicine can cure the original cause of death. Then the body is to be unfrozen, the cure applied, and the patient returned to life. Several bodies are kept in cryogenic facilities around the United States, along with several heads, which are kept for future attachment to cloned bodies. Research the case and discuss how it fits the experimentation model. Here are some references to get you started: issues of *Omni* (October 1986) and *Health* (March 1987); the book *Freezing Point* by L. Kavaler (1970).

THE *CHALLENGER* CASE

Several months before the destruction of the *Challenger*, NASA historian Alex Roland wrote the following in a critical piece about the space shuttle program:

> The American taxpayer bet about $14 billion on the shuttle. NASA bet its reputation. The Air Force bet its reconnaissance capability. The astronauts bet their lives. We all took a chance.
>
> When John Young and Robert Crippen climbed aboard the orbiter *Columbia* on April 12, 1981 for the first shuttle launch, they took a bigger chance than any astronaut before them. Never had Americans been asked to go on a launch vehicle's maiden voyage. Never had astronauts ridden solid propellant rockets. Never had Americans depended on an engine untested in flight (Roland, 1985).

Most of Roland's criticism was directed at the economic and political side of what was supposed to become a self-supporting operation but never gave any indication of being able to reach that goal. Without a national consensus to back it, the shuttle program became a victim of year by year funding politics (Logsdon, 1986).

The *Columbia* and its sister ships, the *Challenger* and *Discovery*, are delta-wing craft with a huge payload bay. Early, sleek designs had to be abandoned to satisfy U.S. Air Force requirements when the latter was ordered to use the NASA shuttle instead of its own expendable rockets for launching satellites and other missions. As shown in Fig. 3-1 each orbiter has three main engines fueled by several million pounds of liquid hydrogen; the fuel is carried in an immense, external, divided fuel tank, which is jettisoned when empty. During lift-off the main engines fire for about 8.5 minutes, although during the first 2 minutes of the launch much of the thrust is provided by two booster rockets. These are of the solid-fuel type, each burning a 1-million-pound load of a mixture of aluminum, potassium chloride, and iron oxide.

The casing of each booster rocket is about 150 feet long and 12 feet in diameter. It consists of cylindrical segments which are assembled at the launch site. The four field joints use seals composed of pairs of O-rings made of vulcanized rubber. The O-rings work in conjunction with a putty barrier of zinc chromide.

The shuttle flights were successful, though not as frequent as had been

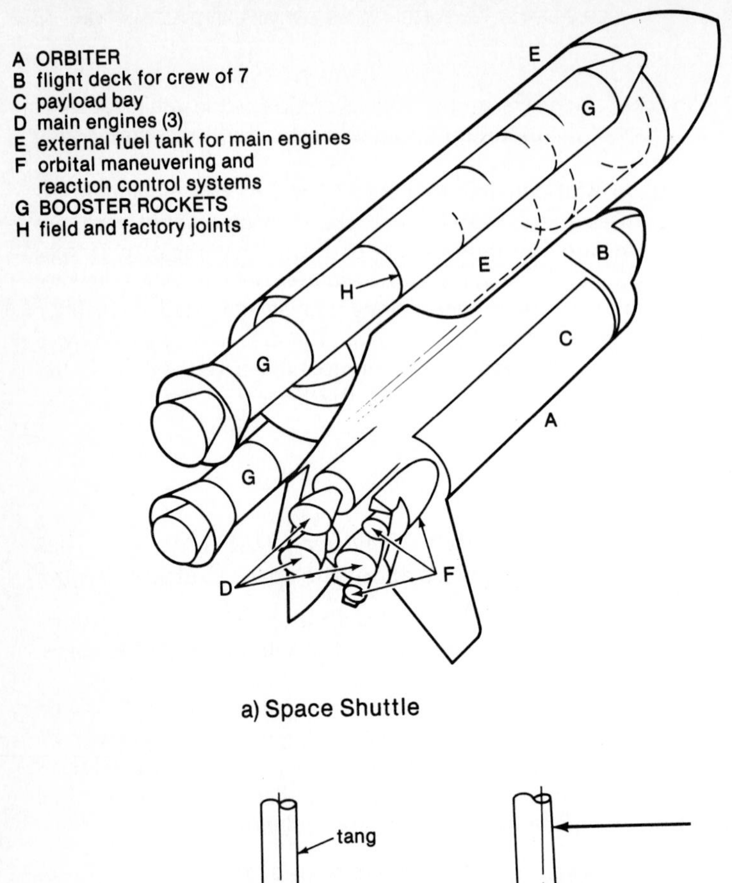

a) Space Shuttle

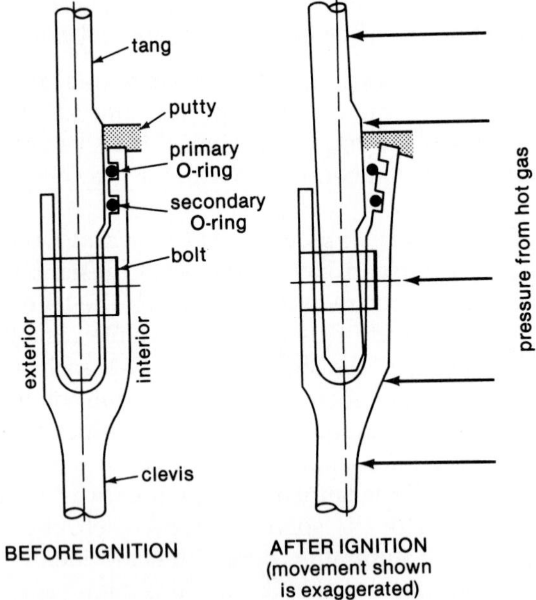

b) Cross-section of Field Joint

FIGURE 3-1
Space shuttle *Challenger.*

hoped. NASA tried hard to portray the shuttle program as an operational system that could pay for itself. Some Reagan administration officials had even suggested that the operations be turned over to an airline. Aerospace engineers intimately involved in designing, manufacturing, assembling, testing, and operating the shuttle still regarded it as an experimental undertaking in 1986. These engineers were employees of manufacturers, such as Rockwell International (orbiter and main rocket) and Morton-Thiokol (booster rockets), or they worked for NASA at one of its several centers: Marshall Space Flight Center, Huntsville, Alabama (responsible for the propulsion system); Kennedy Space Center, Cape Kennedy, Florida (launch operations); Johnson Space Center, Houston, Texas (flight control); and the office of the Chief Engineer, Washington, D.C. (overall responsibility for safety, among other duties).

After embarrassing delays, *Challenger*'s first flight for 1986 was set for Tuesday morning, January 28. But Allan J. McDonald, who represented Morton-Thiokol at Cape Kennedy, was worried about the freezing temperatures predicted for the night. As his company's director of the solid-rocket booster project he knew of difficulties that had been experienced with the field joints on a previous cold-weather launch when the temperature had been mild compared to what was forecast. He therefore arranged a teleconference so that NASA engineers could confer with Morton-Thiokol engineers at their plant in Utah.

Arnold Thompson and Roger Boisjoly, two seal experts at Morton-Thiokol, explained to their own colleagues and managers as well as the NASA representatives how upon launch the booster rocket walls bulge and the combustion gases can blow past one or even both of the O-rings which make up the field joints (see Fig. 3-1). The rings char and erode, as had been observed on many previous flights. In cold weather the problem is aggravated because the rings and the putty packing are less pliable then.

The engineering managers, Bob Lund (V.P. of engineering) and Joe Kilminster (V.P. for booster rockets), agreed that there was a problem with safety. The team from Marshall Space Flight Center was incredulous. Since the specifications called for an operating temperature of the solid fuel prior to combustion of 40 to 90 degrees Fahrenheit, one could surely allow lower or higher outdoor temperatures, notwithstanding Boisjoly's testimony and recommendation that no launch should occur at less than 53 degrees. They were clearly annoyed at facing yet another postponement.

Top executives of Morton-Thiokol were also sitting in on the teleconference. Their concern was the image of the company, which was in the process of negotiating a renewal of the booster rocket contract with NASA. During a recess Senior Vice President Jerry Mason turned to Bob Lund and told him "to take off your engineering hat and put on your management hat." It was a subsequent vote (of the managers only) that produced the company's official finding that the seals could not be shown to be unsafe. The engineers' judgment was not considered sufficiently weighty. At Cape Kennedy, Allan

McDonald refused to sign the formal recommendation to launch; Joe Kilminster had to do that.

Accounts of the *Challenger* disaster (McConnell, 1987; Rogers Commission Report, 1986) tell of the cold Tuesday morning, the high seas which forced the recovery ships to seek coastal shelter, the ice at the launch site, and the concern expressed by Rockwell engineers that the ice might shatter and hit the orbiter or rocket casings. The inability of these engineers to prove that the lift-off would be unsafe was taken by NASA as an approval by Rockwell to launch.

The countdown ended at 11:38 A.M. The temperature had risen to 36 degrees. As the rockets carrying *Challenger* rose from the ground, cameras recorded puffs of smoke which emanated from one of the field joints on the right booster rocket. Soon these turned into a flame which hit the external fuel tank and a strut holding the booster rocket. The hydrogen in the tank caught fire, the booster rocket broke loose, smashed into *Challenger*'s wing, then into the external fuel tank. At 76 seconds into the flight, by the time *Challenger* and its rockets had reached 50,000 feet, it was totally engulfed in a fireball. The crew cabin separated and fell into the ocean, killing all aboard: mission commander Francis (Dick) Scobee; pilot Michael Smith; mission specialists Gregory Jarvis, Ronald McNair, Ellison Onizuka, Judith Resnick; "teacher in space" Christa MacAuliffe.

President Reagan was to give his State of the Union message later that day. He had to change the tone of his prepared remarks on the shuttle flight and its first civilian passenger.

Safety Issues

Unlike the three-stage rockets which carried astronauts to the moon, the space shuttle could be involved in a simultaneous (inadvertent) ignition of all fuel carried aloft. An explosion close to the ground can have catastrophic effects. The crew has no escape mechanism, although McDonnell-Douglas, in a losing shuttle proposal, had provided an abort module with its own thruster. It would have allowed the separation of the orbiter, triggered (among other events) by a field-joint leak. But such a safety measure was rejected as too expensive because of an accompanying reduction in payload.

Working with such constraints, why was safe *operation* not stressed more? First of all we must remember that the shuttle program was indeed still a truly experimental and research undertaking. Next, it is quite clear that the members of the crews knew that they were embarking on dangerous missions. But it has also been revealed that the *Challenger* astronauts were not informed of particular problems such as the field joints. They were not asked for their consent to be launched under circumstances which experienced engineers had claimed to be unsafe.

The reason for the rather cavalier attitude toward safety is revealed in the way NASA assessed the system's reliability. For instance, recovered booster rocket casings had indicated that the field-joint seals had been damaged in many of the earlier flights. The waivers necessary to proceed with launches had become mere gestures. Richard Feynman made the following observations as a member of the Presidential Commission on the Space Shuttle *Challenger* Accident (called the Rogers Commission after its chairman):

> I read all of these [NASA flight readiness] reviews and they agonize whether they can go even though they had some blow-by in the seal or they had a cracked blade in the pump of one of the engines,...and they decide "yes." Then it flies and nothing happens. Then it is suggested...that the risk is no longer so high. For the next flight we can lower our standards a little bit because we got away with it last time....It is a kind of Russian roulette (Rogers Commission Report, 1986).

Since the early days of unmanned space flight, about 1 in every 25 solid-fuel rocket boosters has failed. Given improvements over the years, Feynman thought that 1 in every 50 to 100 might be a reasonable estimate now (Marshall, 1986). Yet NASA counts on only 1 crash in every 100,000 launches. Queried about these figures, NASA Chief Engineer Milton Silveira answered: "We don't use that number as a management tool. We know that the probability of failure is always sitting there..." (Marshall, 1986). So where was this number used? In a risk analysis needed by the Department of Energy to assure everyone that it would be safe to use small atomic reactors as power sources on deep-space probes and to carry both aloft on a space shuttle. As luck would have it, *Challenger* was not to carry the 47.6 pounds of lethal plutonium-238 until its next mission with the *Galileo* probe on board (Grossman, 1986).

Another area of concern was NASA's unwillingness to wait out risky weather. When serving as weather observer, astronaut John Young was dismayed to find his recommendations to postpone launches disregarded several times. Things had not changed much by March 26, 1987, when NASA neglected to heed its monitoring devices for electric storm conditions, launched a Navy communications satellite atop an Atlas-Centaur rocket, and had to destroy the $160 million system when it veered off course after being hit by lightning. The monitors had been installed after a similar event involving an Apollo command module 18 years before had nearly aborted a trip to the moon (Marshall, 1987). Weather, incidentally, could be held partially responsible for the shuttle disaster because a strong wind shear may have contributed to the rupturing of the weakened O-rings (Bell, 1987).

Veteran astronauts were also dismayed at NASA management's decision to land at Cape Kennedy as often as possible despite its unfavorable landing conditions including strong crosswinds and changeable weather. The alternative, Edwards Air Force Base in California, is a better landing place but necessitates a piggyback ride for the shuttle on a Boeing 747 home to Florida. This costs time and money.

In 1982 Albert Flores conducted a study of safety concerns at the Johnson Space Center. He found its engineers to be strongly committed to safety in all aspects of design. When they were asked how managers might further improve safety awareness, there were few concrete suggestions but many comments on how safety concerns were ignored or negatively impacted by man agement. One engineer was quoted as saying, "A small amount of professional safety effort and upper management support can cause a quantum safety improvement with little expense" (Flores, 1982, 79). This points to the important role of management in building a strong sense of responsibility for safety first and to schedules second.

The space shuttle's field joints are designated to be of criticality 1, which means there is no backup. Therefore a leaky field joint will result in failure of the mission and loss of life. There are 700 items of criticality 1 on the shuttle. A problem with any one of them should have been cause enough to do more than launch more shuttles without modification while working on a better system. Improved seal designs had already been developed, but the new rockets would not have been ready for some time. In the meantime the old booster rockets should have been recalled.

At Morton-Thiokol, Roger Boisjoly's personal concern had been heightened by his memory of the DC-10 crash over Paris. That accident had shown him how known defects can be disregarded in a complex organization. For this reason he had started a journal in which he recorded all events associated with the seals (Whitbeck, 1987). But like Dan Applegate in the DC-10 case he probably did not feel that he had the kind of professional backing which would allow him to go beyond his organization directly to the astronauts.

In several respects the ethical issues in the Challenger case resemble those of the DC-10 case. Concern for safety gave way to institutional posturing. Danger signals did not go beyond Convair and Douglas Aircraft in the DC-10 case; they did not go beyond Morton-Thiokol and Marshall Space Flight Center in the *Challenger* case. No effective recall was instituted. There were concerned engineers who spoke out, but ultimately they felt it only proper to submit to management decisions.

The major difference between the cases is found in the late-hour teleconference which Allan McDonald had arranged from the *Challenger* launch site to get knowledgeable engineers to discuss the seal problem from a technical viewpoint. (No similar conference between engineers from different organizations took place in the DC-10 case.) This tense conference did not involve lengthy discussions of ethics, but it revealed the virtues (or lack thereof) which allow us to distinguish between the "right stuff" and the "wrong stuff." This is well described in the following letter to the *Los Angeles Times* by an aerospace engineer.

> In Paul Conrad's cartoon (Feb. 27, 1986), "Autopsy of a Catastrophe," a drawing of the space shuttle *Challenger* is labeled with words like "MONEY," "SCHEDULE," etc. Forty years experience as an engineer in the aerospace industry leads me to believe that Conrad has (uncharacteristically) defused the issue.
>
> He could have used one word, "arrogance." The arrogance that prompts higher-level decision makers to pretend that factors other than engineering judgement should influence flight safety decisions and, more important, the arrogance that rationalizes overruling the engineering judgement of engineers close to the problem by those whose expertise is naive and superficial by comparison.
>
> The flaw is not in the decision-making process; it is in the decision-making

mentality. Consequently it would be of little value to move engineering decisions to a higher level, as has been contemplated by members of the presidential investigating commission" (Moeller, 1986).

Included, surely, is the arrogance of those who reversed NASA's (paraphrased) motto "Don't fly if it cannot be shown to be safe" to "Fly unless it can be shown not to be safe."

At Morton-Thiokol some of the vice presidents in the space division have been demoted. The engineers who were outspoken at the prelaunch teleconference and again before the Rogers Commission kept their jobs at the company because of congressional pressure, but their jobs are of a pro forma nature. In a speech to engineering students at the Massachussetts Institute of Technology a year after the *Challenger* disaster, Roger Boisjoly said: "I have been asked by some if I would testify again if I knew in advance of the potential consequences to me and my career. My answer is always an immediate yes. I couldn't live with any self-respect if I tailored my actions based upon potential personal consequences as a result of my honorable actions..." (Boisjoly, 1987).

Today NASA has a policy which allows aerospace workers with concerns to report them anonymously to the Batelle Memorial Institute in Columbus, Ohio, but open disagreement still invites harrassment (Magnuson, 1988).

Study Questions

1 Read more detailed accounts of the Challenger disaster and then examine if and how the principal actors in this tragedy behaved as responsible experimenters within the framework of the engineering as experimentation model.

2 Chairman Rogers asked Bob Lund: "Why did you change your decision [that the seals would not hold up] when you changed hats?" What might motivate you, as a midlevel manager, to go along with top management when told to "take off your engineering hat and put on your management hat"?

3 Under what conditions would you say it is safe to launch a shuttle without an escape mechanism for the crew?

4 Discuss the role of the astronauts in shuttle safety. To what extent should they (or at least the orbiter commanders) have involved themselves more actively in looking for safety defects in design or operation?

5 Consider the following actions or recommendations and suggest a plan of action to bring about safer designs and operations in a complex organization.

a Lawrence Mulloy represented Marshall Space Flight Center at Cape Kennedy. He did not tell Arnold Aldrich from the National Space Transportation Program at Johnson Space Center about the discussions regarding the field-joint seals even though Aldrich had the responsibility of clearing Challenger for launch. Why? Because the seals were "a Level III issue," and Mulloy was at Level III, while Aldrich was at a higher level (Level II) which ought not to be bothered with such details.

b The Rogers Commission recommended that an independent safety organization directly responsible to the NASA administrator be established. At the end of the *Challenger* case study we mentioned that an anonymous reporting scheme now exists for aerospace industry employees working on NASA projects.

c Tom Peters advises managers to "involve everyone in everything.... Boldly assert that there is no limit to what the average person can accomplish if thoroughly involved" (Peters, 1987).

6 Several Morton-Thiokol engineers were troubled by the seals' poor performance. Long before the *Challenger* disaster Boisjoly wrote in a memo that the result of neglecting the problem "would be a catastrophe of the highest order—loss of human life." By August 1985 a seal task force had been established, but Bob Ebeling sent out this distress message: "HELP! The seal task force is constantly being delayed by every possible means.... This is a red flag." What else could or should these engineers have done in the months before *Challenger's* last flight?

7 On October 4, 1930, the British airship *R 101* crashed about 8 hours into its maiden voyage to India. Of the fifty-four persons aboard, only six survived. Throughout the craft's design and construction, Air Ministry officials and their engineers had been driven by political and competitive forces described by Shute (1954), Higham (1961), Robinson (1973), Meyer (1981), and Squires (1986). Shute, who had worked on the rival, commercial *R 100* wrote in his memoir, *Slide Rule,* that "if just one of [the men at the Air Ministry] had stood up [at a conference with Lord Thomson] and had said, 'This thing won't work, and I'll be no party to it. I'm sorry, gentlemen, but if you do this, I'm resigning'... the disaster would almost certainly have been averted. It was not said, because the men in question put their jobs before their duty" (Shute, 1954, 140). Examine the *R 101* case and compare it with the *Challenger* case.

CODES OF ETHICS

Invoking a code of ethics for engineers might have helped Dan Applegate and Roger Boisjoly with impressing their safety concerns on management. Such a use is one of the most important roles of a code. We shall examine it along with other prominent functions, prominent in terms of both positive and negative consequences. It is suggested that in reading this section the reader examine the sample codes of ethics given in the Appendix as if they were checklists for experimenters.

Roles of Codes

Inspiration and Guidance Codes provide a positive stimulus for ethical conduct and helpful guidance and advice concerning the main obligations of engineers. Often they succeed in inspiring by using language with positive overtones. This can introduce a large element of vagueness, as in phrases like "safeguard the public safety, health, and welfare," a vagueness which may lessen their ability to give concrete guidance. Sometimes lofty ideals and exhortative phrases are gathered into separate documents, such as *Faith of the Engineer,* published by the Accreditation Board for Engineering and Technology (ABET), which succeeded the Engineering Council for Professional Development (ECPD). *Faith of the Engineer* is reprinted in the Appendix along with several other codes or fundamental canons of ethics.

Since codes should be brief to be effective, they offer mostly general guidance. More specific directions may be given in supplementary statements or guidelines. These tell how to apply the code. Further specificity may also be attained by the interpretation of codes. This is done for engineers by the National Society of Professional Engineers. It has established a Board of Ethical Review which applies the Society's code to specific cases and publishes the results in *Professional Engineer* and in periodic volumes entitled *NSPE Opinions of the Board of Ethical Review.*

For inclusion in the Appendix we have selected the codes of the following societies: the Accreditation Board for Engineering and Technology (ABET), the American Association of Engineering Societies (AAES), the National Society of Professional Engineers (NSPE), and the Institute of Electrical and Electronics Engineers (IEEE). The ABET code is accompanied by a set of guidelines which can appear separately or intermeshed with the fundamental canons. The latter format is sometimes followed by the American Society of Civil Engineers (ASCE), which has adopted the code and guidelines of ABET. Among other societies which subscribe to the ABET code and guidelines is the American Society of Mechanical Engineers (ASME).

A number of companies (for example, Bechtel, Hughes Aircraft, McDonnell-Douglas) have instituted their own codes. These tend to concentrate on the moral issues encountered in dealing with vendors and clients, particularly the U.S. federal government.

Support Codes give positive support to those seeking to act ethically. A publicly proclaimed code allows an engineer who is under pressure to act unethically to say: "I am bound by the code of ethics of my profession, which states that. . . ." This by itself gives engineers some group backing in taking stands on moral issues. Moreover, codes can potentially serve as legal support in courts of law for engineers seeking to meet work-related moral obligations.

Deterrence and Discipline Codes can serve as the formal basis for investigating unethical conduct. Where such investigation is possible, a prudential motive for not acting immorally is provided as a deterrent. Such an investigation generally requires paralegal proceedings designed to get at the truth about a given charge without violating the personal rights of those being investigated. In the past, engineering professional societies have been reluctant to undertake such proceedings because they have lacked the appropriate sanctions needed for punishment of misconduct. Unlike the American Bar Association and some other professional groups, engineering societies cannot revoke the right to practice engineering in this country. Yet the American Society of Civil Engineers, for example, does currently suspend or expel members whose professional conduct has been proven unethical, and this alone can be a powerful sanction when combined with the loss of respect from colleagues and the local community that such action is bound to produce.

Education and Mutual Understanding Codes can be used in the classroom and elsewhere to prompt discussion and reflection on moral issues and to encourage a shared understanding among professionals, the public, and government organizations concerning the moral responsibilities of engineers. They can help do this because they are widely circulated and officially approved by professional societies.

Contributing to the Profession's Public Image Codes can present a positive image to the public of an ethically committed profession. Where the image is warranted, it can help engineers more effectively serve the public. It can also win greater powers of self-regulation for the profession itself, while lessening the demand for more government regulation. Where unwarranted, it reduces to a kind of window dressing that ultimately increases public cynicism about the profession.

Protecting the Status Quo Codes establish ethical conventions, which can help promote an agreed upon minimum level of ethical conduct. But it can also stifle dissent within the profession. On occasion this has positively discouraged moral conduct and caused serious harm to those seeking to serve the public. In 1932, for example, two engineers were expelled from the American Society of Civil Engineers for violating a section of its code forbidding public remarks critical of other engineers. Yet the actions of those engineers were essential in uncovering a major bribery scandal related to the construction of a dam for Los Angeles County (Layton, 1980, 17).

Promoting Business Interests Codes can place "restraints of commerce" on business dealings with primary benefit to those within the profession. Basically self-serving items in codes can take on great undue influence. Obviously there is disagreement about which, if any, entries function in these ways. Some engineers believe that in the past the codes were justified in forbidding competitive bidding, while others agree with the decision of the Supreme Court in the case of the National Society of Professional Engineers vs. the United States (April 25, 1978) that such a restriction is inappropriate.

Codes and the Experimental Nature of Engineering

Given that codes may play all these roles, which functions are the most valuable and therefore should be emphasized and encouraged? This is an important question, if only because its answer can greatly influence the very wording of codes. For example, if the disciplinary function is to be emphasized, every effort would have to be made to ensure clear-cut and enforceable rules. This would also tend to make statements of minimal duty predominant, as with standards and laws, rather than statements concerned with higher ideals. By contrast, if the emphasis is to be on inspiration, then statements of high ideals might predominate. Nothing is less inspirational than arid, legalistic wordings, and nothing is less precise than highly emotional exhortations.

The perspective of engineering as social experimentation provides some help in deciding which functions should be primary in engineering codes. It clearly emphasizes those which best enable concerned engineers to express their views freely—especially about safety—to those affected by engineering projects. Only thus can clients and the public be educated adequately enough to make informed decisions about such projects. But as we have already noted and will discuss in more detail later, contemporary working conditions within large corporations do not always encourage this freedom of speech—conditions for which a code of ethics can provide an important counterbalance. Thus the supportive function seems to us of primary importance.

The guidance, inspirational, and educational functions of engineering codes are important also, as is their role in promoting mutual understanding among those affected by them. In seeking to create a common understanding, however, code writers must take every precaution to allow room for reasonable differences between individuals. Wordings in past codes, for example, sometimes used religious language not acceptable to many who did not share that orientation. Codes, we must bear in mind, seek to capture the essential substance of professional ethics; they can hardly be expected to express the full moral perspective of every individual.

The disciplinary function of engineering codes is in our view of secondary importance. There are scoundrels in engineering, as there are everywhere. But when exposed as such, they generally fall subject to the law. Developing elaborate paralegal procedures within professional societies runs the risk of needlessly and at considerable cost duplicating a function better left to the real legal system. At most, enforcement of professional ethics by professional societies should center upon areas that are not covered by law and that can be made explicit and clear-cut, preferably in separate code sections specifically devoted to those areas. In any case, the vast majority of engineers can be counted on to act responsibly in moral situations unless *discouraged* from doing so by outright threats and lack of support on the part of employers.

Probably the worst abuse of engineering codes in the past has been to restrict honest moral effort on the part of individual engineers in the name of preserving the profession's public image and protecting the status quo. Preoccupation with keeping a shiny public image may silence the healthy dialogue and lively criticism needed to ensure the public's right to an open expression. And an excessive interest in protecting the status quo may lead to a distrust of the engineering profession on the part of both government and the public. The best way to *increase* trust is by encouraging and aiding engineers to speak freely and responsibly about the public safety and good as they see it. And this includes a tolerance for criticisms of the codes themselves. Perhaps the worst thing that can happen is for codes to become "sacred documents" that have to be accepted uncritically.

Limitations of Codes

Most codes are limited in several major ways. Those limitations restrict codes to providing only very general guidance, which in turn makes it essential for engineers to exercise a personal moral responsibility in their role as social experimenters rather than to expect codes to solve their moral problems by serving as simple algorithms. The limitations of codes are as follows.

First, as we have already mentioned, codes are restricted to general and vague wording. Because of this they are not straightforwardly applicable to all situations. After all, it is not humanly possible to foresee the full range of moral problems that can arise in a complex profession like engineering. New technical developments and shifting social and organizational structures combine to generate continually new and often unpredictable conditions. And even in the case of foreseeable situations it is not possible to word a code so that it will apply in every instance. Attempting to do so would yield something comparable to the intricate set of laws governing engineering rather than a manageable code.

A sense of responsibility is indispensable for the skillful and at times creative application of code guidelines to concrete situations. It is also the only way certain abuses of codes can be avoided—for example, abuses such as special interpretations being placed on general entries, or legalistic glosses on specific entries, to serve the private gain or convenience of specific individuals or groups.

Second, it is easy for different entries in codes to come into conflict with each other. Usually codes provide no guidance as to which entry should have priority in those cases, thereby creating moral dilemmas.

For example, take the following two former entries from the National Society of Professional Engineers (NSPE) code. Section 1: "The Engineer will be guided in all his professional relations by the highest standards of integrity, and will act in professional matters for each client or employer as a faithful agent or trustee." Section 2: "The Engineer will have proper regard for the safety, health, and welfare of the public in the performance of his professional duties." Which was the more applicable in the DC-10 case mentioned in Chap. 2, where an engineer was told to ignore a situation he believed threatening to the public safety on the basis of a business decision made in the interests of his company?

Recent codes have attempted to address this important area of potential conflict. The NSPE code now states: "Engineers shall hold paramount the safety, health, and welfare of the public in the performance of their professional duties." The word "paramount" means "most important or superior in rank." But even so it is unclear that the provision means engineers should never, under any circumstances, follow a client's or company's directives because they believe those directives might not serve the best interests of the public. This is an issue we will return to in Part III of our book. But here we emphasize again the need for responsible engineers who are able to make reasonable assessments of what "paramount" amounts to in cases where two professional obligations conflict.

A third limitation on codes is that they cannot serve as the final moral authority for professional conduct (Ladd, 1980, 154). To accept the current code of a professional society as the last moral word, however officially endorsed it may be, would be to lapse into a type of Ethical Conventionalism. It will be recalled that Ethical Conventionalism is the view that a particular set of conventions, customs, or laws is self-certifying and not to be questioned simply because it is the set in force at a given time or for a given place. Such a view, of course, rules out the possibility of criticizing that set of conventions from a wider moral framework.

Consider once again the following entry in the pre-1979 versions of the NSPE code: "He [the engineer] shall not solicit or submit engineering proposals on the basis of competitive bidding." This prohibition was felt by the NSPE to best protect the public safety by discouraging cheap engineering proposals which might slight safety costs in order to win a contract. Critics of the prohibition, however, contended that it mostly served the self-interest of engineering firms and actually hurt the public by "preventing" the lower prices that might result from greater competition. In a 1978 decision, National Society of Professional Engineers vs. United States, the Supreme Court ruled that the ban on competitive bidding was unconstitutional and not appropriate in a code of ethics.

The point here is not who holds the correct moral view on this issue—that is a matter of ongoing debate and discussion. And indeed, it is precisely our point that no pronouncement by a code current at any given time should ever be taken as the final word silencing such healthy debates. Codes, after all, represent a compromise between differing judgments, sometimes developed amidst heated committee disagreements. As such, they have a great "signpost" value in suggesting paths through what can be a bewildering terrain of moral decision maker. But equally as such they should never be treated as "sacred canon."

The fourth limitation of codes results from their proliferation. Andrew Oldenquist (a philosopher) and Edward Slowter (an engineer and former NSPE president) point out how the existence of separate codes for different professional engineering societies can give members the feeling that ethical conduct is more "relative" than it is, and how it can convey to the public the view that none of the codes is "really right." But Oldenquist and Slowter have also demonstrated the substantial agreement to be found among the various engineering codes. These authors summarize the core concepts in each and arrange them in order of significance as having to do with (1) the public interest, (2) qualities of truth, honesty, and fairness, and (3) professional performance. They emphasize in their 1979 paper that the time has come for adoption of a unified code (Oldenquist and Slowter, 1979, 8–11). The ABET and AAES codes (see Appendix) are by no means perfect (see Study Question 4), but they are steps in the right direction.

Study Questions

1 Apply a code of ethics taken from the Appendix—or from the collection of Canadian engineering codes cited by Morrison and Hughes (1988) —to the short cases presented as study questions on page 14. Discuss the possible effectiveness of the code(s) as a deterrent to unethical behavior in these cases.

2 Comment on the following passage: "A code only sets the limits beyond which behavior will be condemned, and the moral level is not high when all or most of those who live under it always act within a hairline of those limits. Codes, in fact, are for criminals and competitors, not for professions that want to be known as dedicated" (Barzun, 1978, 67). Specifically, is this true of the engineering codes given in the Appendix?

3 Respond to the following claim: "Even if substantial agreement could be reached on ethical principles and they could be set out in a code, the attempt to impose such principles on others in the guise of ethics contradicts the notion of ethics itself, which presumes that persons are autonomous moral agents" (Ladd, 1980, 154). Is the idea of an officially prescribed, authoritative code of ethics somehow incompatible with an appreciation of the importance of moral autonomy in individuals?

4 Critique the following codes given in the Appendix:

a The AAES Code. Examples of issues for discussion are given by Unger (1986): (*i*) The fundamental principle demands "... concern for the public health and safety." Should "welfare" have been included? (*ii*) Canon 1 restricts activity to "areas of competence and experience." How does an engineer gain experience or deal with new technology? What would be the role of the generalist or manager? (*iii*) Canon 5 might conflict with Canon 6. Which is more binding? (*iv*) Canon 7 omits professional growth of subordinates. Is it important?

b The NSPE code. Consider the following two entries in the 1981 Code of the National Society of Professional Engineers. (*i*) "Engineers shall cooperate in extending the effectiveness of the profession by interchanging information and experience with other engineers and students." (*ii*) "Engineers shall not disclose confidential information concerning the business affairs or technical processes of any present or former client or employer without his consent." Suppose that the two entries come into conflict—for instance, when improving the knowledge and skill of another engineer or student might best be done by passing on confidential information. Which entry should take precedence, and why? Do you think the code should be modified so as to explicitly state which entry should take precedence?

c Other codes. Closely examine the other codes in the Appendix. Are there any entries in them which you think should not be there? Why? Are there any important omissions in the codes?

5 Discuss the Pennwalt advertisement, Fig. 3-2, in light of your understanding of engineering codes of ethics.

A BALANCED OUTLOOK ON LAW

The 1969 Santa Barbara offshore spill of 235,000 gallons of crude oil blackened 30 miles of spectacular beaches, damaged wildlife, and hurt the local tourist trade. Predictably, the disaster prompted demands for new laws and tighter controls to prevent such occurrences in the future (Lawless, 1977, 233–247). A group of Southern Californians staged a burning of gasoline credit cards issued by the offending oil company, Union Oil, only to be taken to task by a local newspaper for taking the wrong aim. The newspaper ar-

A code of ethics isn't something you post on the bulletin board.

It's something you live every day.

Suddenly everybody seems to be rediscovering ethics.

In the business community, in Congress, on the campus and in the pulpit.

We think the trend is healthy. And needed. So we'd like to disclose a discovery of our own on this subject.

We found a long time ago that when it comes to any sort of corporate decree, the more you reduce it to writing the more you reduce participation.

It's much better, we learned, to create a working environment in which communication is a two-way process. And corporate goals are shared.

So that your code of ethics is expressed not in a news release, but in the release of appropriate thought and action.

Nobody's perfect, but it seems to work.

As our chairman put it: "The character of this company is simply a reflection of how Pennwalt people think and act. *That's* our code of ethics."

And so it is.

Admittedly, it's an approach that places more stress on the integrity and good judgment of our people than on manuals from Personnel. (A *lot* more stress.)

But it pays off. In pride. In performance. In a belief that the work we do is important. And in the enhancement of our worldwide reputation.

You might say it's the difference between a bulletin that goes up on the board, and the life that goes on every day.

(We have a brief booklet on corporate citizenship which we believe covers this subject. If you'd like one, just write our Director of Corporate Communications.)

Pennwalt Corporation, Three Parkway, Philadelphia, Pa. 19102.

For 126 years we've been making things people need—including profits.

Courtesy of Penwalt Corporation.

FIGURE 3-2
A statement on codes of ethics (see Study Question 5).

gued that the gas station operators, who would suffer the most from a boycott, were not at fault. The real offenders were the federal authorities who required less stringent safeguards in offshore drilling than did California state authorities, it was claimed.

Yet we might well ask, who would be involved in drafting safety regulations for offshore drilling? Obviously experienced petroleum engineers, geologists, and well drillers, members of the same group which had prepared the state regulations and who—in their capacity as oil company employees—had also conducted the drilling off the coast of Santa Barbara. If expert knowledge was available, then why was it not applied, law or no law?

It is worth noting that some safeguards were indeed required by federal law. Following the Santa Barbara incident, then Secretary of the Interior Walter Hickel ordered an inspection of the thousands of offshore oil wells, mostly in the Gulf of Mexico. The inspection showed that hundreds lacked mandatory safety chokes. Hickel ordered prosecutions and later justified his tough approach to pollution with what has been called "Hickel's law": "You've got to hit them [i.e., polluters] with a two-by-four to make them believe you" (Rosenbaum, 1977, 129).

Is it really necessary to burden engineering practice with ever more—and increasingly restrictive—rules? Earlier we discussed the bases for responsible action. Here we shall examine the role of formal rules and their ethical implications. The model of engineering as social experimentation will assist us again as we consider the interaction of rules with the engineering process. The problem of product liability and safe design will be postponed until we take up a more detailed analysis of risk in Chap. 4.

A Regulated Society

In order to live, work, and play together in harmony as a society, we need to carefully balance individual needs and desires against collective needs and desires. Ethical conduct, which by definition includes a strong element of altruism, provides such a balance. Unfortunately people all too frequently disagree on what constitutes right action in specific instances, even when they agree on ultimate goals. At such times we need to negotiate, and if a compromise can be agreed upon, it should be recorded for repeated reference and use.

Engineers can play an active role in establishing or changing rules as well as in enforcing them. Indeed, some people would say that the engineer's ethical duties should be limited to just such activities—in addition to following accepted rules of conduct, of course (Florman, 1978, 30–33).

At various times in history, and in various countries, engineers have had less say in how rules affecting their work were made or carried out, except perhaps for a few who were among a ruler's trusted advisors; often engineers were merely subject to those rules. We assume that the time of Hammurabi fits this description.

1758 B.C.: Babylon's Building Code Hammurabi as King of Babylon was concerned with strict order in his realm, and he decided that the builders of his time should also be governed by his laws. Thus he decided as follows:

> If a builder has built a house for a man and has not made his work sound, and the house which he has built has fallen down and so caused the death of the householder, that builder shall be put to death. If it causes the death of the householder's son, they shall put that builder's son to death. If it causes the death of the householder's slave, he shall give slave for slave to the householder. If it destroys property he shall replace anything it has destroyed; and because he has not made sound the house which he has built and it has fallen down, he shall rebuild the house which has fallen down from his own property. If a builder has built a house for a man and does not make his work perfect and the wall bulges, that builder shall put that wall into sound condition at his own cost (Hammurabi; also quoted by Firmage, 1980, 7).

The substantive or normative part of Babylon's building code is admirably succinct. The procedural aspects would find little approval today, although we cannot help but wistfully reflect on how small a bureaucracy it probably took to maintain standards. One can imagine how builders passed on their carefully drawn rules for sound design from generation to generation. There was indeed a powerful incentive for self-regulation! In other words, the law was broad and the specifics of how to comply with it were left to those presumably best able to formulate them for each application—the builders themselves. We might note that this was by no means a simple matter, for "the Babylonians found only deep alluvium in their flood plains between the Tigris and the Euphrates, which settled under the weight of their cities" (Sowers, 1970, 389).

Let us turn to another example, some four millennia later. In this case, procedural aspects and regulations in detail took the form of ready-made standards.

A.D. 1852: The U.S. Steamboat Code Early steam engines were large and cumbersome. So to make them more practical for use, James Watt, and later Oliver Evans and Richard Trevethick, increased steam pressure, did away with the condenser in some models, and thus ushered in the age of compact, portable sources of motive power. In spite of these pioneers' careful calculations and guidelines, however, boiler explosions were frequent, particularly on steamboats. Nowhere was the problem as acute as in the United States, where riverboats were vying for trade on the great midwestern rivers. Races were common, boilers were stressed beyond their limits, and safety valves were disabled to keep steam pressure up; 233 explosions contributed to a total of 2563 persons killed and 2097 injured during the period 1816 to 1848. One explosion alone, on the *Moselle* in Cincinnati in 1838, claimed 151 lives (Burke).

Demands for safety rules finally moved Congress to exert its river and interstate regulatory powers. Steamboat interests objected. It was argued that the prudent self-interest of steamboat owners and operators would in itself dictate caution. Cumbersome rules and an unyielding bureaucracy were predicted. But self-regulatory commitments by owners and operators were clearly not in evidence. So in 1838 a law was passed which provided for the inspection of the safety features of ships and their boilers and engines. The occurrence of an explosion was to be taken as prima facie evidence of negligence and any loss of life was to be considered manslaughter.

But the 1838 law turned out to be ineffective. Shipowners could find corruptible inspectors. Even honest inspectors were not much help. They had no training and the law did not specify how a safety check should be conducted. Nevertheless, after a safety check was carried out, a shipowner could claim to be blameless. Boiler explosions continued unabated.

Among those who were troubled by the situation was Alfred Guthrie, an engineer from Illinois. Guthrie had inspected, at his own expense, about 200 steamboats to learn the causes of boiler explosions and written a report on his findings. His recommendations were published by a Senator Shields of Illinois and included in Senate documents. By 1852, when a new steamboat bill came before Congress, the groundwork had been carefully laid, and an effective law was passed. Guthrie was made the first supervisor of the regulatory agency established by the law.

Congress was able to intervene as it did because of its powers to regulate interstate shipping. But even then it was left to ad hoc associations, insurance companies, and later the American Society of Mechanical Engineers to promulgate the standards which would govern the manufacture of steam boilers and their operation in mines, factories, and railroads. In France, boiler safety standards were earlier and more rapidly promulgated under the more centralized state authority of the Napoleonic code. Between 1823 and 1830 a committee of engineers, assisted by prominent scientists of the time, developed accurate steam tables, stress values for metals, and design standards which called for hemispherical end plates and initial testing of boilers at three times their expected operating pressure. France had very few boiler explosions thereafter—nor did the United States after 1852.

The Trend Toward Greater Detail

In Hammurabi's time one could let the law take care of building failures *after* the structure had failed. While many houses may have crumbled, there were probably not many casualties associated with any one occasion (unless an earthquake had struck). However, when 150 passengers and crew members can be killed all at once by a boiler explosion and the ship is likely to sink, there will be demands for rules which prevent such accidents from happening in the first place. As technology's machines became more complex, simplicity in rule-making appeared to be doomed. The 1852 steamboat law even had to regulate the qualifications of steamboat inspectors.

But lawmakers cannot be expected always to keep up with technological development. Nor would we necessarily want to see laws changed with each new innovation. What is needed are regulating agencies and commissions—the Food and Drug Administration (FDA), Federal Aviation Agency (FAA), and the Environmental Protection Agency (EPA) are examples of these in the United States—to fill the void. These agencies employ experts who can set up precise regulations. And even though they are independent and belong to neither the judicial nor the executive branches of government, their rules have, for all practical purposes, the effect of law.

Industry tends to complain that excessive restrictions are imposed on it by regulatory agencies. But one needs to reflect on why regulations may have been necessary in the first place. Take, for example, the U.S. Consumer Product Safety Commission's rule for baby cribs which specifies that "the distance between components (such as slats, spindles, crib rods, and corner posts) shall not be greater than 2⅜ inches at any point." This rule came about because some manufacturers of baby furniture had neglected to consider the danger of babies strangling in cribs or had neglected to measure the size of babies' heads (Lowrance, 1976, 134).

Again, why must regulations be so specific when broad statements would appear to make more sense? When the EPA adopted rules for asbestos emissions in 1971, it was recognized that strict numerical standards would be impossible to promulgate. Asbestos dispersal and intake, for example, are difficult to measure. So, being reasonable, EPA specified a set of work practices to keep emissions to a minimum—that asbestos should be wetted down before handling, for example, and disposed of carefully.

> [A wrecking company], after promising repeatedly to comply with the rules, came along and demolished buildings without taking any of the precautions—thereby endangering its workers and the surrounding community. The violations were so blatant, EPA felt, and civil procedures so inadequate under the Clean Air Act, that the agency asked for and received a criminal indictment.... The U.S. Supreme Court overruled the Court of Appeals... and threw out the charges. Thanks to the High Court's technical illiteracy, EPA might now be justified in attempting to prescribe voluminous measurement techniques covering all possible asbestos-generating situations since its reasonableness led to an all but unenforceable rule. The engineering community would then snicker and joke about EPA's foolishness. Wouldn't it be better for the construction industry to police itself, and for demolition instructions with regard to asbestos to be clearly specified by contract? (S. Ross, 1978, 6) [Modifications in the Clean Air Act eventually permitted EPA to issue enforceable rules on work practices, and now the Occupational Safety and Health Administration is also involved.]

Industrial Standards

There is one area in which industry usually welcomes greater specificity, and that is in regard to standards. Standards facilitate the interchange of components, they serve as ready-made substitutes for lengthy design specifications, and they decrease production costs.

Standards consist of explicit specifications which, when followed with care, assure that stated criteria for interchangeability and quality will be attained.

Examples range from automobile tire sizes and load ratings to computer languages. Table 3-1 lists purposes of standards and gives some examples to illustrate those purposes.

Standards are established by companies for in-house use, are adopted by professional associations and trade associations for industrywide use, and may also be prescribed as parts of laws and official regulations. The latter would be examples of mandatory standards, which frequently arise from lack of adherence to voluntary standards.

Standards do not help the manufacturers only; they also benefit the client and the public. They preserve some competitiveness in industry by reducing overemphasis on name brands and giving the smaller manufacturer a chance to compete. They assure a measure of quality and thus facilitate more realistic trade-off decisions.

Standards can also be a hindrance at times. For many years they were mostly descriptive, specifying, for instance, how many joists of what size should support a given type of floor. Clearly such standards tended to stifle innovation. The move to performance standards, which in the case of a floor may merely specify the required load-bearing capacity, has alleviated that problem somewhat. But other difficulties can arise when special interests (e.g., manufacturers, trade unions, exporters and importers) manage to impose unnecessary provisions on standards, or remove important provisions from them, to secure their own narrow self-interest. Requiring metal conduits for home wiring is one example of this problem. Modern conductor coverings have eliminated the need for metal conduit in numerous applications, but many localities still require it. Its use sells more conduit and labor time for installation.

There are standards nowadays for practically everything, it seems, and

TABLE 3-1
TYPES OF STANDARDS

Criterion	Purpose	Selected examples
Uniformity of physical properties and functions	Accuracy in measurement; interchangeability; ease of handling	Standards of weights; screw thread dimensions; standard time; film size
Safety and reliability	Prevention of injury, death, and loss of income or property	National Electric Code; boiler code; methods of handling toxic wastes
Quality of product	Fair value for price	Plywood grades; lamp life
Quality of personnel and service	Competence in carrying out tasks	Accreditation of schools; professional licenses
Use of accepted procedures	Sound design; ease of communications	Drawing symbols; test procedures
Separability	Freedom from interference	Highway lane markings; radio frequency bands

consequently we frequently assume that stricter regulation exists than may actually be the case. The public tends to trust implicitly the National Electrical Code in all matters related to power distribution and wiring, but how many people realize that this code, issued by the National Fire Protection Association, is primarily oriented toward *fire* hazards? Only recently have its provisions against electric shock begun to be strengthened. Few consumers know that an Underwriter Laboratories seal prominently affixed to the cord of an electrical appliance may pertain to the cord only and not to the rest of the device. In a similar vein, a patent notation inscribed on the handle of a product may refer just to the handle, and then possibly only to the design of the handle's appearance.

Sometimes standards are thought to apply when in actuality there is no standard at all. An example can be found in the widely varying worth and quality of academic degrees—doctorates are even available from mail order houses. Appearances can be misleading in this respect. Years ago when competing foreign firms were attempting to corner the South American market for electrical fixtures and appliances, one manufacturing company had a shrewd idea. It equipped its light bulbs with extra-long bases and threads. These would fit into the competitors' lamp sockets and its own deep sockets. But the competitors' bulbs would not fit into the deeper sockets of its own fixtures (see sketch in Fig. 3-3). Yet so far as the unsuspecting consumer was concerned, all the light bulbs and sockets continued to look alike.

Problems with the Law in Engineering

The legal regulations which apply to engineering and other professions are becoming more numerous and more specific all the time. We hear many complaints about this trend, and a major effort to "deregulate" various spheres of our lives is currently underway. Nevertheless, we continue to hear cries of "there ought to be a law" whenever a crisis occurs or a special interest is threatened.

FIGURE 3-3
The light bulb story. (*a*) Long base, deep socket: firm contact. (*b*) Short base, deep socket: no contact. (*c*) Long base, shallow socket: firm contact.

(*a*)

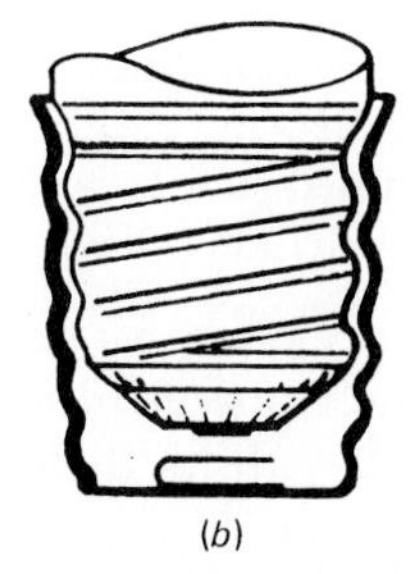
(*b*)

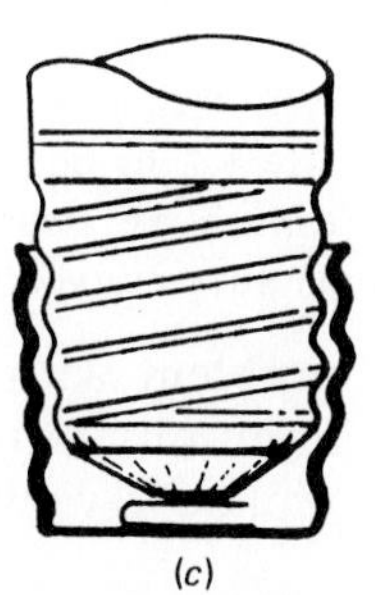
(*c*)

This is not surprising. We pride ourselves on being a nation that lives under the rule of law. We even delegate many of our decisions on ethical issues to an interpretation of laws. And yet this emphasis on law can cause problems in regard to ethical conduct quite aside from the more practical issues usually cited by those who favor deregulation.

For example, one of the greatest moral problems in engineering, and one fostered by the very existence of minutely detailed rules, is that of *minimal compliance*. Companies or individuals hunt around for loopholes in the law that will allow them to keep to its letter while violating its spirit. Or hard-pressed engineers sometimes find it convenient to refer to standards with specifications already prepared as a substitute for original thought, perpetuating the "handbook mentality" and the repetition of mistakes.

Minimal compliance led to the tragedy of the *Titanic* (Wade, 1980, 68): Why should that ship have been equipped with enough lifeboats to accommodate all its passengers and crew when British regulations in effect at the time did not require it? Or why should the Tampa Bay Bridge have been designed with possible ship collisions in mind when the code required that only wind loads, not impact loads, be considered in the calculation of horizontal forces?

On the other hand, remedying the situation by continually updating laws or regulations with further specifications may also be counterproductive. Not only will the law inevitably lag behind changes in technology, leading to a judicial vacuum; there is also the danger of overburdening the rules and the regulators. As Robert Kates puts it:

> If cooperation is not forthcoming—if the manufacturer, for example, falsifies or fails to conduct safety tests—there is something akin to the law of infinite regress in which the regulator must intrude more and more expensively into the data collection and evaluation process. In the end, the magnitude of the task overwhelms the regulators (Kates, 1977, 32).

The public is frequently lulled into a sense of security by the passage of new laws. Yet many laws are "nonlaws"—that is, laws without enforceable sanctions. These merely serve as window dressing—a false display of caring. Or a law may be burdened intentionally by its opponents with so many unreasonable provisions that a repeal will not be far off. Thus there is a need for the critical examination of many laws—and of their sources. Even Adam Smith was moved to make the following observation:

> The proposal of any new law or regulation of commerce which comes from the capitalist class ought always to be listened to with great skepticism, and ought never to be adopted until it has been examined, not only with the most scrupulous, but with the most suspicious attention (Jenkins, 1948, 156).

And still another problem with and occasion for frustration with the law is the apparent immunity with which powerful interests, including the government, can violate laws when they think they can get away with it by inviting would-be challengers to face them in lengthy and costly court proceedings.

The Proper Role of Laws in Engineering

Society's attempts at regulation have indeed often failed, and in various ways. But it would be wrong to write off rule making and rule following as futile. Good laws, effectively enforced, clearly produce benefits. They authoritatively establish reasonable minimal standards of professional conduct and provide at least a self-interested motive for most people and corporations to comply with those standards. Moreover, they serve as a powerful support and defense for those who wish to act ethically in situations where ethical conduct might be less than welcome. By being able to point to a law, one can feel freer to act as a responsible engineer.

We contend that to view engineering as social experimentation can provide engineers with a proper perspective on laws and regulations. And the rules which govern engineering practice should not be devised or construed as rules of a game but as rules of responsible experimentation.

Such a view places great responsibility on the engineer, who is intimately connected will his or her "experiment" and responsible for its safe conduct; moreover, it suggests the following conclusions: Precise rules and enforceable sanctions are appropriate in cases of ethical misconduct which involve violations of well-established and regularly reexamined procedures that have as their purpose the safety and well-being of the public. Little of an experimental nature is probably occurring in such standard activities, and the type of professional conduct required is most likely very clear-cut. In areas where experimentation is involved more substantially, however, rules must not attempt to cover all possible outcomes of an experiment, nor must they force the engineer to adopt a rigidly specified course of action. It is here that regulations should be broad, but so written as to hold the engineer accountable for his or her decisions.

Consider genetic "engineering," for example. One can foresee the time when genetic manipulation will be carried out in a routine manner under strict sets of guidelines. At present, however, the field is still so new that any rules will invariably leave uncovered some very important safety aspects. Rather than provide unintentional loopholes through such omissions, or convey a false sense of security to laboratory personnel, it would be better to issue only very general guidelines. The gist of these guidelines would be to place responsibility and accountability for unforeseen consequences on the experimenter.

Study Questions

1 How do the roles of standards, regulations, and laws differ with respect to engineering products and practice?

2 A growth in regulatory practices in the United States during the 1970s was followed by a reversal during the 1980s. Where should regulations go from here? Discuss the relevant factors from the standpoint of the engineer, the lawyer, and the (government) regulator. Discuss their roles in rule making. Consider the influences of rapidly changing technology. You may discuss these issues in generic terms or pick a particular industry and its regulator as an example (e.g., air transport and FAA; chemicals and EPA; electronic media and FCC; consumer products and FTC).

3 In 1975, Hydrolevel Corporation brought suit against the American Society of Mechanical Engineers (ASME), charging that two ASME volunteers, acting as agents of ASME, had conspired to interpret a section of ASME's Boiler and Pressure Vessel Code in such a manner that Hydrolevel's low-water fuel-cutoff for boilers could not compete with the devices built by the employers of the two volunteers. On May 17, 1982, the Supreme Court upheld the lower courts which had found ASME guilty of violating antitrust provisions and had opened the way for awarding of treble damages. (A U.S. District Court's award of $7.5 million had been found excessive by the Court of Appeals.) Writing on behalf of the 6 to 3 majority. Justice Harry A. Blackmun said: "When ASME's agents act in its name, they are able to affect the lives of large numbers of people and the competitive fortunes of businesses throughout the country. By holding ASME liable under the antitrust laws for the antitrust violations of its agents committed with apparent authority, we recognize the important role of ASME and its agents in the economy, and we help to ensure that standard-setting organizations will act with care when they permit their agents to speak for them." Acquaint yourself with the particulars of this case and discuss it as an illustration of the possible misuses of standards.

4 On February 26, 1972, the Buffalo Creek dam near Lorado, West Virginia, collapsed, "unleashing a wall of water that killed 118 persons and swept away four communities." A U.S. Senate labor subcommittee investigating the damage found that "lack of adequate design and construction measures as well as the poor planning and operation make all similar dams presently in use a serious hazard. . . . The safety factor slipped between the cracks of responsibility." Regulations of the U.S. Bureau of Mines called for inspections which had not been carried out. But, stated the Bureau's director, "Even if a bureau coalmine inspector had been at the dam site as the waters rose, his authority would have been limited to the issuance of an imminent danger order, withdrawing the mine workers on the mine property." It would not, he said, have "prevented the retaining dam from failing nor would it have been applicable to persons off the mine property in the path of the flood." The West Virginia Public Service Commission denied responsibility because it certifies dams for safety only at the time that a builder applies for a permit to build a dam. The Commission claims to have no jurisdiction over dams once they have been built (based on an Associated Press report in the *Los Angeles Times*, 1 June 1972).

A Governor's Ad Hoc Committee found that the dam had been built by a non-engineer, that inspectors should have been aware of problems, and that the engineering profession should have sounded a warning since some of its members were aware of the substandard construction. The registration system had failed in this instance, because "the specialty required by any engineer designing and constructing such a dam as that which failed, is not covered in any of the categories mentioned by the West Virginia State Registration Board. Moreover, since the technology of building such dams as this had not been developed, there was no way of

judging any competence in the persons constructing such dams" (from *The West Virginia Engineer*, December 1972, courtesy of Robert D. Miles, Purdue University).
Write an essay on the Buffalo Creek flood in which you touch upon the issues we have covered so far in this book. More technical information can be found in the book on dam failures by R. B. Jansen.

5 Should owners of passenger cars be protected against extensive front-end damage to their cars when they or other authorized drivers back-end trucks or high-riding off-road vehicles which have incompatible (or no) bumpers? Are there standards governing bumper location? What do they say, and are they enforced?

SUMMARY

Engineering is an inherently risky activity, usually conducted with only a partial knowledge of the underlying scientific laws about nature and society and often producing uncertain results and side effects. It lends itself to being viewed as an experiment on a societal scale involving human subjects. While it differs from standard experiments using control groups, it nevertheless imposes the same moral requirements on engineers that are imposed on researchers in other experimental areas involving human subjects. Most important, it requires the following: imaginative forecasting of possible bad side effects, and with this the development of an attitude of "defensive engineering"; careful monitoring of projects; respect for people's rights to give informed consent; and in general that engineers act as *responsible agents.*

Responsible agency, understood as a moral virtue, involves several features: (1) conscientious committment to live by moral values, (2) a disposition to maintain a comprehensive perspective on the context and possible consequences of one's actions, (3) autonomous, personal involvement in one's activities, and (4) an acceptance of accountability for the results of one's conduct.

There are many contemporary threats to efforts by engineers to act responsibly, as well as obstacles placed in the way of their respecting the public's right to have the knowledge needed for making informed decisions about engineering products and projects. Those threats and obstacles include the pressures caused by time schedules and organizational rules restricting free speech; the narrow division of labor which tends to cause moral "tunnel vision"; a preoccupation with legalities in a time of proliferating malpractice lawsuits; and the human tendency to divorce oneself from one's actions by placing all responsibility on an "authority" such as one's employer.

Codes of ethics promulgated by professional societies play a variety of roles: (1) inspiration, (2) guidance, (3) support for responsible conduct, (4) deterring and disciplining unethical professional conduct, (5) education and promotion of mutual understanding, (6) contributing to a positive public image of the profession, (7) protecting the status quo and suppressing dissent within the profession, and (8) promoting business interests through restraint of trade.

From the perspective of engineering as social experimentation, the emphasis of codes should be on support of responsible conduct, general guidance, and promotion of mutual understanding rather than on punishment; and the roles of protecting the status quo and promoting business should be avoided altogether. On the other hand, it should be kept in mind that codes are only a small part of engineering ethics. Their brevity renders them overly general and vague, so that some provisions occasionally contradict others. They also represent compromises between many differing viewpoints, the expression and discussion of which must never be stifled. Codes are anything but sacred writ, and should always be viewed as open to critical examination.

A balanced outlook on laws emphasizes both the necessity of laws and regulations and their limitations in governing engineering practice. Laws are necessary because people are not fully responsible and because the competitive nature of our free enterprise system does not always encourage the requisite moral initiative on the part of corporations. Their effects are limited because they encourage minimal compliance with their provisions and tend toward the kind of detailed regulation which can harm productivity and sometimes actually promote violations of the spirit of the law. Moreover, laws inevitably lag behind technological development.

The model of engineering as social experimentation allows for the importance of clear laws, effectively enforced. But it places equal emphasis on the moral responsibility of engineers—an emphasis that goes beyond merely following laws and is especially vital for those working at the frontiers of technological development.

CHAPTER 4

THE ENGINEER'S CONCERN FOR SAFETY

Pilot Dan Gellert was flying an Eastern Airlines Lockheed L-1011, cruising at an altitude of 10,000 feet, when he inadvertently dropped his flight plan. Being on autopilot control, he casually leaned down to pick it up. In doing so, he bumped the control stick. This should not have mattered, but immediately the plane went into a steep dive, scaring the 230 passengers no end. Badly shaken himself, Gellert was nevertheless able to grab the control stick and ease the plane back onto course. Though much altitude had been lost, the altimeter still read a stable 10,000 feet.

Not long before this incident, one of Gellert's colleagues had been in a flight trainer when the autopilot and the flight trainer disengaged, producing a crash on an automatic landing approach. Fortunately it all happened on simulation. But just a short time later, an Eastern Airlines L-1011 actually crashed on approach to Miami. On that flight there seemed to have been some problem with the landing gear, so the plane had been placed on autopilot at 2000 feet while the crew investigated the trouble. Four minutes later, after apparently losing altitude without warning while the crew was distracted, it crashed in the Everglades, killing 103 people.

A year later Gellert was again flying an L-1011 and the autopilot disengaged once more when it should not have done so. The plane was supposedly at 500 feet and on the proper glide slope to landing as it broke through a cloud cover. Suddenly realizing it was only at 200 feet and above a densely populated area, the crew had to engage the plane's full takeoff power to make the runway safely.

The L-1011 incidents point out how vulnerable our intricate machines and control systems can be, how failures in their working can be caused by unanticipated circumstances, and how important it is to design for proper human-machine interactions whenever human safety is involved. The reader who is curious to find out what happened in the course of Gellert's frustrating efforts to seek corrective action is referred to his account, "Insisting on Safety in the Skies" (Westin, 1981, 17–30). Here we turn to a more general discussion of the role of safety as seen by the public and the engineer.

Members of the public are "active consumers" when they use appliances to mow the lawn, wash clothes, or toast bread. The same persons are "passive consumers" of gasoline, water, and electricity because they have less control over or power of selection with regard to the latter commodities or services. Finally, they are mere "bystanders" when they are exposed to pollution from sources beyond their control. The "engineers" are those members of the engineering and allied professions who are knowledgeable in the design, manufacture, application, and operation of a specific engineered product. They may act as individual entrepreneurs or employees who produce and sell engineered products, buy and operate them, or educate persons to perform those activities. Gellert fit the operator category of engineer, while his passengers were passive consumers.

Thus typically several groups of people are involved in safety matters, each with its own interests at stake. If we now consider that within each group there are differences of opinion regarding what is safe and what is not, it becomes obvious that "safety" can be an elusive term. It behooves us, therefore, to decide upon a working definition of safety. And to help us understand the subject even more fully, we will discuss it in conjunction with the term "risk." Following a look at these basic concepts, we will then turn to safety and risk assessment and methods of reducing risk (increasing safety). Finally, by way of examining the nuclear power plant accidents at Three Mile Island and Chernobyl, we will consider the implications of an ever-growing complexity in engineered systems and the ultimate need for "safe exits."

SAFETY AND RISK

We demand safe products and services because we do not wish to be threatened by potential harm, but we also realize that we may have to pay for this safety. To complicate matters, what may be safe enough for one person may not be so for someone else—either because of different perceptions about what is safe or because of different predispositions to harm. For example, a power saw in the hands of a child will never be as safe as it can be in the hands of an adult. And a sick adult is more prone to suffer ill effects from air pollution than is a healthy adult.

Absolute safety, in the sense of a degree of safety which satisfies all individuals or groups under all conditions, is neither attainable nor affordable.

Yet it is important for our discussion here that we come to some understanding of what we mean by safety.

The Concept of Safety

One approach to defining "safety" would be to render the notion thoroughly subjective by defining it in terms of whatever risks a person judges to be acceptable. Such a definition was given by William W. Lowrance: "*A thing is safe if its risks are judged to be acceptable*" (Lowrance, 1976, 8). This approach helps underscore the notion that judgments about safety are tacitly value judgments about what is acceptable in the way of risk to a given person or group. Differences in appraisals of safety are thus correctly seen as reflecting differences in values.

Lowrance's definition, however, needs to be modified, for it departs too far from our common understanding of safety. This can be shown if we consider three types of situations that can arise. Imagine, first, a case where we seriously *underestimate* the risks of something—say of using a toaster we see at a garage sale. On the basis of that mistaken view, we judge it to be very safe and buy it. On taking it home and trying to make toast with it, however, it sends us to the hospital with a severe electric shock and burn. Using the ordinary notion of safety, we conclude we were wrong in our earlier judgment: The toaster was not safe at all! Given our values and our needs, its risks should not have been judged acceptable earlier. Yet by Lowrance's definition we would be forced to say that prior to the accident the toaster was entirely safe since, after all, at that time we had judged the risks to be acceptable.

Consider, second, the case where we grossly *overestimate* the risks of something. For example, we irrationally think fluoride in drinking water will kill a third of the populace. According to Lowrance's definition, the fluoridated water is unsafe, since we judge its risks to be unacceptable. It would, moreover, be impossible for someone to reason with us to prove that the water is in reality safe. For again, according to his definition, the water became unsafe the moment we judged the risks of using it to be unacceptable for us. But of course, our ordinary concept of safety allows us to say the water has been perfectly safe all along, in spite of such irrational judgments.

Third, there is the situation in which a group makes no judgment at all about whether the risks of a thing are acceptable or not—they simply do not think about it. By Lowrance's definition this means the thing is neither safe nor unsafe with respect to that group. Yet this is somewhat paradoxical, given our ordinary ways of thinking about safety. For example, we normally say that some cars are safe and others unsafe, even though many people may never even think about the safety of the cars they drive.

The point is that there must be at least some objective point of reference outside ourselves which allows us to decide whether our judgments about safety are correct or not. An adequate definition should capture this element, without omitting the insight already noted that safety judgments are relative to people's value perspectives (Council for Science and Society, 1977, 13).

We propose to adopt as our working definition a modified version of Lowrance's definition:

> A thing is safe if, were its risks fully known, those risks would be judged acceptable in light of settled value principles.

More fully,

> A thing is safe (to a certain degree) with respect to a given person or group at a given time if, were they fully aware of its risks and expressing their most settled values, they would judge those risks to be acceptable (to a certain degree).

The objections to Lowrance's definition raised by the examples given above are met by the new definition's "knowledge" condition. And the further condition that a judgment about safety express "settled value principles" helps to rule out as irrelevant many other types of judgments that could be problematic: For example, judgments made while heavily intoxicated would not count.

Thus in our view safety is a matter of how people *would* find risks acceptable or unacceptable *if* they knew the risks and were basing their judgments on their most settled value perspectives. To this extent safety is an *objective* matter. It is a *subjective* matter to the extent that value perspectives differ. In what follows we will usually speak of safety simply as acceptable risk. But this is merely for convenience, and should be interpreted as an endorsement of Lowrance's definition only as we have qualified it.

Safety is frequently thought of in terms of degrees and comparisons. We speak of something as "fairly safe" or "relatively safe" (compared with similar things). Using our definition, this translates as the degree to which a person or group, judging on the basis of their settled values, would decide that the risks of something are more or less acceptable in comparison with the risks of some other thing. For example, when we say that airplane travel is safer than automobile travel, we mean that for each mile traveled it leads to fewer deaths and injuries—the risky elements which our settled values lead us to avoid.

We interpret "things" to include products as well as services, institutional processes, and disaster protection. The definition could therefore be extended to medicine, finance, and international affairs, to mention just a few of the "things" and "services" organized by people. And for engineers the definition would extend to the safe operation of systems and the prevention of natural or people-caused disasters.

Risks

We say a thing is "not safe" if it exposes us to unacceptable danger or hazard. What is meant by "risk"? *A risk is the potential that something unwanted and harmful may occur*. We take a risk when we undertake something or use a product or substance that is not safe. Rowe refers to the "potential for the realization of unwanted consequences from impending events" (Rowe, 1977, 24). Thus a future, possible, occurrence of harm is postulated.

Risk, like harm, is a broad concept covering many different types of unwanted occurrences. In regard to technology, it can equally well include dangers of bodily harm, of economic loss, or of environmental degradation. These in turn can be caused by delayed job completion, faulty products or systems, and economically or environmentally injurious solutions to technological problems.

Good engineering practice has always been concerned with safety. But as technology's influence on society has grown, so has public concern about technological risks increased. In addition to measurable and identifiable hazards arising from the use of consumer products and from production processes in factories, some of the less obvious effects of technology are now also making their way to public consciousness. And while the latter are often referred to as "new risks," many of them have existed for some time. They are new only in the sense that (1) they are now identifiable (because of changes in magnitude of the risks they present, having passed a certain threshold of accumulation in our environment, or because of a change in measuring techniques, allowing detection of hitherto unnoticeable traces), or (2) the public's perception of them has changed (because of education, experience, or media attention, or because of a reduction in other hitherto dominant and masking risks).

Meanwhile, natural hazards continue to threaten human populations. Technology has greatly reduced the scope of some of these, such as floods, but at the same time it has increased our vulnerability to others as they affect our ever greater concentrations of population and cause greater damage to our finely tuned technological networks.

A word here should be said about disasters. A disaster does not take place until a seriously disruptive event coincides with a state of insufficient preparedness (Dynes, 1970). Hence the *Titanic*'s collision with an iceberg did not in itself constitute a disaster, but rather an emergency. The real disaster in terms of lives lost came about because emergency preparedness was inadequate: There were too few lifeboats, and there had been no lifeboat drills worth mentioning.

And if a disaster emerges from a combination of factors, so too does a risk—in the latter case from a combination of probability and consequence. The probabilistic aspects arise out of uncertainties over the event and who its victims will be, and the severity of the risk is judged by its nature and possible consequences (Rowe, 1977, 28). All this, of course, is related to the notion of experiment, for we are speaking of the "experimental" risks connected with the introduction of new technology, the risks associated with new applications of familiar technology, and the risks arising from attempts at disaster control. So again we shall find our paradigm of engineering as social experimentation to be of some use as we unravel the ethical implications of safety and risk in regard to engineered products.

Acceptability of Risk

Having adopted a modified version of Lowrance's definition of safety as acceptable risk, we need to examine the idea of acceptability more closely. William D. Rowe says that "a risk is acceptable when those affected are generally no longer (or not) apprehensive about it" (Rowe, 1979, 328). Apprehensiveness depends to a large extent on how the risk is perceived. This is influenced by such factors as whether or not the risk is assumed voluntarily; the effects of knowledge on how the probabilities of harm (or benefit) are perceived; job-related or other pressures that cause people to be aware of or (alternatively) to overlook risks; whether or not the effects of a risky activity or situation are immediately noticeable or are close at hand; and whether or not the potential victims are identifiable beforehand. Let us illustrate these elements of risk perception by means of some examples.

Voluntarism and Control John and Ann Smith and their children enjoy riding motorcycles over rough terrain for amusement. They take *voluntary* risks—that is part of being engaged in such a potentially dangerous sport. They do not expect the manufacturer of their dirt bikes to adhere to the same standards of safety as they would the makers of a passenger car used for daily commuting. The bikes should be sturdy, but guards covering exposed parts of the engine, padded instrument panels, collapsible steering mechanisms, or emergency brakes are clearly unnecessary, if not inappropriate.

In discussing dirt bikes and the like we do not include the all-terrain three-wheel vehicles. Those represent hazards of greater magnitude because of the false sense of security they give the rider. They tip over easily. During the 5 years before they were forbidden in the United States, they were responsible for nearly 900 deaths and 300,000 injuries. About half of the casualties were children under 16.

John and Ann live near a chemical plant. It is the only area in which they can afford to live, and it is near the shipyard where they both work. At home they suffer from some air pollution, and there are some toxic wastes in the ground. Official inspectors tell them not to worry. Nevertheless they do, and they think they have reason to complain—they do not care to be exposed to risks from a chemical plant with which they have no relationship except on an involuntary basis. Any beneficial link to the plant through consumer products or other possible connections is very remote and, moreover, subject to choice.

John and Ann behave as most of us would under the circumstances: We are much less apprehensive about the risks to which we expose ourselves voluntarily than those to which we are exposed involuntarily. "We are loath to let others do unto us what we happily do to ourselves" (Starr, quoted in Lowrance, 1976, 87). In terms of our "engineering as social experimentation" paradigm, people are more willing to be the subjects of their own experiments (social or not) than of someone else's.

Intimately connected with this notion of voluntarism is the matter of *control.* The Smiths choose where and when they will ride their bikes. They have selected their machines and they are proud of how well they can control them (or think they can). They are aware of accident figures, but they tell themselves those apply to other riders, not to them. In this manner they may well display the characteristically unrealistic confidence of most people when they believe hazards to be under their control (Slovic, Fischhoff, and Lichtenstein, April 1979, June 1980, and May 1979). But still, riding motorbikes cross-country, skiing, hanggliding, horseback riding, boxing, and other hazardous sports are usually carried out under the implied control of the participants, which is a good part of why they are engaged in voluntarily at all and why their enthusiasts worry less about their risks than the dangers of, say, air pollution or airline safety. Another reason for not worrying so much about the consequences of these sports is that rarely does any one accident injure any appreciable number of innocent bystanders.

Effect of Information on Risk Assessments The manner in which information necessary for decision making is presented can greatly influence how risks are perceived. The Smiths are careless about using seat belts in their car. They know that the probability of their having an accident on any one trip is infinitesimally small. Had they been told, however, that in the course of 50 years of driving, at 800 trips per year, there is a probability of 1 in 3 that they will receive at least one disabling injury, their seat belt habits (and their attitude about seat belt laws) would likely be different (Arnould, 1981, 35).

Studies have verified that a change in the manner in which information about a danger is presented can lead to a striking reversal of preferences about how to deal with that danger. Consider, for example, an experiment in which two groups of around 150 people each were told about the strategies available for combatting a disease. The first group was given the following description:

> Imagine that the U.S. is preparing for the outbreak of an unusual Asian disease, which is expected to kill 600 people. Two alternative programs to combat the disease have been proposed. Assume that the exact scientific estimate of the consequences of the programs are as follows:
>
> If Program A is adopted, 200 people will be saved. . . .
>
> If Program B is adopted, there is ⅓ probability that 600 people will be saved, and ⅔ probability that no people will be saved. . . .
>
> Which of the two programs would you favor? (Tversky and Kahneman, 1981, 453)

The researchers reported that 72 percent of the respondents selected Program A, and only 28 percent selected Program B. Evidently the vivid prospect of saving 200 people led many of them to feel averse to taking a risk on possibly saving all 600 lives.

The second group was given the same problem and the same two options, but the options were worded differently:

> If Program C is adopted 400 people will die. . . .
>
> If Program D is adopted there is ⅓ probability that nobody will die, and ⅔ probability that 600 people will die. . . .
>
> Which of the two programs would you favor? (Tversky and Kahneman, 1981, 453)

This time only 22 percent chose Program C, which is the same as Program A. 78 percent chose Program D, which is identical to Program B.

One conclusion that we draw from the experiment is that options perceived as yielding firm gains will tend to be preferred over those from which gains are perceived as risky or only probable. A second conclusion is that options emphasizing firm losses will tend to be avoided in favor of those whose chances of success are perceived as probable. In short, people tend to be more willing to take risks in order to avoid perceived firm losses than they are to win only possible gains.

The difference in perception of probable gain and probable loss is illustrated graphically in Fig. 4-1. The typical risk-benefit value function shown there drops more steeply on the loss portion than it rises on the gain portion. We have included on this graph a loss side threshold, as does Rowe (Rowe, 1979, 331). The threshold is ascribable to the human habit of ignoring smaller hazards in order to avoid anxiety overload and means that no value is attached to a first small amount of loss or that no effort is expended to overcome the loss. We have added a similar, though smaller, threshold on the gain side to account for the normal human inertia and a certain amount of inherent generosity that often restrain people in how they set about seeking their own gain.

The thresholds are significant because they remind us that different people have different tolerances for specific conditions. Someone with a respiratory illness, for example, will react to the smallest amount of air pollution; thus the threshold for pollution is near zero for that person. An entire population living near an oil refinery, however, will have a fairly high tolerance (i.e., a large threshold) as far as automobile emissions alone are concerned.

Job-Related Risks John Smith's work in the shipyard has in the past exposed him to asbestos. He is aware now of the high percentage of asbestosis cases among his coworkers, and after consulting his own physician was told that he was slightly affected himself. Even Ann, who works in a clerical position at the shipyard, has shown symptoms of asbestosis as a result of handling her husband's clothes. Earlier John saw no point to "all the fuss stirred

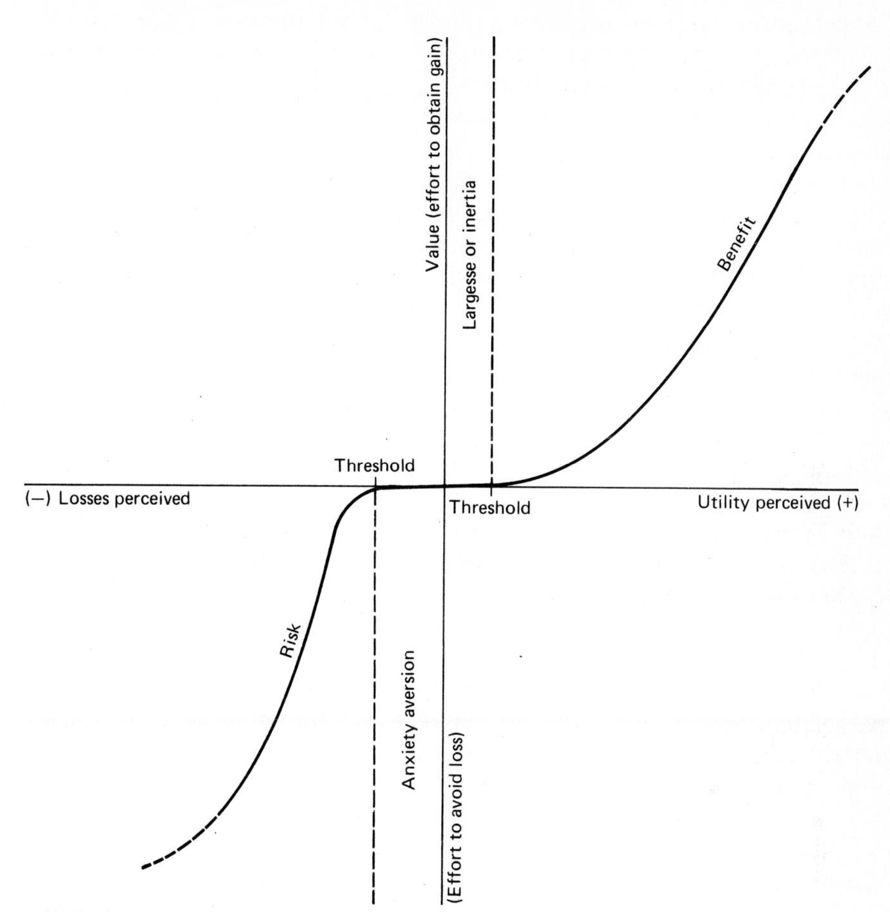

FIGURE 4-1
Typical risk-benefit value function.

up by some do-gooders." He figured that he was being paid to do a job, he felt the masks which were handed out occasionally gave him sufficient protection, and he thought the company physician was giving him a clean bill of health.

In this regard John's thinking is similar to that of many workers who take risks on their jobs in stride, and sometimes even approach them with a bravado attitude. Of course exposure to risks on a job is in a sense voluntary since one can always refuse to submit oneself to them, and workers perhaps even have some control over how their work is carried out. But often employees have little choice other than to stick with what is for them the only available job and to do as they are told. What they are often not told about is their hidden exposure to toxic substances and other dangers. Unions and occupational health and safety regulations (such as right to know rules regarding toxics) can correct the worst situations, but standards regulating conditions in the workplace (its air quality, for instance) are generally still far below those which regulate conditions in our general (public) environment. It may be argued that the "public" encompasses many people of only marginal health whose low thresholds for pollution demand a fairly clean environment. On the other hand factory workers are seldom carefully screened for their work. And in all but the most severe environments (those conducive to black lung or brown lung, for instance), unions display little desire for change lest necessary modifications of the workplace force employers out of business altogether.

Engineers who design and equip work stations must take into account the cavalier attitude toward safety shown by many workers, especially when their pay is on a piecework basis. And when one worker complains about unsafe conditions, but others do not, the complaint should not be dismissed as coming from a crackpot. Any report regarding unsafe conditions merits serious attention.

Pilot Dan Gellert's reports about problems with the L-1011s should have been looked into promptly. Later attempts to discredit him through psychiatric examinations may have been used partly as a legal ploy, but management might also have found it hard to believe that anyone could be so particular about safety. And yet Gellert was properly concerned about his responsibilities as a pilot—and all the more so because he had an airliner full of passengers and crew relying on his skill and the integrity of the plane rather than just himself to take into consideration. Which brings us to yet another factor that colors our perception of risks: their magnitude and proximity.

Magnitude and Proximity Our reaction to risk is affected by the *dread* of a possible mishap, both in terms of its magnitude and of the personal identification or relationship we may have with the potential victims. A single major airplane crash, the specter of a child we know trapped in a cave-in—these affect us more acutely than the ongoing but anonymous carnage on the highways, at least until someone close to us is killed in a car accident.

In terms of numbers alone we feel much more keenly about a potential risk if one of us out of a group of 20 intimate friends is likely to be subjected to great harm than if it might affect, say, 50 strangers out of a larger group of 1000. This proximity effect is noticeable in the time domain as well. A future risk is easily dismissed by various rationalizations including (1) the attitude of "out of sight, out of mind," (2) the assumption that predictions for the future must be discounted by using lower probabilities, or (3) the belief that a countermeasure will be found in time.

The numbers game can easily make us overlook losses which are far greater than the numbers would reveal by themselves. Consider the 75 men lost when the unfinished Quebec Bridge collapsed in 1907. As William Starna relates,

> Of those 75 men, no fewer than 35 were Mohawk Indians from the Caughnawaga Reserve in Quebec. Their deaths had a devastating effect on this small Indian community, altering drastically its demographic profile, its economic base, and its social fabric. Mohawk steelworkers would never again work in such large crews, opting instead to work in small groups on several jobs. Today, Mohawk high-steelworkers remain among the highest regarded and most skilled in their field (Starna, 1986).

Two other examples which involved large-scale disruptions of communities were mentioned earlier: the Buffalo Creek flood (Study Question 4, p. 102.) and Bhopal (discussed in Chap. 7). The forceful evacuation of Pripyat next to the Chernobyl nuclear power plant is part of the story of the reactor failure discussed later in this chapter.

Lessons for the Engineer

Engineers in their work face two problems in regard to public conceptions of safety. On the one hand there is the overoptimistic attitude that things which are familiar, which have not hurt us before, and over which we have some control, present no real risks. On the other hand is the dread people feel when accidents kill or maim in large numbers or harm those we know, even though statistically speaking such accidents might occur infrequently.

Leaders of industry are sometimes heard to proclaim that those who fear the effects of air pollution, toxic wastes, or nuclear power are emotional and irrational or politically motivated. This in our view is a misperception of legitimate concerns expressed publicly by thoughtful citizens. It is important that engineers recognize as part of their work reality such widely held perceptions of risk and take them into account in their designs. It is not wise to proceed under the assumption that "education" will quickly change the public's underestimation or overestimation of risk. As Paul Slovic, Baruch Fischhoff, and Sarah Lichtenstein point out:

> Another barrier to educational attempts is that people's beliefs change slowly and are extraordinarily resistant to new information. Research in social psychology has often demonstrated that once formed, people's initial impressions tend to structure the way they interpret subsequent information. They give full weight to evidence that is consistent with their initial beliefs while dismissing contrary evidence as unreliable, erroneous, or unrepresentative. Whereas opponents of nuclear power believe the accident at Three Mile Island "proved" how dangerous reactors are, proponents felt that it confirmed their faith in the effectiveness of the multiple safety and containment system (Slovic, Fischhoff, and Lichtenstein, 1980, 48).

And in regard to professionals they continue:

> Since even well-informed citizens have difficulty in judging risk accurately, and the cognitive functioning of experts appears to be basically like that of everyone else, it seems clear that no one person or profession knows how to get the right answers. The best we can hope to do is to keep the particular kinds of mistakes to

> which each of us is prone to a minimum by being more aware of our tendency to make mistakes (Slovic, Fischhoff, and Lichtenstein, 1980, 48).

Finally, in regard to wisdom over which no one holds a monopoly, Slovic writes:

> Perhaps the most important message from this research [on risk perception] is that there is wisdom as well as error in public attitudes and perceptions. Lay people sometimes lack certain information about hazards. However, their basic conceptualization of risk is much richer than that of experts and reflects legitimate concerns that are typically omitted from expert risk assessments. As a result, risk communication and risk management efforts are destined to fail unless they are structured as a two-way process. (Slovic, 1987)

Study Questions

1 Describe a real or imagined traffic problem in your neighborhood involving children and elderly people who find it difficult to cross a busy street. Put yourself in the position of (*a*) a commuter traveling to work on that street, (*b*) the parent of a child, or the relative of an older person, who has to cross that street on occasion, (*c*) a police officer assigned to keep the traffic moving on that street, and (*d*) the town's traffic engineer working under a tight budget.

Describe how in these various roles you might react to (*i*) complaints about conditions dangerous to pedestrians at that crossing and (*ii*) requests for a pedestrian crossing protected by traffic lights.

2 In some technologically advanced nations, a number of industries which have found themselves restricted by safety regulations have resorted to dumping their products on, or moving their production processes to, less-developed countries where higher risks are tolerated. Examples are the dumping of unsafe or ineffective drugs on the Third World by pharmaceutical companies from Western Europe, communist bloc countries, Japan, and the United States (Silverman, Lee, and Lydecker, 1981) and the transfer of asbestos processing from the United States to Mexico (Shue, 1981, 586). To what extent do differences in perception of risk justify the transfer of such merchandise and production processes to other countries? Is this an activity that can or should be regulated?

3 The industrial accident described below and illustrated in Fig. 4-2 raises several issues of ethical import. Identify and discuss them.

The following news story is based on the Nassau edition of Newsday, *the Long Island, N.Y., newspaper, April 24, 1981.*

Inadequate safety precautions and an accident inside an empty water tank caused the deaths of two workmen in New Jersey on April 23. At 4 p.m., a scaffold inside the tank collapsed and caused the two men painting the tank to fall to the bottom. Stranded there, they were overcome by paint fumes and eventually lost consciousness. John Bakalopoulos, 34, of Brooklyn, N.Y. and Leslie Salomon, 31, also of Brooklyn, were not wearing oxygen masks. The Suffolk County Water Authority's contract for the painting job specified that workmen wear "air hoods," masks connected to air compressors. The masks were available, but Bakalopoulos and Salomon had decided not to wear them because they

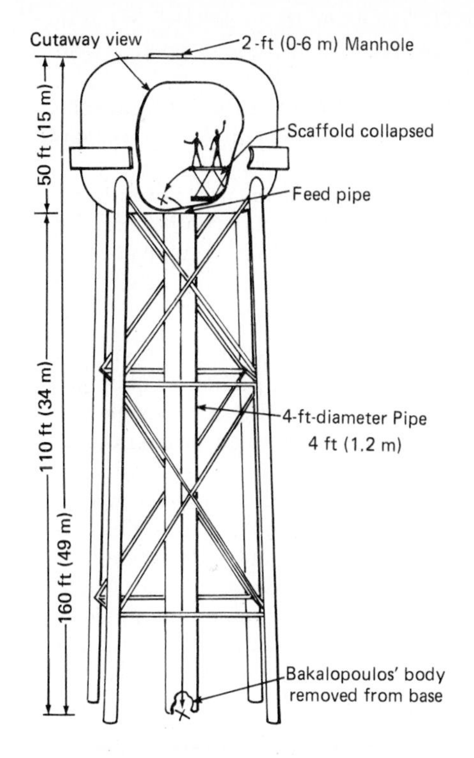

FIGURE 4-2
Accident in a water tank. (*Source:* Opflow, *American Water Works Association, vol. 7, no. 6, June 1981, p. 3, using material from* Newsday, *Nassau edition, Long Island, New York, April 24, 1981.*)

were unwieldy. Instead, Bakalopoulos wore a thin gauze mask designed to filter out dust and paint particles. Salomon wore no mask.

Peter Koustas, the safety man who was handling the compressor and paint feed outside the tank, asked a nearby resident to call firemen as soon as he realized the scaffold had collapsed. Then he rushed into the tank with no oxygen mask, and he, too, was overcome by the fumes and lost consciousness.

The men lay unconscious for hours as rescue efforts of more than 100 policemen, firemen, and volunteers were hampered by bad weather. Intense fog, rain, and high winds made climbing the tank difficult and restricted the use of machinery. Several men collapsed from fatigue.

Inside the tank, conditions were worse. Because of the heavy fumes, rescuers used only hand-held, battery-powered lights, fearing that sparks from electric lights might cause an explosion. Lt. Larry Viverito, 38, a Centereach, N.Y. volunteer fireman, was overcome by fumes 65 ft (20 m) above the floor of the tank. Fellow rescuers had to pull him out.

Rescuer John Flynn, a veteran mountain climber, said he hoped he would never have to go through anything like that night again. For five hours he set up block-and-tackle pulleys, tied knots, adjusted straps on stretchers, and attached safety lines and double safety lines. The interior of the tank was as blindingly white as an Alpine blizzard—completely and nauseatingly disorienting. Fans that had been set up to pull fresh air into the tank caused deafening noise.

When Flynn first reached the tank floor, he stepped into the wet paint and began to slide toward the uncovered 4-ft (1.2-m) opening to the feeder pipe in the center of the floor. Flynn was able to stop sliding, but John Bakalopoulos wasn't as fortunate.

As rescuers watched helplessly, Bakalopoulos, still out of reach, stirred, rolled over, and in the slippery paint slid into the feeder pipe. He plunged 110 ft (34 m) to the bottom.

Bakalopoulos was dead on arrival at the University Hospital in Stony Brook, N.Y. Peter Koustas, rescued at 1:45 a.m. and suffering from hypothermia, died the following morning when his heart failed and he could not be revived. Only Leslie Salomon survived. (*Quoted with permission from* Opflow, *Am. Water Works Assoc., vol. 7, no. 6, June 1981, p. 3, and from* Newsday.)

4 Grain dust is pound for pound more explosive than coal dust or gunpowder. Ignited by an electrostatic discharge or other causes, it has ripped apart grain silos and killed or wounded many workers over the years. When fifty-four people were killed during Christmas week 1977, grain handlers and the U.S. government finally decided to combat dust accumulation (Marshall, 1983). Ten years, 59 deaths, and 317 serious injuries later, a compromise standard has been agreed upon which designates dust accumulation of ⅛ inch or more as dangerous and impermissible. Use grain facility explosions for a case study of workplace safety and rule making.

5 The oil rig *Alexander L. Kielland* collapsed during a storm, taking 122 men to their deaths. The structure was weak because of a faulty weld. So many men died because it was difficult to launch the lifeboats. Search the literature (you may start with a chapter on the *Kielland* in Bignell and Fortune, 1984), then assess the hazards of working and living on an oil rig in the North Sea, with special emphasis on the opportunities for safe escape.

ASSESSMENT OF SAFETY AND RISK

Absolute safety is not attainable, and any improvement in safety as it relates to an engineered product is often accompanied by an increase in cost of that product. On the other hand, products that are not safe incur secondary costs to the manufacturer beyond the primary (production) costs that must also be taken into account—costs associated with warranty expenses, loss of customer goodwill and even loss of customers due to injuries sustained from use of the product, litigation, possible downtime in the manufacturing process, and so forth. It is therefore important for manufacturers and users alike to reach some understanding of the risks connected with any given product and of what it might cost to reduce those risks or not reduce them.

As Fig. 4-3 indicates, an emphasis on high safety (low risks) leads to high primary costs, but secondary costs are low. At the other extreme of high risks (low safety), one saves on primary costs but pays dearly because of high secondary costs. In between, where the slopes of the primary and secondary cost curves are equal, is the point of minimum total cost. If all costs were quantifiable, that optimum point would be the goal to reach for. But before we crank up our computers to home in on such an optimal design, we must be clear about how to determine risk (to be discussed in this section) and how to compare losses with benefits (to be covered in the next section).

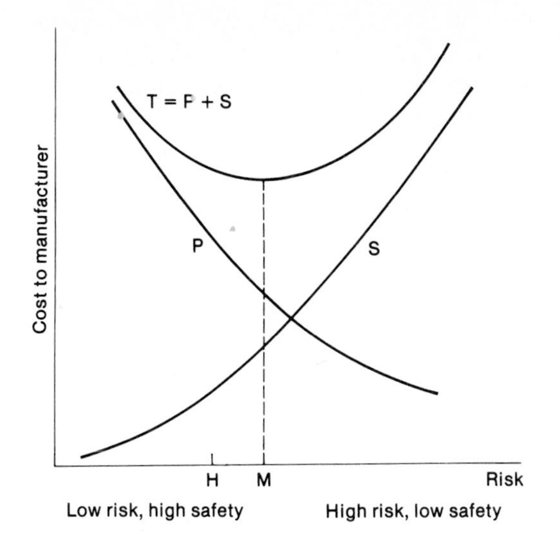

FIGURE 4-3
Why both low-risk and high-risk products are costly. *P* = primary cost of product, including cost of safety measures involved; *S* = secondary costs, including warranties, loss of customer goodwill, litigation costs, costs of down time, and other secondary costs. *T* = total cost. Minimum total cost occurs at *M*, where incremental savings in primary cost (slope of *P*) are offset by an equal incremental increase in secondary cost (slope of *S*). Highest acceptable risk (*H*) may fall below risk at least cost (*M*), in which case *H* and its higher cost must be selected as design or operating point.

Knowledge of Risk

One would think that experience and historical data would provide good information about the safety of standard products. Much has been collected and published; gaps remain, however, because (1) there are some industries where information is not freely shared, and (2) there are always new applications of old technology which render the available information less useful.

Engineers are by nature inclined to share information freely. It is in this spirit, according to R. R. Whyte in *Engineering Progress through Trouble,* that Robert Stephenson, famous bridge builder during the first half of the nineteenth century, took the following position upon reviewing a technical paper:

> ...he hoped that all the casualties and accidents, which had occurred during their progress, would be noticed in revising the paper; for nothing was so instructive to the younger Members of the Profession, as records of accidents in large works, and of the means employed in repairing the damage. A faithful account of those accidents, and of the means by which the consequences were met, was really more valuable than a description of the most successful works. The older Engineers derived their most useful store of experience from the observations of those casualties which had occurred to their own and to other works, and it was most important that they should be faithfully recorded in the archives of the Institution (R. R. Whyte, 1975, v).

We also take the following account from pp. 54–57 of the same book: In 1950 the chief engineer for a British manufacturer of large generators invited his competitors to study the failure of a rotor endbell during overspeed tests. The endbell is a retaining sleeve which holds in place the endturns of the rotor winding so they will not fly apart at the high speed of the turbine generator. Because the endbell has to be made of nonmagnetic steel, it presents severe metallurgical problems. This engineer's action led to a similar frank divulgence of information on the occasion of another, similar, failure in Canada some years later.

Such examples are noteworthy because they are so rare—which is regrettable. Too many companies believe that releasing technical information might hurt their competitive position, if not place them at a disadvantage in case of litigation. So it is that new engineers and new companies usually have to learn from scratch, although sometimes past experience is used effectively to educate beginners (see, for example, the textbook on soil mechanics and foundations, *Introductory Soil Mechanics and Foundations*, by George B. Sowers and George F. Sowers, 1970).

Uncertainties in Design

Risk is seldom intentionally designed into a product. It arises because of the many uncertainties faced by the design engineer, the manufacturing engineer, and even the sales and applications engineer.

To start with, there is the purpose of a design. Let us consider an airliner. Is it meant to maximize profits for the airline, or is it intended to give the highest possible return on investment? The answer to that question is important to the company because on it hinge different decisions and their outcomes and the possibility of the airline's economic success or ruin. Investing $50 million in a jumbo jet to bring in maximum profits of, say, $10 million during a given time period involves a lower return on investment than spending $24 million on a medium-sized jet to bring in a return of $6 million in that same period.

Regarding applications, designs which do quite well under static loads may fail under dynamic loading. A famous example is the wooden bridge that collapsed when a contingent of Napoleon's army crossed it marching in step. This even affected one of Robert Stephenson's steel bridges, which shook violently under a contingent of marching British troops. Ever since then, soldiers are under orders to fall out of step when crossing a bridge. Wind can also cause severe vibrations: The Tacoma Narrows Bridge, which collapsed some years ago, and a high tension power line across the Bosphorus that broke after a short circuit caused by swinging cables are but two examples.

Apart from uncertainties about applications of a product, there are uncertainties regarding the materials of which it is made and the quality of skill that goes into designing and manufacturing it. For example, changing economic realities or hitherto unfamiliar environmental conditions such as extremely low temperatures may affect how a product is to be designed. A typical "handbook engineer" who extrapolates tabulated values without regard to their implied limits under different conditions will not fare well under such circumstances. Even a careful analyst will face difficulties when con-

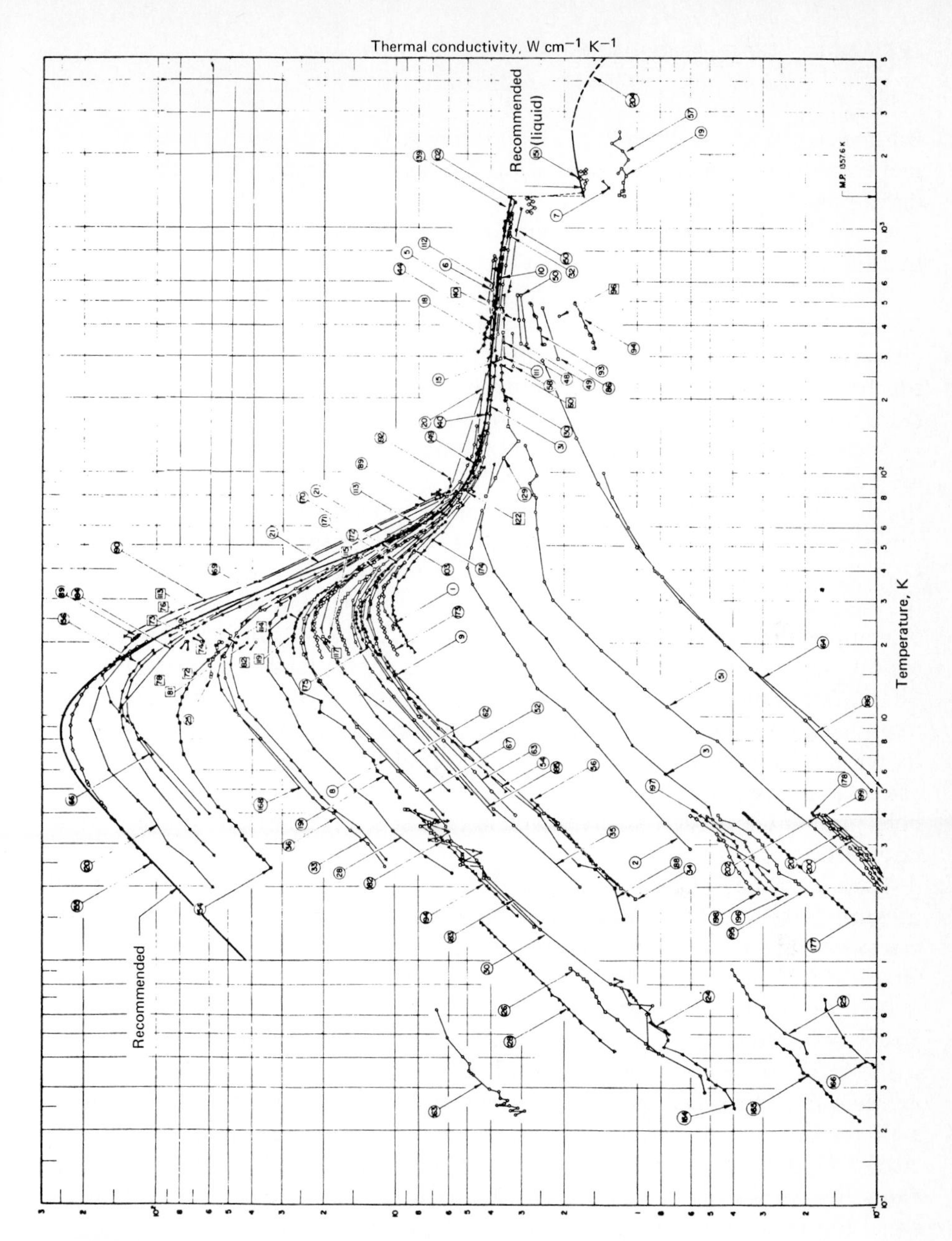

FIGURE 4-4
Thermal conductivity of copper over wide ranges of temperature as observed by different investigators—an example of the diversity in test results that can affect engineering decisions about safety. (*From D. R. Lide, Jr., "Critical Data for Critical Needs,"* Science, *vol. 212, June 19, 1981, p. 1344.*)

fronted with data such as those illustrated in Fig. 4-4, which gives the thermal conductivity of copper over wide ranges of temperature as observed by different investigators.

Caution is required even with standard materials specified for normal use. In 1981 a new bridge that had just replaced an old and trusted ferry service across the Mississippi at Praire du Chien, Wisconsin, had to be closed because 11 of the 16 flange sections in both tie girders were found to have been fabricated from excessively brittle steel (ENR, 1981). In the meantime, the ferries had disappeared! While strength tests are routinely carried out on concrete, the strength of steel is all too often taken for granted.

Such drastic variations from the standard quality of a given grade of steel are rather exceptional. More typically the variations are small. Nevertheless the design engineer should realize that the supplier's data on items like steel, resistors, insulation, optical glass, and so forth apply to statistical averages only. Individual components can vary considerably from the mean.

Engineers traditionally have coped with such uncertainties about materials or components, as well as incomplete knowledge about the actual operating conditions of their product, by introducing a comfortable "factor of safety." That factor is intended to protect against problems arising when stresses due to anticipated loads (duty) and stresses the product as designed is supposed to withstand (strength or capability) depart from their expected values. Stresses can be of a mechanical or any other nature—for example, an electric field gradient to which an insulator is exposed, or the traffic density at an intersection.

A product may be said to be safe if its capability exceeds its duty. But this presupposes exact knowledge of actual capability and actual duty. In reality the stress calculated by the engineer for a given condition of loading and the stress which ultimately materializes at the loading may vary quite a bit. This is because each component in an assembly has been allowed certain tolerances in its physical dimensions and properties—otherwise the production cost would be prohibitive. The result is that the assembly's capability as a whole cannot be given by a single numerical value but must be expressed as a probability density which can be graphically depicted as a "capability" curve (see Fig. 4-5*a* and *b*). For a given point on a capability curve, the value along the vertical axis gives the probability that the capability, or strength, is equal to the corresponding value along the horizontal axis.

A similar curve can be constructed for the duty which the assembly will actually experience. The stress exposure varies because of differences in loads, environmental conditions, or the manner in which the product is used. Associated with the capability and duty curves are nominal or, statistically speaking, expected values C and D. We often think and act only in terms of nominal or expected values. And with such a deterministic frame of mind, we may find it difficult to conceive of engineering as involving experimentation. The "safety factor" C/D rests comfortably with our consciences. But how sure can we be that our materials are truly close to their specified

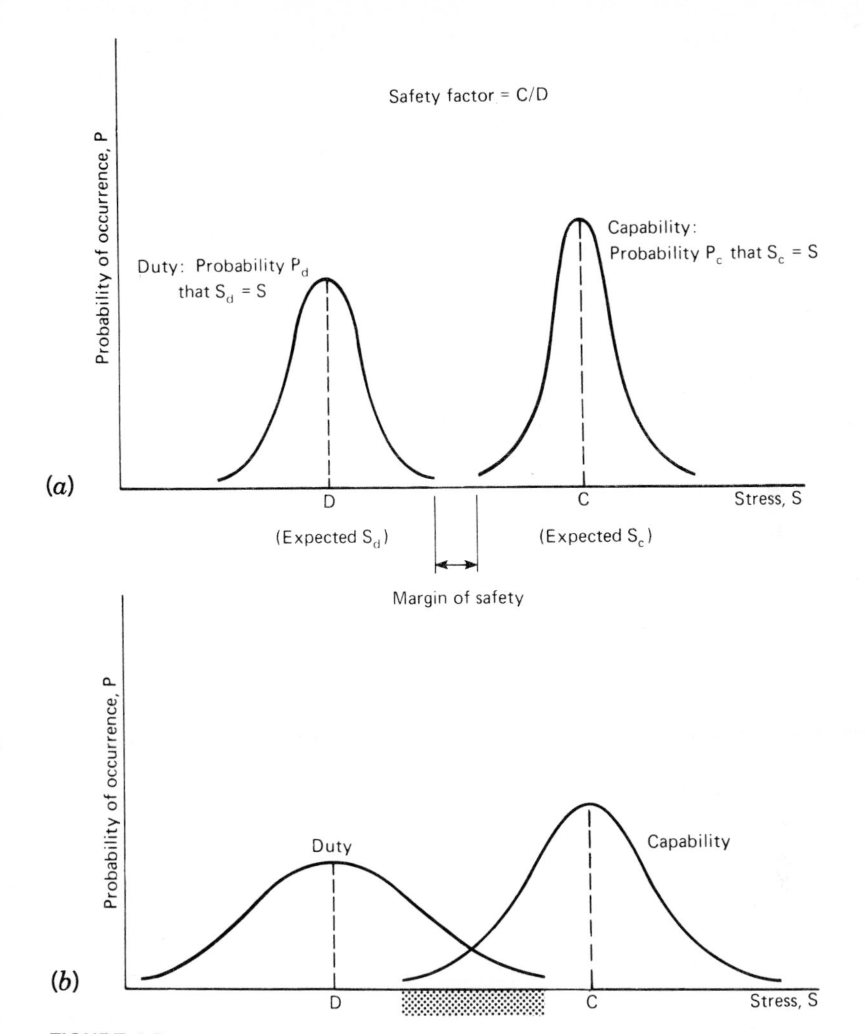

FIGURE 4-5
Probability density curves for stress in an engineered system (*a*) Variability of stresses in a relatively safe case. (*b*) An unsafe case (danger is due to some probable overlap in duty and capability stresses).

nominal properties? Or that the loads will not vary too widely from their anticipated values or occur in environments hostile to the proper functioning of the materials? It is entirely conceivable that the capability and duty curves will assume flatter shapes because of increased variances (see Fig. 4-5*b*) than they would have under normal conditions, as in Fig. 4-5*a*. And Fig. 4-5*b* shows how it is probable that load stress may exceed design stress along the shaded region of stress. Mathematical treatment of this topic is offered by, among others, Edward B. Haugen, who reminds his readers how "the safety factor concept completely ignores the facts of variability that result in different reliabilities for the same safety factor" (Haugen, 1968, 5).

A more appropriate measure of safety would be the "margin of safety," which is shown in Fig. 4-5*a*. If it is difficult to compute such a margin of safety for ordinary loads used every day, imagine the added difficulties that arise when repeatedly changing loads have to be considered. As F. Nixon, N. E. Frost, and K. J. March point out:

> The use of a general safety margin on the static strength of a material (or structure) is quite unsuitable for cyclic loadings. For example, local static stresses are usually unimportant in ductile materials because regions which become overstressed may yield, with the result that the stress becomes more uniformly distributed. However, when the loading is cyclic, yield can affect only the mean value of the local stresses to which any local discontinuities give rise. Thus, the merit of a structure subjected to dynamic loadings depends to a great extent on the skill of the designer in avoiding unnecessary concentrations of stress (Nixon, Frost, and March, 1975, 137).

Testing for Safety

The widely proclaimed "safety factor" thus obviously has some serious conceptual flaws. So what can the engineer do to assure safety? "Rely on experience" was mentioned at the outset of this discussion. But it was also pointed out that experience gained by one engineer is all too often not passed on to others, especially if the experience was professionally embarrassing. Bad news travels fast, we agree, but usually it travels unaccompanied by hard facts.

Another way of gaining experience is through tests. Under certain conditions this can be a valuable source of information, especially if the testing of materials or a product is carried out to destruction. An example of such testing was that performed on a Comet aircraft fuselage after two of those early jets had crashed in service in 1954. One conclusion emerged from initial investigations:

> ...fatigue of the pressure cabin. The only possible objection was that fatigue so early in the lives of the two aircraft seemed at that time incredible. These arguments provided the justification and the objective for the experimental attack that then followed.
>
> The fatigue-testing of a complete full-size pressure cabin and fuselage went beyond anything ever before attempted. For safety, water was used as the testing medium, and the whole fuselage was immersed in a 250,000 gal. tank.
>
> It was clearly necessary to introduce every conceivable adverse factor in addition to pressure. In particular, the wings had to be included in the test and given the appropriate load fluctuations. Total immersion of the wings was considered, but rejected in favor of their projecting through the sides of the tank, the water being held back by flexible sleeves that allowed the wings to articulate.
>
> With [special apparatus] the cabin was made to breathe in and out as the wings

> were forced to bend up and down realistically. The test went on night and day until a piece of pressure cabin 20 ft^2 in area was pushed out almost instantaneously. The failure was initiated by a fatigue crack at a corner of one of the forward windows. In the air, of course, this would have meant an explosion (De Havilland and Walker, 1975, 53).

The reader should note that this test was carried out *after* real accidents had occurred. The more usual procedure is to subject prototypes to testing. Yet there are severe limitations to relying on prototype tests, as R. R. Whyte points out:

> ...successful prototype testing and prolonged test-bed running also do not ensure a reliable machine....Like the designer, the development engineer is seldom allowed sufficient time. In many cases all he has time for is a functional test too short to reveal potential weaknesses in the design such as a fatal resonant frequency, for example. In others all that is stipulated even by customers who should know better, is that a single prototype should satisfactorily pass a "type-approval test"; maybe a few hundred operations or hours of running.
>
> Such a test can do no more than demonstrate that one individual product—seldom representative of large quantities—is capable of passing the stipulated test. It can do little to demonstrate the ultimate capability of the design. It can do nothing to indicate the variability in capability which is likely to exist when numbers of similar products are to be produced or the long-term effects such as corrosion or fatigue (R. R. Whyte, 1975, 139).

In the case of the space shuttle *Challenger* the flights occurred outside the field joints' test envelope; extrapolation to performance at lower temperatures was based more on handbook material specifications than on available engineering judgment.

Even prototype tests and routine quality assurance tests are frequently not carried out properly. Suppliers of rifles and ammunition to the Armed Forces have been found to have committed fraud in the testing of their products; the Alaska pipeline was plagued with poor welds and inadequate testing; the Ford Motor Co. at one time was found to have falsified emission test data; the B. F. Goodrich Co. delivered an aircraft brake for test flights on a new Air Force plane although the brake failed to meet Air Force specifications even according to a common sense interpretation of those specifications; and bogus parts, indistinguishable from the originals in appearance but much lower in quality, are flooding the U.S. market.

In short, we cannot trust testing procedures uncritically. Time pressure, as we have said, is one factor contributing to shoddy testing. The boredom of routine that tempts one to quickly duplicate test data of a repetitive nature can be another. At times there is pressure from management to "fudge the data for now" since "by the time we get into production we will have ironed out the problem." And not to be overlooked is the problem of outright fraud, as when testers are bribed to pass faulty items or when no tester was on the job although testing was claimed to have been undertaken. Conscientious engineers had therefore better make occasional spot checks on their own unless the organization they work for operates an independent testing service free of production pressures.

When Testing Is Inappropriate

Not all products can be tested to destruction. In such cases a simulation which traces the outcome of one or more hypothetical, risky events should be applied. A common approach is *scenario analysis,* in which one starts from a given event, then studies the different consequences which might evolve from it. Another approach, known as *failure modes and effects analysis,* systematically examines the failure modes of each component, without, however, focusing on causes or interrelationships among the elements of a complex system. In contrast to this is the *fault-tree analysis* method, in which one proposes a system failure and then traces the events back to possible causes at the component level.

Of these several techniques, the fault-tree method can perhaps most effectively illustrate the disciplined approach required to capture as much as possible of everything that affects the proper functioning and safety of a complex system. Please note, however, that no safety analysis should be attempted without a thorough understanding of the physical aspects of the system under study; the mere use of ready-made computer programs does not suffice except to facilitate intermediate calculations.

To illustrate the use of the fault-tree technique we resort to a rather simple example, a water system without filtration plant, depicted in Fig. 4-6. A fault tree for that water system is shown in Fig. 4-7. We start with the system failure at the top and work down to failures in various subsystems, components, and outside factors or events which could have given rise to the problem. Each level in the tree lists events that could have caused the problem listed in the level above it. "Could have" implies that one *or* more events could be the cause for the event at the next higher level. Sometimes one *and* another event—perhaps several events—must all occur for that next event to happen. Thus there are two types of logical statements that appear on the chart, OR and AND.

The fault tree in Fig. 4-7 has not been completed. There are several further levels which could be indicated. There are also possible omissions at the levels shown. But even though incomplete, this fault tree can give us a good qualitative sense of the types of risks to which the water system is exposed. Some analysts proceed further and attach probability figures to each event. The accuracy of such figures is always problematic, however, particularly when there is the chance of common-mode failures. An earthquake, for example, could damage not only the reservoir—it could cause any of the events in Fig. 4-7 trailed by the circled E, eventually affecting delivery of water. Say an earthquake causes crumbling of riverbanks and other damage leading to silting of the pump's source of water. If not properly filtered out, the silt

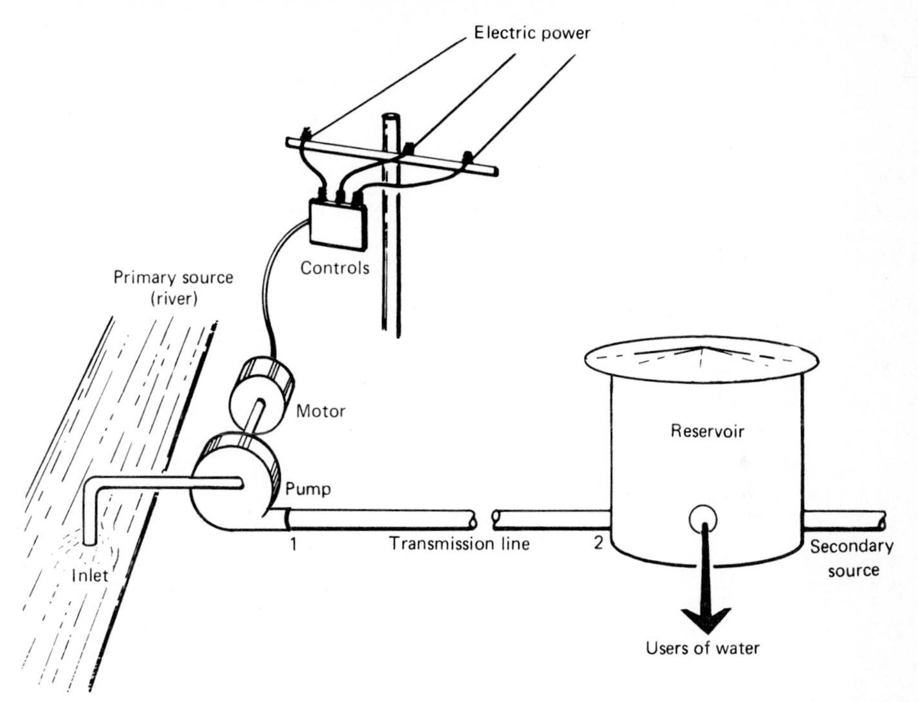

FIGURE 4-6
A simple water system.

could cause the impeller of the pump to wear out in no time at all. Alternatively the silting could cause the inlet to clog, thus stopping the inflow of water necessary to the pump's operation.

In the disaster that befell the passengers and the crew of the *Titanic,* the fact that the accident occurred at night certainly played its part as a "common-mode event" in producing the tragedy (Machol, 1975, 53). The iceberg could not be seen easily at night, the only radio operator on a nearby ship was asleep, and abandoning ship was more difficult at night than it would have been during the day. Thus because of the difficulty in foreseeing all common-mode events, one should treat the results of quantitative risk assessments with caution (see Study Question 4 below for a numerical example of this problem).

The strength of a fault-tree analysis lies in its qualitative aspects. It assists in the exposure of hitherto unforeseen situations. In a real-life water system analysis we would test for the availability of potable water and its usefulness. Water which is contaminated is not useful, even when available in large quantities at the householder's tap. Even high-quality water may not be useful when the sewer system is inoperative. On the other hand, even when no

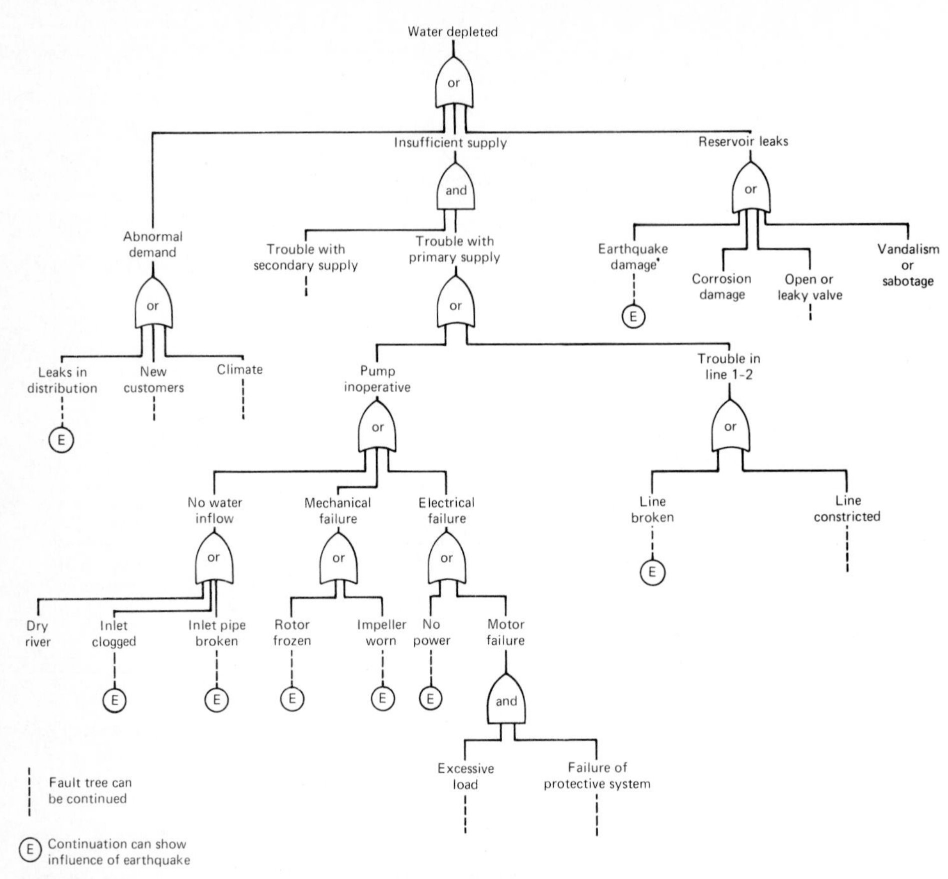

FIGURE 4-7
Fault tree for the water system shown in Fig. 4-6.

water comes out of the tap, water may still be available: from reserve tanks or from emergency delivery trucks. Thus we note that the fault-tree examination can be extended to embrace emergency measures as well.

Study Questions

1 "It is a sobering but uncertain possibility that our ability to respond to *unknown* hazards is diminished by the prevailing emphasis on control of the *known* and the specific" (Kates, 1977, *Preface,* emphasis added). Describe situations where this statement applies (for example, the past emphasis on fire hazards and the neglect of shock hazards in electrical wiring).

2 Testing is a critical step in product development and manufacture. It is not free from external influences, as the case described below reveals. Discuss how you would proceed with the testing program.

XYZ Aircraft is developing a new airplane. FAA procedures require a simulated emergency evacuation test in which the maximum number of people expected to occupy the aircraft at one time must be evacuated from it in less than 5 minutes. The test is conducted on a prototype plane in a dark hangar. It will be very expensive (insurance costs alone are considerable), and it has attracted a great deal of attention from the FAA, airline clients, and the media. The problem is that the emergency door and inflatable slide system have already revealed some serious design problems. Attempts to solve those problems have also already caused the development budget to be exceeded. Only 2 months remain until the schedule date of the test. If the plane fails it, XYZ will be greatly embarrassed, the FAA will be alerted to the problems, and an even larger financial burden will be placed on XYZ, what with the need to rerun the tests on top of the necessity still to solve the design problems themselves.

The possible actions that XYZ can take include the following:

a Expect the test to be successful. Construction of the test aircraft can proceed while design changes are made on the emergency systems. The financial savings will be great if the plane passes the test.

b Postpone the test. Allow more time for redesign. Avoid a possible rerun of the test at a high cost.

3 Draw a fault-tree diagram for the event "automobile passenger falls out of a car during accident."

4 This is a problem which involves some manipulation of logic operations and probabilistic data. It is introduced here to demonstrate that common-cause events can drastically reduce a system's (a brake's) calculated reliability.

Examine first the fault tree shown in Fig. 4-8*a*. Let P_J be the probability that event J occurs; let $\bar{P}_J$ be the probability that event J does not occur, with $\bar{P}_J = 1 - P_J$. For simplicity, let $P_D = P_E = P_F = P_G = 10^{-6}$. Then $P_B = 1 - \bar{P}_D\bar{P}_E = P_D + P_E - P_D P_E \Supset 2 \cdot 10^{-6}$. Similarly $P_C \Supset 2 \cdot 10^{-6}$. *The top event A* will then occur with probability $P_A = P_B P_C = 4 \cdot 10^{-12}$.

Now assume that the failure of the plunger in the rear half of the cylinder is not independent of the failure of the plunger in the front half. For instance, both failures could have originated from a mismatch in the properties of the plunger rubber and the brake fluid, thereby weakening the rubber. If such is the case, a different fault tree needs to be drawn, as shown in Fig. 4-8*b*. What is the value of P_A now?

5 List major failures of a particular type of structure. Describe the failures and discuss probable causes (faulty design, materials, construction, maintenance) and frequency of occurrence. You may consult earlier case studies and study questions and the following entries in the Bibliography. General: S. Ross (1984). b) Buildings: Hayward (1981); Klein et al (1982); McKaig (1962); McQuade (1979); Ransom (1981). c) Bridges: Fisher (1984); Kardos (1969); Petroski (1985). d) Dams: Jansen (1980). e) Airframes, ships, rails, etc.: Consult current periodical indexes.

RISK-BENEFIT ANALYSES AND INDUCEMENTS TO REDUCE RISK

Many large projects, especially public works, are justified on the basis of a risk-benefit analysis. The questions answered by such a study are the following: Is the product worth the risks connected with its use? What are the benefits? Do they outweigh the risks? We are willing to take on certain levels of

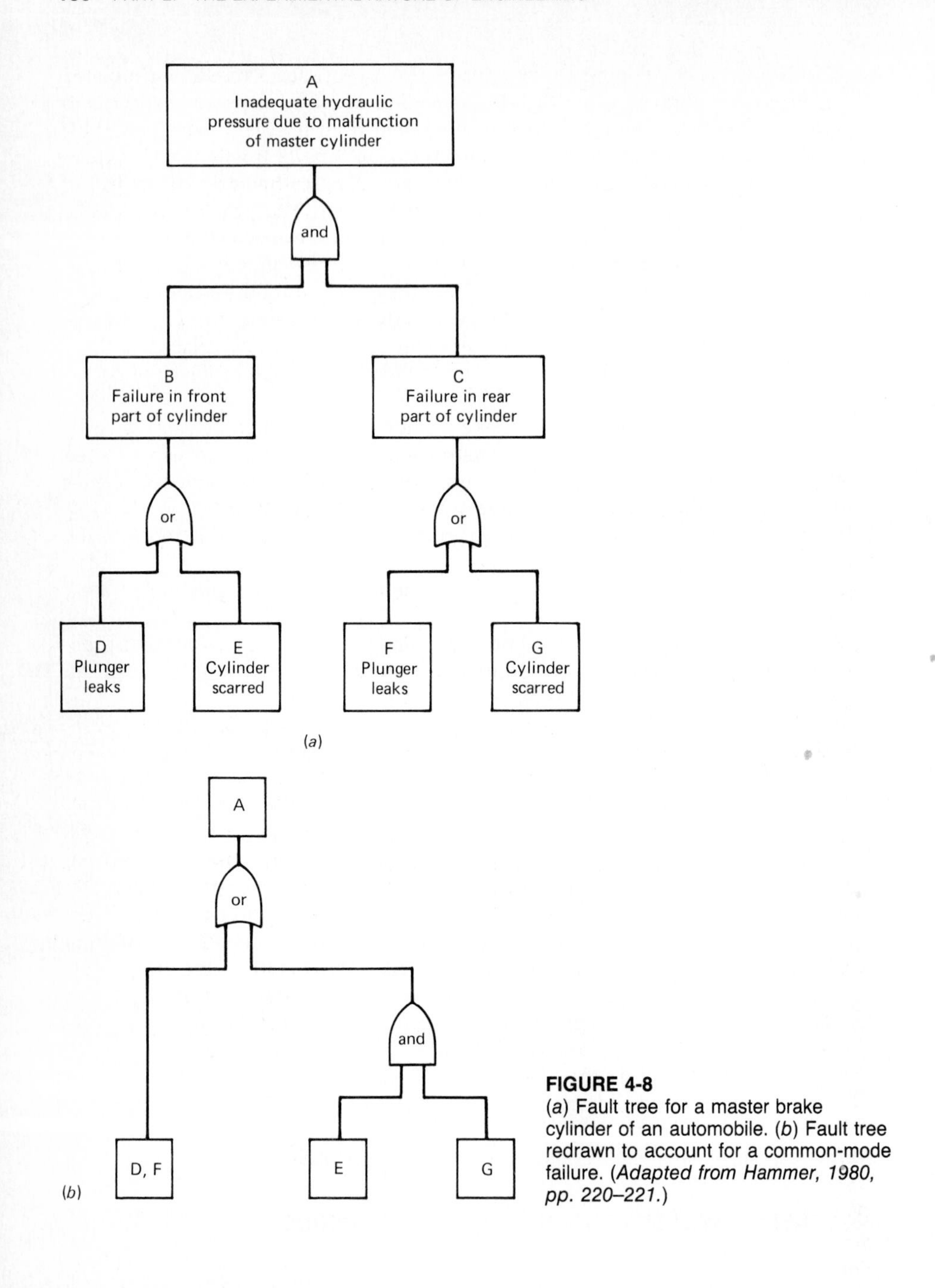

FIGURE 4-8
(*a*) Fault tree for a master brake cylinder of an automobile. (*b*) Fault tree redrawn to account for a common-mode failure. (*Adapted from Hammer, 1980, pp. 220–221.*)

risk as long as the project (the product, the system, or the activity that is risky) promises sufficient benefit or gain. If risk and benefit can both be readily expressed in a common set of units (say, dollars), it is relatively easy to carry out a risk-benefit analysis and to determine whether or not we can expect to come out on the benefit side. For example, an inoculation program may produce some deaths, but it is worth the risk if many more lives are saved by suppressing an imminent epidemic.

A closer examination of risk-benefit analyses will reveal some conceptual difficulties. Both risks and benefits lie in the future. Since there is some uncertainty associated with them, we should deal with their expected values; in other words, we should multiply the magnitude of the potential loss by the probability of its occurrence, and similarly with the gain. But who establishes these values, and how? If the benefits are about to be realized in the near future but the risks are far off (or vice versa), how is the future to be discounted in terms of, say, an interest rate so we can compare present values? What if the benefits accrue to one party and the risks are incurred by another party?

The matter of delayed effects presents particular difficulties when an analysis is carried out during a period of high interest rates. Under such circumstances the future is discounted too heavily because the very low present values of cost or benefit do not give a true picture of what a future generation will face.

How should one proceed when risks or benefits are composites of ingredients which cannot be added in a common set of units, as for instance in assessing effects on health plus aesthetics plus reliability? At most one can compare designs that satisfy some constraints in the form of "dollars not to exceed *X*, health not to drop below *Y*" and attempt to compare aesthetic values with those constraints. Or when the risks can be expressed and measured in one set of units (say, deaths on the highway) and benefits in another (speed of travel), we can employ the ratio of risks to benefits for different designs when comparing the designs.

It should be noted that risk-benefit analysis, like cost-benefit analysis, is concerned with the advisability of undertaking a project. When we judge the relative merits of different designs, however, we move away from this concern. Instead we are dealing with something similar to cost-effectiveness analysis, which asks what design has the greater merit given that the project is actually to be carried out. Sometimes the shift from one type of consideration to the other is so subtle that it passes unnoticed. Nevertheless, engineers should be aware of the differences so that they do not unknowingly carry the assumptions behind one kind of concern into their deliberations over the other.

These difficulties notwithstanding, there is a need in today's technological society for some commonly agreed upon process—or at least a process open to scrutiny and open to modification as needed—for judging the acceptability of potentially risky projects. What we must keep in mind is the following eth-

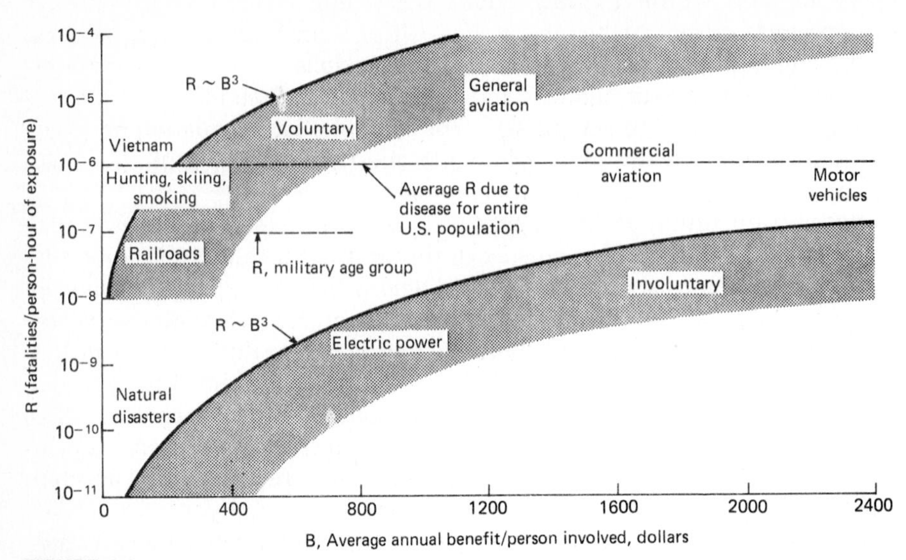

FIGURE 4-9
Willingness to assume voluntary risks as opposed to involuntary ones correlated to benefits those risks produce. (*After Starr, 1969, p. 1234.*)

ical question: *"Under what conditions, if any, is someone in society entitled to impose a risk on someone else on behalf of a supposed benefit to yet others?"* (Council for Science and Society, 1977, 37). In examining this problem further, we can trace our steps back to an observation on risk perception made earlier: A risk to a known person (or to identifiable individuals) is perceived differently by people than statistical risks merely read or heard about. Engineers do not affect just an amorphous public—their decisions have a direct impact on people who feel the impact acutely, and that fact should be taken into account equally as seriously as studies of statistical risk.

Personal Risk

Given sufficient information, an individual is able to decide whether or not to participate in (or consent to be exposed to) a risky activity (an experiment). Chauncey Starr has prepared some widely used figures which indicate that individuals are more ready to assume voluntary risks than they are to be subjected to involuntary risks (or activities over which they have no control), even when the voluntary risks are 1000 times more likely to produce a fatality than the involuntary ones. We show this graphically in Fig. 4-9.

The difficulty in assessing personal risks arises when we consider those that are involuntary. Take John and Ann Smith and their discomfort over living near a refinery. Assume the general public was all in favor of building a new refinery at that location, and assume the Smiths already lived in the area. Would they and others in their situation have been justified in trying to veto its construction? Would they have been entitled to compensation if the plant was built over their objections anyway? If so, how much compensation would have been adequate? These questions arise in numerous instances. Nuclear power plant siting is another example. Indeed, Fig. 4-9 was produced in the context of nuclear safety studies.

The problem of quantification alone raises innumerable problems in assessing personal safety and risk, as was alluded to earlier. How, for instance, is one to assess the dollar value of an individual's life? This question is as difficult as deciding whose life is worth saving, should such choice ever have to be made.

Some would advocate that the marketplace should decide, assuming market values can come into play. But today there is no over-the-counter trade in lives. Nor are even more mundane gains and losses easily priced. If the market is being manipulated, or if there is a wide difference between "product" cost and sales price, it matters under what conditions the buying or selling takes place. For example, if one buys a loaf of bread, it can matter whether it is just one additional daily loaf among others one buys regularly or whether it is the first loaf available in weeks. Or if you are compensated for a risk by an amount based on the exposure tolerance of the "average" person, yet your tolerance of a condition or your propensity to be harmed is much greater than average, the compensation is apt to be inadequate.

The result of these difficulties in assessing personal risk is that analysts employ whatever quantitative measures are ready at hand. In regard to voluntary activities one could possibly make judgments on the basis of the amount of life insurance taken out by an individual. Is that individual going to offer the same amount to a kidnapper to be freed? Or is there likely to be a difference between future events (requiring insurance) and present events (demand for ransom)? In assessing a hazardous job one might look at the increased wages a worker demands to carry out the task. Faced with the wide range of variables possible in such assessments, one can only suggest that an open procedure, overseen by trained arbiters, be employed in each case as it arises. On the other hand, for people taken in a population-at-large context, it is much easier to use statistical averages without giving offense to anyone in particular. The ethical implications of that practice will be addressed in the following subsections. Other aspects were discussed earlier in the context of giving valid consent to participation in an experiment (see Chap. 3, under "Engineering as Experimentation").

Public Risk and Public Acceptance

Risks and benefits to the public at large are more easily determined because individual differences tend to even out as larger numbers of people are considered. The contrast between costs of a disability viewed from the stand-

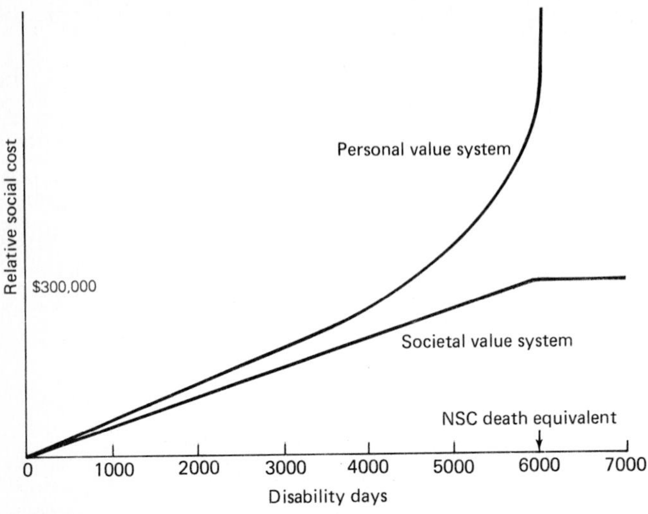

FIGURE 4-10
Value systems for social costs of disability. Using the National Safety Council equivalent of 6000 disability days for death and L. A. Sagan's 1972 assumed rate of $50 per day of disability (Sagan, 488) yields a "death equivalent" of $300,000—valid for societal value analysis only. *(After Starr, Rundman, and Whipple, 1976, p. 637.)*

point of a private value system and from that of a societal value system, for example, is vividly illustrated in Fig. 4-10. Also, assessment studies relating to technological safety can be conducted more readily in the detached manner of a macroscopic view as statistical parameters take on greater significance. In that context, the National Highway Traffic Safety Administration (NHTSA) has proposed a value for human life based on loss of future income and other costs associated with an accident. Intended for study purposes only, NHTSA's "blue-book value" amounted to $200,725 in 1972 dollars. This is certainly a more convenient measure than sorting out the latest figures from court cases. (On April 23, 1981, for example, the *Los Angeles Times* reported settlements for relatives of victims of a 1979 DC-10 crash in Chicago at $2,287,257 for a 36-year-old promising executive, $750,000 for a telephone company employee, and $275,000 for a stewardess on duty.)

A recent study by Shulamit Kahn gives a labor market value of life in the amount of $8 million. This does not include the value people place on other persons' lives, which would be 115 to 230 percent higher. "Yet even the $8 million figure is higher than is typically used in policy analysis. The unavoidable implication... is therefore that policy analysts do not evaluate the risk of their subjects' lives as highly as people evaluate risks to their own (and others') lives. Consequently, too many risks are taken" (Kahn, 1986).

NHTSA, incidentally, emphasized that "placing a value on a human life can be nothing more than a play with figures. We have provided an estimate of some of the quantifiable losses in social welfare resulting from a fatality and can only hope that this estimate is not construed as some type of basis for determining the 'optimal' (or even worse, the 'maximum') amount of expenditure to be allocated to saving lives" (O'Neill and Kelley, 1975, 30).

Accounting Publicly for Benefits and Risks

The conclusions of risk assessment and cost-benefit studies are increasingly being challenged by special interest groups. And though engineers are generally reluctant to face the rough and tumble of the political and legal arenas, they are often called upon as expert witnesses in such cases. When testifying, they will find that they are treated with far more respect than perhaps equally well versed lay people who, as interested citizens, have volunteered their time to study a crucial issue. But if formal expertise bestows a certain power on engineers, it behooves them not to abuse that power. There is a noblesse oblige which demands that they remain as objective as humanly possible in their investigations and the conclusions they reach and that, in order to place their testimony in the proper perspective, they state at the outset any personal biases they may have about the subject at issue.

No expert—or even group of experts—can be expected to be omniscient. Hence the public processes designed to establish safeguards and reasonable regulations in relation to technology themselves suffer from the already mentioned problem of incomplete knowledge that engineering is subject to. Moreover, in the view of a judge who has heard many cases involving new technologies and who has written searchingly on the subject, there is yet another problem affecting public accountability for risk:

> The other kind of uncertainty that infects risk regulation comes from a *refusal to face the hard questions created by lack of knowledge.* It is uncertainty produced by scientists and regulators who assure the public that there are no risks, but know that the answers are not at hand. Perhaps more important, it is a false sense of security because the hard questions have never been asked in the first place.
>
> In the early days of nuclear plant licensing, for example, the problem of long-term waste disposal was never even an issue. Only after extensive prodding by environmental and citizens' groups did the industry and regulators show any awareness of waste disposal as a problem at all. Judges like myself became troubled when those charged with ensuring nuclear safety refused even to recognize the seriousness of the waste disposal issue, much less to propose a solution (Bazelon, 1979, 279, emphasis added).

A willingness to admit uncertainty and bias and to reveal methodology and sources is particularly important when numerical data and statistics are presented. Special caution is required when stating probabilities of rare events. We have already mentioned how conceptions of risk can vary—even be turned around—depending on how the facts are presented. How presentations of data interpretation (even the best intended) can be misleading is pointed out by W. Hammer in his discussion of the tabular material we give

TABLE 4-1
INDIVIDUAL RISK OF ACUTE FATALITY BY VARIOUS CAUSES*

Accident type	Total number for 1969	Approximate individual risk of acute fatality, probability/yr.†
Motor vehicle	55,791	3×10^{-4}
Falls	17,827	9×10^{-5}
Fires and hot substance	7,451	4×10^{-5}
Drowning	6,181	3×10^{-5}
Poison	4,516	2×10^{-5}
Firearms	2,309	1×10^{-5}
Machinery (1968)	2,054	1×10^{-5}
Water transport	1,743	9×10^{-6}
Air travel	1,778	9×10^{-6}
Falling objects	1,271	6×10^{-6}
Electrocution	1,148	6×10^{-6}
Railway	884	4×10^{-6}
Lightning	160	5×10^{-7}
Tornadoes	91	4×10^{-7}
Hurricanes	93	4×10^{-7}
All others	8,695	4×10^{-5}
All accidents		6×10^{-4}

*Use Figures with caution. See text.
†Based on total U.S. population.
Source: Rasmussen, p. 230.

here in Table 4-1. The table presents statistics on accident fatalities for the U.S. population. The approximate individual risk entries were calculated as part of a study for the U.S. Atomic Energy Commission in 1974 (Rasmussen). Hammer writes:

> The values [in Table 4-1] illustrate one problem with the use of quantitative assessments. All the fatality risk values shown are based on total U.S. population in 1969, which may not be valid in some cases. For example, if half the people of the United States do not travel by air, the probability of any one of them being killed (or of having been killed in 1969) in a plane crash is zero. The probability of the average air traveler being killed is increased to 1.8 times 10^{-5}. The probability of a person who travels more than the average (which isn't specified) is even higher. This method of risk assessment becomes even more invalid when the operation considered is one in which few persons participate. Assume that there are 160 persons killed in hang-gliding accidents in a year (the same as . . . [the number] killed by lightning). When the total U.S. population is used as the base for determining the risk, the probability of a fatality is 5 times 10^{-7}, as shown; a relatively safe operation. However, if there are only 20,000 enthusiasts who participate in hang gliding and they suffer a 160-person loss each year, the fatality risk is 8 times 10^{-3}. To make correct and comparable risk assessments it is therefore necessary to base them on correct and acceptable assumptions and data (Hammer, 1980, 246).

A particularly controversial study was performed by Inhaber for the Atomic Energy Control Board of Canada (Inhaber, 1979, 1982). He claimed that nonconventional energy sources, such as methanol and solar, were riskier in terms of deaths and disabling injuries (measured in workdays lost) than nuclear energy. The questions raised by the study's critics concern the lack of differentiation between injuries, the reliability of downtime data, and in connection with the latter, the casualties ascribable to replacement sources such as coal. The use of historical data (for instance, on heavy coal mine losses) and projected data (on atomic energy) also creates difficulty if one considers that coal mine safety is improving while nuclear plant safety has gone down with the bigger plants. Inhaber's report and the objections to it will make one appreciate the difficulty of preparing an objective study. (Many of the objections are quoted and rebutted in his 1982 book; other objections were raised by Shrader-Frechette, 1986). It appears that as many questions are left unanswered as are answered and the observer must conclude that decisions are made on sociopolitical grounds after all. Engineers can provide background material to support or rebut various positions, but unless they are willing to enter the debate, their contributions to the final outcome may be small. Engineers and scientists who find themselves in such situations might prefer the model of a science court, but its time has not yet arrived.

Engineers are usually asked for numbers when assessing safety and risk; therefore they should insist on meaningful numbers. This means that they should regard statistics with caution, whatever source may have issued them. Engineers should also recognize the previously mentioned difficulties with measuring risks and benefits on a cardinal scale (that is, in absolute terms), and should instead employ ordinal rankings. One of the difficulties with risk-benefit and cost-benefit analyses, again, is the matter of who does the assessing. The parties that will be affected by a project are rarely polled, especially when they are not represented by an influential lobby or trade organization (Nelson and Peterson, 1981, 2).

But difficulties in publicly accounting for risks and benefits are not related only to methods of quantification. There is also the question of justice, which involves *qualitative* value judgments:

> There are things which are wrong to do regardless of the benefits of the consequences. This is contrary to the basic assumptions in cost-benefit analysis. Without going so far as to say that consequences are never important, we can say that they are not as important as the [analyses] would imply. . . . The type of action one does should be morally evaluated regardless of its consequences; if it is wrong to violate certain rights, then figuring out the benefits of the consequences of doing so is irrelevant (Nelson and Peterson, 1981, 4).

Paying compensation, for example, may be an efficient and bureaucratically pleasing method for trying to make restitution when harm has been done, and indeed it is an improvement over earlier government and business procedures. But efficiency in itself does not promote ethics, as much as one may be tempted to equate the two. As an example we may cite the attempts of the city of San Francisco in the 1960s to demolish in one fell swoop a large area slated for urban renewal. Even if the residents being displaced could have afforded to live in the new housing units once built, they would have scattered and the neighborhood would have been destroyed by the time the new dwellings were completed. The response by those residents was a near revolt which caused the city to rethink its cost-benefit formula and to set about rebuilding the area on a block by block basis, with accommodations for the temporarily displaced being provided by the city. A more expensive solution, certainly, but also more humane. And engineers need to be sensitive to such considerations.

Incentives to Reduce Risk

The engineer is faced with the formidable tasks of designing and manufacturing safe products, of giving a fair accounting of benefits and risks in regard to those products, and of meeting production schedules and helping his or her company maintain profits all at the same time. Of these objectives, product safety should command top priority. Yet this is not often so in practice, partly because of some commonly held misconceptions which militate against application of the extra thought and effort required to make a product safe.

Among the popular though faulty assumptions about safety are the following (adapted from Hammer, 1980, 52).

Assumption: Operator error and negligence are the principal causes of all accidents.

Reality: Accidents are caused by dangerous conditions that can be corrected.

For example, introduction of automatic couplers drastically reduced the number of deaths and injuries suffered by train workers. Dangerous design characteristics of products cause more accidents than failures (by fatigue, etc.) of components.

Assumption: Making a product safe invariably increases costs.

Reality: Initial costs need not be higher if safety is built into a product from the beginning. It is design changes at a later date that are costly. Even then life-cycle costs can be lower for the redesigned, safe product.

Assumption: We learn about safety after a product has been completed and tested.

Reality: If safety is not built into the original design, people can be hurt during the testing stage.

Reluctance to change a design may mean safety features will not be incorporated into the product.

Assumption: Warnings about hazards are adequate; insurance coverage is cheaper than planning for safety.

Reality: A warning merely indicates that a hazard is known to exist, and thus provides only minimal protection against harm. Insurance rates are skyrocketing. Recall actions can add to costs, even when no accidents have occurred.

Engineers should recognize that reducing risk is not an impossible task, even under financial and time constraints. All it takes in many cases is a different perspective on the design problem: a recognition from the outset that one is embarking on an experiment in which safety is an important factor.

Some Examples of Improved Safety

This is not a book on design; therefore only a few simple examples will be given to show that safety need not rest on elaborate contingency features.

The first example is the magnetic door catch introduced on refrigerators to prevent death by asphyxiation of children accidentally trapped in them. The new catch permits the door to be opened from the inside without major effort. It also happens to be cheaper than the older types of latches.

The second example is the dead-man handle used by the engineer (engine driver) to control a train's speed. The train is powered only as long as some pressure is exerted on the handle. If the engineer becomes incapacitated and lets go of the handle, the train stops automatically. Perhaps cruise controls for newer model automobiles should come equipped with a similar feature.

Railroads provide the third example as well. Old-fashioned semaphores actuated by cable usually indicated STOP when the arm was lowered. This was the position the arm assumed all by itself if the cable snapped accidentally. Here we have an early instance of a fail-safe design dating back more than a hundred years.

The motor-reversing system shown diagrammatically in Fig. 4-11 gives still another example of a situation in which the introduction of a safety feature involves merely the proper arrangement of functions at no additional expense. As the mechanism is designed in Fig. 4-11*a*, sticky contacts could cause battery B to be shorted, thus making it unavailable for further use even after the contacts are coaxed loose. A simple reconnection of wires as shown in Fig. 4-11*b* removes that problem altogether.

As a final example we mention the Volkswagen safety belt. A simple attachment on the door ensures that the belt automatically goes into place whenever one enters the car. Forgetting to strap oneself in is no longer a problem.

In the rush to bring a product onto the market, safety considerations are frequently slighted. This would not be so much the case if the venture were regarded as an experiment—an experiment which is about to enter its active phase as the product comes into the hands of the user. Space flights were carried out with such an attitude, but everyday ventures involve less obvious dangers and therefore less attention is usually paid to safety. If moral concerns alone do not sway engineers and their employers to be more heedful of

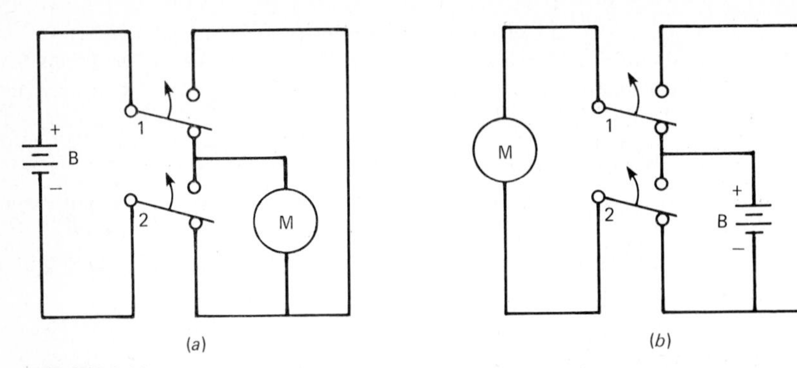

FIGURE 4-11
Motor-reversing system. (*a*) Arms 1 and 2 of the switch are both raised by a solenoid (not shown). If either one does not move—say, a contact sticks—while the other does, there is a short across the battery. The battery will discharge and be useless even after the trouble is detected. (*b*) By exchanging the positions of battery and motor, a stuck switch will cause no harm to the battery. (The motor can be shorted without harm.)

potential risks, then recent trends in product liability law should certainly do so.

Liability

The last two decades have seen a drastic change in legal protection for the consumer. Richard C. Vaughn, in his informative book *Legal Aspects of Engineering*, describes the changes as follows:

> Early English social and legal philosophy reflected the manufacturing nature of the economy. Producers of goods and services were held in high esteem. Their success meant success of the nation. The legal climate fostered their growth. Both logic and social philosophy supported the legal defense available if someone complained about a product—*caveat emptor* (the buyer beware). The logic was simple: one should examine what he is to receive before he buys it. If he is so negligent that he does not examine before he buys, then he should live with his bad bargain. Legal support of an action to recover for a bad product would be, in effect, support of buyer negligence and the law usually will not aid those who are negligent. But then, of course, the products produced in those days were somewhat more easily examined than what we buy today....
>
> Another defense in the producers' armament was "privity of contract"—the idea that one who is not a party of a contract should have no rights arising from it. In other words, if one was injured by a product but he did not buy it directly from the manufacturer, he could not act against that manufacturer to recover for his injury. The producer or manufacturer only needed to interpose a middleman—a wholesaler or retailer—as an insulator. Then, if the injured person could prove he was the buyer of the product, he might sue the middleman, but he could not reach "the deep pocket" (Vaughn, 1977, 41).

A turning point in the reversal of these attitudes came in 1916 when Judge Cordoza found the Buick Motor Company responsible for the injuries suffered by one McPherson when a wheel on a new Buick collapsed. Gradually thereafter, over a period of half a century, the notion took hold that a manufacturer can be held liable for injuries resulting from negligence in a product's design or manufacture. Still, such negligence had to be proven. Then, in 1963, the concept of strict liability was established in California by the case Greenman vs. Yuba Power Products, a concept soon thereafter incorporated into tort law by most states. *Strict liability* means it is sufficient for a product to have been defective as sold for the manufacturer to be held liable for any harm that results to users. Negligence is not at issue. What matters is that the product has a defect not obvious to users.

Despite the responsibilities implied by the doctrine of strict liability, negligence certainly remains a more grievous offense, as does breach of warranty. The latter is interpreted as a breach of contract and usually covers only the purchaser and members of the purchaser's household, not just any user. The warranty is established by advertising, labels, and other information that causes the buyer to expect a serviceable and safe product.

Engineers—and students of engineering—need to be aware of strict liability. As Richard Moll writes: "The fact that proof of negligence is not essential to impose liability is a frightening prospect for most manufacturers. . . . The significance of the strict liability doctrine, as far as engineers are concerned, is that although in many cases it is impossible to test every product, the engineer must weigh the chances of a defect causing serious injury against the cost of eliminating or minimizing defects in the product" (Moll, 1976, 331). Adhering to accepted practices and observing standards is not sufficient. We have labeled such behavior *minimal compliance.* It neither guarantees a safe product nor provides a valid excuse in court. As Seiden impresses on managers of engineering firms, "[standards] are excellent starting points . . . and they are excellent as checklists. But they must be used creatively and judgementally. They must be only the beginning, not the end, of the design process" (Seiden, 1984, 195).

While most engineers strive to be responsible in their work and have the safety of the public in mind, conditions imposed by employers may circumscribe these professional goals. On the other hand, engineers can also be sued individually, even when acting according to guidelines set by their employers. This may happen when an injured party is frustrated by laws which shield the employer or limit the employer's liability, as is the case with government agencies (the U.S. government is not liable for nuclear test fall-out). Thus it happened that one county highway engineer was sued for damages exceeding $2 million. As Dean M. Dilley reports, "The plaintiff suffered permanent injuries in an automobile accident caused by a washed out section of highway, and the suit argued that the engineer's failure to repair the road gave rise to an action for negligence against him personally" (Dilley, 1979, 12).

Some companies and government agencies protect their engineers by allowing themselves to be sued for money damages over harm arising from activities of employees working within the scope of their professional duties. Engineers not so protected can resort to malpractice insurance. Independent engineers who contract for their work can minimize the effects of adverse legal judgments by writing liability limits into their contracts. This, in turn, can reduce their malpractice insurance rates. As has been mentioned before, engineering practice should be preventive or defensive in approach. A knowledge of liability is therefore well advised. Good introductory material is offered in books by Richard C. Vaughn (already mentioned) and by James F. Thorpe and William H. Middendorf (see the Bibliography).

On a larger scale, liability limitation was offered to electric utilities through the Price-Anderson Act in order to entice them into nuclear power ventures during the Atoms for Peace Program of the Eisenhower years. The $400 million limit seemed inadequate even then (it is to be raised to $7 billion), but it was probably assumed that the government would step in with disaster aid in case of a serious accident. Unfortunately the expectation of bailouts with such aid has often resulted in inadequate planning by public agencies for potential disasters of many different types, from floods to droughts, from earthquakes to conflagrations. While the Three Mile Island nuclear power plant incident of 1979 did not constitute a disaster in the usual sense, it serves as an example of the need to take safety seriously in large-scale engineering design, so it will be discussed in greater detail in the next section.

Study Questions

1 A 250,000-ton tanker (capable of holding 200,000 tons of oil) is cruising at 16 knots when engine power (25,000 hp) is lost. Control over the rudder is lost as well.

- **a** Estimate how long it will take in time and distance before the ship comes to a stop. Assume it had been on a straight course before the mishap. (Under normal circumstances, with reversing power available, it will take 3 miles or 22 minutes.)
- **b** What might be the consequences of the tanker running aground or colliding with another vessel?
- **c** How many tankers were lost at sea during the last decade? You may consult an almanac.
- **d** What are the benefits of operating a supertanker? Some are larger than 500,000 tons. Some 1-million-ton tankers have been planned. (An interesting source on this topic is Noel Mostert's book *Supership*.)

2 The following appeared in the July 10, 1981, issue of *Science:*

> The Supreme Court, in the cotton dust decision on 17 June, says explicitly that OSHA must ignore the results of any cost-benefit comparison when setting a standard for worker exposure to a hazardous substance. Justice William Brennan, writing for the court's five-person majority, said that "Congress *itself* decided the basic relationship between costs and benefits by placing the 'benefit' of the worker's health above all other considerations" when it wrote the law in 1970. Yet the agency cannot require exposure controls that are impossible to

achieve, nor can it bankrupt an entire industry, Brennan wrote. He concluded that consideration of anything besides these questions would be inconsistent with Congress's direction (R. J. Smith, 1981, 185).

Discuss this opinion of the court and how you think it might affect efforts to combat byssinosis (brown lung disease suffered by textile workers). What interpretations do you think OSHA, the textile industry, and the textile workers' unions would give this court decision? If you were an engineer in a textile plant, how would you react? Discuss your reactions in terms of their ethical foundations.

3 The following excerpt is from an article by T. W. Lockhart, "Safety Engineering and the Value of Life." Attempt to answer the question posed by Lockhart. Try again after you have finished reading this entire chapter.

> ...there is an honored tradition in moral philosophy, associated primarily with Immanuel Kant, according to which human beings have a worth that is not commensurable with that of mere objects. According to this view, because of this incommensurability we must recognize and respect the liberty and dignity of each person and refrain from treating him merely as a means to some end. Human beings may not be used in order to achieve some higher good, for there is no higher good. Let us call this view the Incommensurability Principle.
>
> The Incommensurability Principle has had a powerful appeal for many. This has been true mainly because it has been felt that unless it, or something like it, is accepted it is not possible to account for such fundamental human rights as the right not to be killed, the right not to have one's liberty abridged without just cause, and the right to be treated fairly and honestly. The Incommensurability Principle is clearly incompatible with an attempt to place a monetary value on human life or to justify actions on the basis of such a valuation. There is thus further reason for doubting the wisdom of any such attempt....
>
> Is it possible to reconcile the Incommmensurability Principle with the commonsense view that considerations of safety must be weighed against economic costs? (Lockhart, 1981, 3)

4 During the early stages of its development, crashworthiness tests revealed that the Pinto, a Ford automobile, could not sustain a front-end collision without the windshield breaking. A quick-fix solution was adopted: The drive train was moved backward. As a result the differential was moved very close to the gas tank (Camps, 1981, 119-129). Thus many gas tanks collapsed and exploded upon rear-end collisions once the Pinto became available to the public. Extensive lawsuits followed (Strobel, 1980; Cullen, 1987). Study the Pinto story and write it up as a case study. You may use the Predicasts F&S (Funk and Scott) Index of Corporations and Industries (Predicasts, Inc., Cleveland, Ohio) to find additional references.

5 The U.S. Consumer Product Safety Commission states the following in its 1981 annual report:

> Each year, an estimated 36 million Americans are injured and 30,000 are killed in accidents involving consumer products. The cost of these accidents is staggering—over $12.5 billion annually in medical costs and lost earnings.

These figures have been repeatedly cited in the literature to impress on engineers the need to design safe products. Review the data presented by the Commission in its annual and monthly reports and estimate what portion of the figures can be rightfully ascribed to poor design and what portion to consumer carelessness.

6 Only about 15 years ago 600 women reportedly lost hands (one, not both, we hope) in top-loading washing machines every year (Papanek, 75). If you had been in the business of manufacturing washing machines then, how do you think you might have reacted and should have reacted to such statistics?

7 A worker accepts a dangerous job after being offered an annual bonus of $2000. The probability that the worker may be killed in any one year is 1/10,000. This is known to the worker. The bonus may therefore be interpreted as a self-assessment of life with a value equal to $2000 divided by 1/10,000, or $20 million. Is the worker more or less likely to accept the job if presented with the statistically identical figures of a $100,000 bonus over 50 years (neglecting interest) and a 1/200 probability of a fatal accident during that period?

8 As emerging technologies have reduced risks from some sources, other risks have frequently become prominent. For instance, efforts to combat famine by drawing and transporting water for irrigation brought early benefits coupled with lasting problems to such diverse areas as Mesopotamia and the Andes in the past, or to the Punjab and the Sahel in the present. We also observe that the decrease in infectious disease was accompanied by an increase in chronic disease. Give another similar example and discuss it as a problem in risk-benefit analysis.

9 The owner of a television set brought legal action against its manufacturer (Admiral) seeking compensatory and punitive damages for severe burns and other injuries suffered when the set burst into flames. It was revealed at the trial that other sets made by the manufacturer had also gone up in smoke and flames. Prepare a case study for use in discussing the legal and ethical implications of the manufacturer's actions (or lack thereof). The case is Zora H. Gillham vs. The Admiral Corporation. It was brought to the authors' attention by Donald E. Wilson. An early report appeared in the *Federal Reporter,* 2d series, vol. 253, F.2d, West Publishing Co., St. Paul, Minnesota, 1976. Consult the current literature.

10 "Airless" paint spray guns do not need an external source of compressed air connected to the gun by a heavy hose (although they do need a cord to attach them to a power source) because they have incorporated into them a small electric motor and pump. One common design uses an induction motor which does not cause sparking since it does not require a commutator and brushes (which are sources of sparking). Nevertheless the gun carries a label, warning users that electrical devices operated in paint spray environments pose special dangers. Another type of gun which, like the first, also requires only a power cord, is designed to weigh less by using a high-speed universal motor and a disk-type pump. The universal motor *does* require a commutator and brushes, which cause sparking. This second kind of spray gun carries a warning similar to that attached to the first, but it states in addition that the gun should never be used with paints which employ highly volatile and flammable thinners such as naphtha. The instruction booklet is quite detailed in its warnings.

A painter had been lent one of the latter types of spray guns and was operating it while it was partially filled with paint thinner in order to clean it out. It caught fire, and the painter was severely burned as the fire spread. The instruction booklet pointing out the spray gun's dangers was in the cardboard box in which the gun was kept, but it had not been read by the painter, who was a recent immigrant and did not read English very well. Do you see any ethical problems in continuing over-the-counter sales of this second type of spray gun? What should the manufacturer of this novel, lightweight device do?

In answering these questions, consider the fact that courts have ruled that hidden design defects are not excused by mere warnings attached to the defective products or posted in salesrooms. Informed consent must rest on a more thorough understanding than can be transmitted to buyers by simple warning labels.

11 Discuss the ethical aspects of various methods of measuring benefits and costs described by Bailey (1980) and/or, Shrader-Frechette (1985, 2 books). This may take the form of book reports.

12 Industry generally maintains that restrictive regulations on potentially toxic substances should be enacted only after it has been proven by rigorous scientific methods that a link exists between health effects and a pollutant. The opposition to this viewpoint argues that "waiting for firm evidence of human health effects amounts to using the nation's people as guinea pigs, and that is morally unacceptable. It proposes that far from overestimating the risks from toxic substances, conventional risk assessments underestimate them, for there may be effects from chemicals in combination that are greater than would be expected from the sum effects of all chemicals acting independently" (Ruckelshaus, 1987, 27). What are the risk management problems when *several* pollutants from *several* sources contribute to an areawide pollution problem but are regulated by *several* different agencies, each of which "has statutory authority to regulate the use and emission of *some* of the substances from *some* of the sources, in *some* of the pathways, for the purpose of protecting *some* of the population under *some* circumstances"? (Baram, 1976)

THREE MILE ISLAND AND CHERNOBYL: THE NEED FOR SAFE EXITS

As our engineering systems grow more complex, it becomes more difficult to operate them. As Charles Perrow (see Bibliography) argues, our traditional systems tended to incorporate sufficient slack, which allowed system aberrations to be corrected in a timely manner. Nowadays, he points out, subsystems are so tightly coupled within more complex total systems that it is not possible to alter a course safely unless it can be done quickly and correctly. Frequently the supposedly corrective action taken by operators may make matters worse because they do not know what the problem is. At Three Mile Island, for instance, so many alarms had to be recorded by a printer that it fell behind by as much as 2½ hours in reporting the events.

Designers hope to ensure greater safety during emergencies by taking human operators out of the loop and mechanizing their functions. The control policy would be based on predetermined rules. This in itself creates problems because (1) not all eventualities are foreseeable, and (2) even those that can be predicted will be programmed by an error-prone human designer (Senders, 1980). In addition, another problem arises when (3) the mechanized system fails and a human operator has to replace the computer in an operation that demands many rapid decisions. The proposed air traffic control system allowing for 10-second separation between planes would fall into this category.

Operator errors were the main causes of the nuclear reactor accidents at Three Mile Island and Chernobyl. Beyond these errors a major deficiency was revealed at both installations: inadequate provisions for evacuation of nearby populations. This lack of "safe exit" is found in too many of our otherwise amazingly complex systems. After examining the reactor incidents we shall discuss this wider issue.

Three Mile Island

Walter Creitz, president of Metropolitan Edison, the power company in the Susquehanna Basin, was obviously annoyed by a series of articles in the *Record*. This local, daily newspaper of York, Pennsylvania, had cited unsafe conditions at Metropolitan Edison's Three Mile Island nuclear power plant Unit 2. Creitz dismissed the stories as "something less than a patriotic act—comparable in recklessness...to shouting 'Fire!' in a crowded theater." A few days later a minor malfunction in the plant set off a series of events which made "Three Mile Island" into household words across the world (Rogovin, 1980, 3).

Briefly, this is what happened (for details see Kemeny, 1979; Rogovin, 1980; Ford, 1982; Mason, 1979; Moss and Sills, 1981; Keisling, 1980; D. Martin, 1980). At 4 A.M. on March 28, 1979, Unit TMI-2 was operating under full automatic control at 97 percent of its rated power output. For 11 hours a maintenance crew had been working on a recurring minor problem. Resin beads are used in several demineralizers (labeled 14 in the schematic diagram shown as Fig. 4-12) to clean or "polish" the water on its way from the steam condenser (12) back to the steam generator (3). Some beads had clogged the resin pipe from a demineralizer to a tank in which the resin is regenerated. In flushing the pipe with water, perhaps a cupful of water backed up into an air line which provides air for fluffing the resin in its regeneration tank. But that air line is connected to the air system which also serves the control mechanisms of the large valves at the outlet of the demineralizers. Thus it happened that these valves closed unexpectedly.

With water flow interrupted in the secondary loop (26), all but one of the condensate booster pumps turned off. That caused the main feedwater pumps (23) and the turbine (10) to shut down as well. In turn, an automatic emergency system started up the auxiliary feedwater pumps (25). But with the turbines inoperative, there was little outlet for the heat generated by the fission process in the reactor core. The pressure in the reactor rose to over 2200 pounds per square inch, opening a pressure-relief valve (7) and signaling a SCRAM, in which control rods are lowered into the reactor core to stop the main fission process.

The open valve succeeded in lowering the pressure, and the valve was readied to be closed. Its solenoid was deenergized and the operators were so

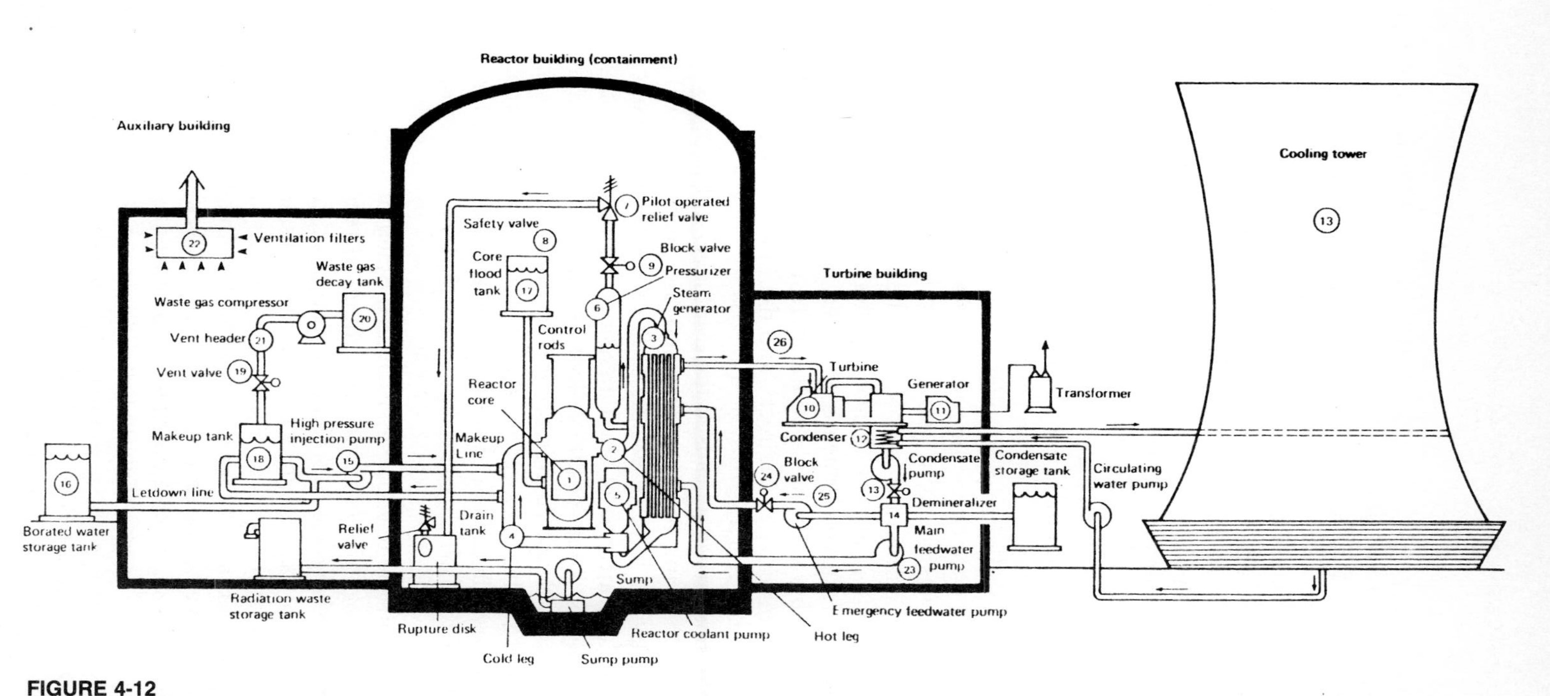

FIGURE 4-12
Schematic diagram of Three Mile Island nuclear power plant Unit 2. Pressurized water reactor system: Heat from reactor core (1) is carried away by water in a primary loop (1, 2, 3, 5, 4). In steam generator (3) the heat is transferred to water in a secondary loop (26) at lower pressure. The secondary-loop water turns to steam in the steam generator or boiler (3), drives the turbine (10), turns into water in the condenser (12), and is circulated back to (3) by means of pumps (13, and 23 and 25). (*Adapted from* IEEE Spectrum, *November 1979, copyright ©1979 by the Institute of Electrical and Electronics Engineers, Inc., and from Rogovin and Frampton, 1980.*)

informed by their control-panel lights. But something went wrong: The valve remained open, contrary to what the control panel indicated. Apart from this failure everything else had proceeded automatically as it was supposed to. Everything, that is, except for one other serious omission: The pumps (25) could not supply the auxiliary feed water because block valves 24 had inadvertently been left closed after maintenance work done on them two days earlier. Without feedwater in loop 26 the steam generator (3) boiled dry. Now there was practically no heat removal from the reactor, except through the relief valve. Water was pouring out through it at the rate of 220 gallons per minute. The reactor had not yet cooled down, and even with the control rods shutting off the main fission reaction there would still be considerable heat produced by the continuing radioactive decay of waste products.

Loss of water in the reactor caused one of a group of pumps, positioned at 15, to start automatically; another pump was started by the operators to rapidly replenish the water supply for the reactor core. Soon thereafter the full emergency core-cooling system went into operation in response to low reactor pressure. Low reactor pressure can promote the formation of steam bubbles which reduce the effectiveness of heat transfer from the nuclear fuel to the water. There is a pressurizer which is designed to keep the reactor water under pressure. (The relief valve sits atop this pressurizer.) The fluid level in the pressurizer is also used as an indirect—and the only—means of measuring the water level in the reactor.

The steam in the reactor vessel caused the fluid level in the pressurizer to rise. The operators, thinking they had resolved the problem and that they now had too much water in the reactor, shut down the emergency core-cooling system and all but one of the emergency pumps. Then they proceeded to drain water at a rate of 160 gallons per minute from the reactor, causing the pressure to drop. At this point they were still unaware of the water escaping through the open relief valve. Actually they assumed some leakage, which occurred because of poor valve seating even under normal circumstances. It was this which made them disregard the high-temperature readings in the pipes beyond 7.

The steam bubbles in the reactor water covered much of the fuel, and the tops of the fuel rods began to crumble. The chemical reaction between the steam and zircaloy covering the fuel elements produced hydrogen, some of which was released into the containment structure, where it exploded.

The situation was beginning to get really serious when, 2 hours after the initial event, the next shift arrived for duty. With some fresh insights into the situation, the relief valve was deduced to be open. Blocking valve 9 in the relief line was then closed. Soon thereafter, with radiation levels in the containment building rising, a general alarm was sounded. While there had been telephone contact with the Nuclear Regulatory Commission (NRC), as well as with Babcock & Wilcox (B&W), who had constructed the reactor facility, no one answered at NRC's regional office and a message had to be left with an answering service. The fire chief of nearby Middletown was to hear about the emergency on the evening news.

In the meantime, a pump was transferring the drained water from the main containment building to the adjacent auxiliary building, but not into a holding tank as intended; because of a blown rupture disk, the water landed on the floor. Eventually there was to be sufficient airborne radiation in the control room to force evacuation of all but essential personnel. Those left behind wore respirators, making communication difficult.

Eventually the operators decided to turn the high-pressure injection pumps back on again, as the automatic system had been set to do all along. The core was covered once more with water, though there were still some steam and hydrogen bubbles on the loose. Thirteen and one-half hours after the start of the episode there was finally hope of getting the reactor under control. Confusion over the actual state of affairs, however, continued for several days.

Nationwide the public watched television coverage in disbelief as responsible agencies displayed their lack of emergency preparedness at both the reactor site and evacuation-planning centers. More than 8 years later, the decommissioning of the reactor was still not complete. Its radioactive water had been decontaminated, but only one-half of the 300,000 pounds of core debris had been gingerly removed. The cleanup alone was expected to cost over a billion dollars. Three Mile Island was a financial disaster.

Prior Warnings

Apart from the technical lessons learned at the Three Mile Island "laboratory," the experience has offered lessons in the need for disaster planning and openmindedness. "Openmindedness" refers, once again, to not allowing a preoccupation with rules to prevent close examination of safety problems which may not be covered by rules. It also refers to a willingness on the part of management to take seriously the safety concerns expressed by engineers within or outside the organization. A lack of concern in this direction is particularly troubling, as exemplified by the experience of a number of engineers who reported dangerous behavior of B&W reactors of the type used at Three Mile Island well before the accident. A selection follows.

Stephen Hanauer, a government nuclear safety expert, communicated in 1969 to then Atomic Energy Commission head Glenn Seaborg his concern regarding common-mode failure possibilities at nuclear power plants. Later he became particularly worried about the poor performance record of valves in nuclear power plants. He also decried the lack of proper analysis of field reports which would help identify weak spots in design (Ford, 1982, 53-61).

As it turned out, the largest single failure at TMI *was* of a common-mode type, although not of a physical nature, as is more usual. It lay in the operators' inadequate training. No one on duty at the plant that early morning of March 28 was a nuclear engineer, and no one was even a college graduate; no one in charge was trained to handle complex reactor emergencies. [Interestingly enough, the 1974 Rasmussen Report on nuclear power plant safety (see the Bibliography) had singled out human response as the weakest link in plant reliability.]

Moreover, Hanauer's concern over valves was well founded. Problems with valves have not been restricted to the nuclear industry, and the particular type of relief valve that malfunctioned at TMI had done so earlier at other nuclear power plants. In fact, since 1970 eleven pilot-operated relief valves had stuck open at other such plants (Lombardo, 1980, 55).

2 John O'Leary, Deputy Secretary of Energy when he presided at the opening ceremonies of TMI's Unit 2, had earlier prepared a memo for incoming President Carter. Then a deputy director for licensing at the Atomic Energy Commission, he had written that "the frequency of serious and potentially catastrophic nuclear incidents supports the conclusion that sooner or later a major disaster will occur at a major generating facility" (Ford, 1982, 15). Yet the probability of serious accidents continued to be given such a low value that it was not thought necessary to plan for orderly emergency measures involving off-site populations and agencies.

3 James Creswell was a reactor inspector for the NRC assigned to the startup of Toledo Edison's Davis-Besse nuclear generating station, which also used a reactor built by B&W. Reading a B&W report on some strange behavior of its unit at Rancho Seco in California following the accidental dropping of a small light bulb into the main control panel, Creswell noticed startling similarities to an incident which had occurred at Davis-Besse during low-power tests in September 1977. A severe and sudden increase in heat had taken place in both cases, at Rancho Seco because of faulty control signals produced by difficulties with the control panel and at Davis-Besse because of failure of the main feedwater system. At both plants the instruments had not given the operators adequate indication of the reactors' true operating conditions. At Davis-Besse the pressure-relief valve had stuck open and the operator had misinterpreted the level of water in the reactor based on indications of the level in the pressurizer. This had led them to mistakenly shut off the emergency core-cooling pumps, even at Davis-Besse. The only differences were that Davis-Besse had been operating at 9 percent of rated capacity (as against 96 percent at TMI) and the failure of the relief valve to close had been detected after 22 minutes (instead of more than 2 hours as at TMI).

Creswell worried about this and, from early 1978 on, tried to communicate his concerns to various parties at the NRC, the utility, and B&W. Eventually he took a day off and flew at his own expense to Bethesda, Maryland. There he met with two NRC commissioners who were willing to listen. The result was a memo to the NRC staff requesting answers to some of Creswell's questions. The memo was delivered the day after the TMI incident.

While the similarities between what happened at Davis-Besse and at Rancho Seco are of interest, the near duplication at TMI of the occurrences at Davis-Besse makes a strong case for the free exchange of information and expeditious correction of faults, unfettered by organizational expediencies or short-range interests. And an ironic twist to the story points out clearly what approaches to safety are usually the more valued: Creswell eventually received a special NRC award of $4000 for his efforts. Two other officials, who had earlier been unresponsive to Creswell's pleas, were awarded $10,000 and $20,000 each for their efforts during the ensuing dramatic emergency at TMI (Ford, 1982, 72-82).

4 Also late in 1977, Carlyle Michelson, a senior nuclear engineer with the Tennessee Valley Authority (TVA), had been questioning safety aspects of a new B&W reactor for TVA's Bellefonte Nuclear Plant. Included in his analysis was the possible formation of a steam bubble in the reactor's cooling-water system, something which had also been mentioned by Creswell. Michelson's study, which additionally mentioned the dangers of misreading the reactor coolant level by looking at the pressurizer, received some support on its way to the NRC's Division of System Safety. If handled according to protocol, the memo should have been sent on to the Division of Operating Reactors, whence reactor users, particularly at Davis-Besse, would have been alerted to the problems. The assistant director of the Division of System Safety later said that the Davis-Besse analysis was the responsibility of the Office of Inspection and Enforcement. He thought they would have taken action if any was needed. Ten days after the TMI incident, the Advisory Committee on Reactor Safeguards asked the NRC to carry out some of Michelson's recommendations (Ford, 1982, 82-84).

5 At B&W, meanwhile, the 1977 occurrence at Davis-Besse caused concern. Before the TMI incident the company did not have any formal procedure to analyze ongoing problems at B&W-equipped plants or to study reports filed by its customers with the NRC. B&W was busy building and shipping new reactors. Nevertheless, the Davis-Besse incident was sufficiently unusual that B&W engineers were sent out to investigate. Some individuals maintained their interest and tried—in vain—to get word out to B&W customers: John Kelly, a plant-integration engineer; Bert Dunn, manager of emergency-cooling-system analysis; and Donald Hallman, a customer service employee. Internal B&W memos later revealed the company's awareness of reactor defects. The company denied NRC charges that it had failed to notify the Commission, but it instituted measures to make sure that this would not occur again (Ford, 1982, 86-92).

Chernobyl

The nuclear power plant complex at Chernobyl, near Kiev (Ukraine, U.S.S.R.) had four reactors in place by 1986. With the planned addition of Units 5 and 6, for which foundation work was underway, the site would be the world's second largest electric power plant park, with an output of 6000 megawatts (electrical). The reactors are of a type referred to as RBMK; they are graphite-moderated and use boiling-water pressure tubes. Chernobyl and the Russian nuclear power program were prominently featured in 1985 issues of the English language periodical *Soviet Life*. The articles featured the safety of atomic energy and the low risk of accidents and radiation exposure. For readings on the accident and its aftermath we refer the reader to Hawkes et al. (1986), Marples (1986), Edwards (1987), and Ahearne (incl. disc., 1987).

On April 25, 1986, a test was underway on reactor 4 to determine how long the mechanical inertia of the turbine-generator's rotating mass could keep the generator turning and producing electric power after the steam supply was shut off. This was of interest because reactor coolant pumps and other vital electric machinery have to continue functioning though the generators may have had to be disconnected suddenly from a malfunctioning power grid. Special diesel generators will eventually start to provide emergency power for the plant, but diesel units cannot always be relied upon to come up promptly. This test was undertaken as part of a scheduled plant shutdown for general maintenance purposes.

It requires 3600 megawatts of thermal power in the reactor to produce 1200 megawatts at the generator output. Unit 4 had been gradually reduced from 3200 megawatts (thermal) to 1600 megawatts and was to be slowly taken down to between 1000 and 700 megawatts, but at 2 P.M. the power dispatch controller at Kiev requested that output be maintained to satisfy an unexpected demand. This meant a postponement of the test. In preparation for the test the reactor operators had disconnected the emergency core-cooling system so its power consumption would not affect the test results. This was to be the first of many safety violations. Another error occurred when a control device was not properly reprogrammed to maintain power at the 700- to 1000-megawatt level. When at 11:10 P.M. the plant was authorized to reduce power, its output dropped all the way to 30 megawatts, where the reactor is difficult to control. Instead of shutting down the reactor, the operators tried to keep the test going by raising the control rods to increase power. Instead of leaving fifteen controls inserted as required, the operators raised almost all control rods because at the low power level the fuel had become poisoned by a buildup of xenon-135, which absorbs neutrons.

The power output stayed steady at 200 megawatts (thermal)—still below what the test called for—but the test was continued. In accord with the test protocol, two additional circulating pumps were turned on to join the six already in operation. Under normal levels of power output this would have contributed to the safety of the reactor, but at 200 megawatts it required many manual adjustments to maintain the balance of steam and water. "The operators at this point recognized that because of the instabilities in this reactor and the way xenon poisoning builds up, once the reactor is shut down, they would have to wait a long time before starting it up again" (Ahearne,

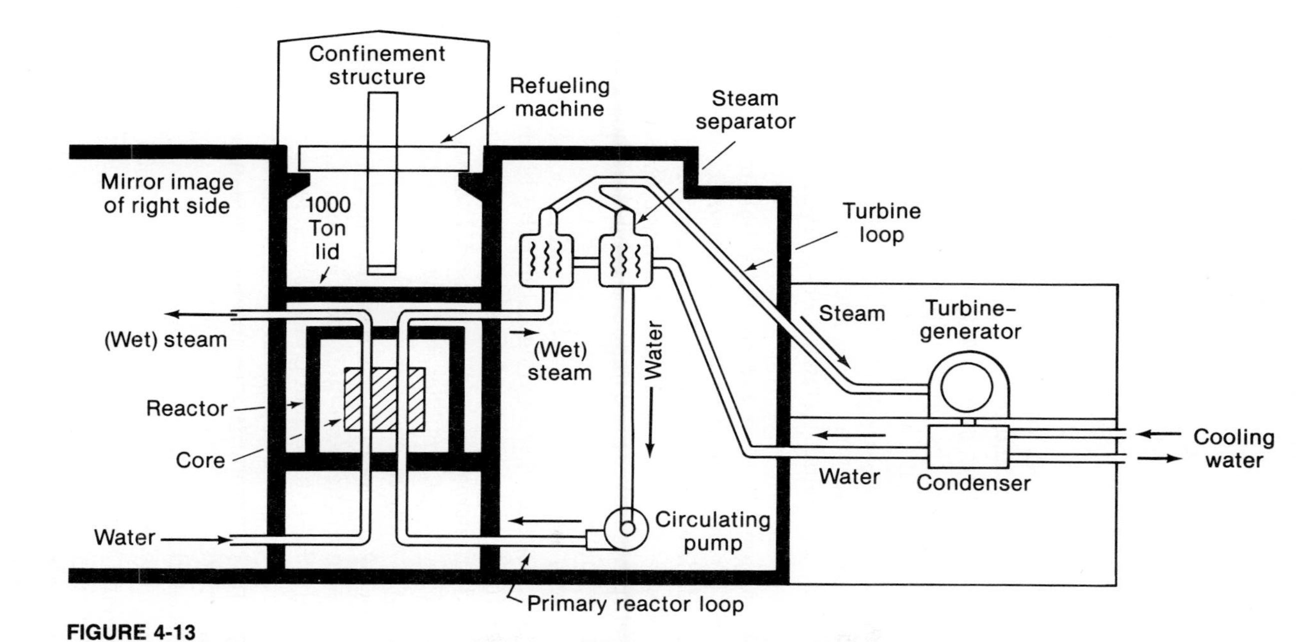

FIGURE 4-13
Schematic diagram of Reactor 4 at Chernobyl. This RBMK-type reactor produces steam for two 500-megawatt steam turbine generators, only one of which is shown. (*After Ahearne, 1987, p. 674.*)

1987). So, deciding to proceed with the test, the operators blocked the emergency signals and automatic shutdown controls because they would have gone into action upon removal of the electrical load.

"The reactor was now running free, isolated from the outside world, its control rods out, and its safety system disconnected." As Legasov, the U.S.S.R. representative to the International Atomic Energy Commission, told the conference reviewing the accident in Vienna: "The reactor was free to do as it wished" (Hawkes et al., 1986, 102).

At 1:23 A.M. the test began. When the steam valves were closed and its load was effectively removed, the reactor's power and temperature rose sharply. Unlike water-moderated reactors, the graphite-moderated RBMK reactor uses water only as a heat-transfer medium, not as a moderator. As the core becomes hotter it allows fission to increase. This positive feedback effect produced a surge of power in Chernobyl's reactor 4, from 7 percent to hundreds of times its rated thermal output. "The effect was the equivalent of ½ ton of TNT exploding in the core. . . . The fuel did not have time to melt . . . it simply shattered into fragments" (Hawkes et al., 1986). The fuel, bereft of its cladding, came in contact with the water. A second explosion occurred (very likely a steam explosion). It lifted and shifted a 1000-ton concrete floor pad separating the reactor from the refueling area above it. The zirconium cladding of the fuel rods interacted with the circulating water to form hydrogen. This produced a spectacular display of fireworks. A shower of glowing graphite and fuel spewed over the compound while a radioactive plume was driven sky high by the heat.

What followed was as inexcusable as what had caused the accident. While valiant firefighters lost their lives extinguishing the blaze, it took hours to warn the surrounding communities. Only when alert nuclear plant operators in Sweden detected an increase in radioactivity did Moscow learn that something was amiss. The Soviet republics and the rest of Europe did not know how to handle such a grave event, especially not the radioactive fallout. Many blamed Moscow for not notifying them but had no monitoring devices of their own, not even to check on their local nuclear plants. Instructions on what to do about drinking milk, eating vegetables, letting children play outside, and other concerns of the populations of Europe depended more on the political leanings and the pronuclear or antinuclear stance of the health minister issuing a directive.

Acute radiation sickness, combined with burns, severely affected about 200 Chernobyl plant workers, of whom 31 died. The 1000 families living in a workers' settlement 1 mile from the plant were evacuated 12 hours after the explosion, but the plant had no responsibility for, nor direct link with, the communities beyond a 1.5-mile radius. The evacuation of nearby Pripyat and 71 villages within 18 miles of the plant started the next day. About 120,000 people had to be moved by buses and trucks. Numerous new villages were constructed to house the displaced. The near- and long-term effects of radiation on the people and fauna of Europe will be widely discussed for many years.

It took one week to contain the fire by covering the reactor with a mix of sand, clay, and dolomite deposited by helicopters. Tunnels were dug underneath the reactor to install cooling pipes carrying liquid nitrogen. The tunnels also served to lay down a concrete layer to prevent leakage of radioactive water to the aquifer. Eventually the entire plant was completely entombed in concrete.

Three Mile Island, Chernobyl, and a Hushed-up Forerunner

There are similarities and dissimilarities between the events at Three Mile Island and Chernobyl, but the lessons to be learned do not differ much.

Pressurized-water reactors (PWR) as used at TMI have strong containment structures. Thus the radioactive products of the accident at TMI Unit 2 were fairly well contained. The RBMK reactors at Chernobyl have a much weaker containment system relative to the space into which gases and steam can expand during an accident. It should be noted that many reactors in the United States also do not have a sturdy containment. Examples of such less well protected types are the ones producing weapons-grade plutonium for the U.S. Department of Energy and the earlier versions of boiling-water reactors. Such units depend on special cooling methods to limit pressure rises and to keep radioactive gases within confinement structures (which are smaller and weaker than containment structures).

Both reactor types are sensitive to perturbations. Three Mile Island's PWR has a once-through reactor-cooling system with a rather small amount of water and an undersized pressurizer. The RMBK exhibits a positive temperature-power feedback which at low power levels is not sufficiently offset by the negative fuel temperature coefficient.

Also common to both plants was the complacency shown by management and operators, largely created by the absence of prior, major accidents at their respective sites. What happened elsewhere was either "out of sight, out of mind" or greeted by "It can't happen here." This is how the engineers at Chernobyl had felt about TMI, and how more recently the Chernobyl episode was received elsewhere. But serious accidents can happen, and when they do, they usually occur in ways not foreseen—which is what makes them serious. The physical layout of systems may be different from plant to plant and country to country, but managers and operators are never so different in their behavior. At TMI, operating procedures were not continuously and thoroughly reviewed by experts. At Chernobyl, the test protocol had not even been discussed with plant designers and nuclear engineers or physicists. At neither plant were the operators fully conversant with the operating principles of the plant equipment.

Discussions regarding dangers to the public at the time of the events at

TMI and Chernobyl became "mass-mediated" events (a term used by some in connection with Bhopal), while engineers, physicists, physicians, health officials, and regulators were unable to issue authoritative status reports and offer professionally sound advice. The official reports that eventually came from the Soviet Union were refreshingly candid. In the past, moreover, secrecy had not been a monopoly of the U.S.S.R. In the United States, the former Atomic Energy Commission had kept information about embarrassing events close to its chest; so did the atomic energy establishment in the United Kingdom, where the Windscale nuclear plant had emitted so much radioactive material that its name was changed to Sellafield to deflect attention.

Not only did Windscale discharge ¼ ton of plutonium into the Irish sea and experience several leaks, it also had a reactor fire in 1957, with graphite and uranium fuel cladding ablaze for 42 hours. Efforts at extinguishing the fire with the carbon dioxide system provided for this purpose had failed. Only by gambling on extinction by a "tidal wave" of water to forestall a steam explosion during its application was the plant saved. Fortunately most of the potential radioactive fallout was trapped by special filters installed at the insistence of Sir John Cockcroft, who had worked on atomic bombs. Before the accident the filters had been jokingly called Cockcroft's Folly because they were felt to be superfluous. Even with those filters enough fallout escaped to require eventual disposal of 2 million liters of milk in a 500-square-mile area. The reactor was smaller than the Chernobyl unit and not all of it was demolished. Yet it took 10 years before dismantling could begin, and for over two decades the official reports of an inquiry were not released to the public to protect the nuclear industry.

Central Europe has the greatest concentration of atomic power in the world, with 388 plants in operation or in some phase of construction and planning. Electricité de France (EdF) is often cited as a model of nuclear plant operation. By concentrating on standardized designs early on (a gamble, because the standards could have turned out to be poor), and insisting on highly trained personnel, an excellent safety record has been accumulated (except for nuclear-fuel-reprocessing at La Hague). Yet, to the people on the other side of the Rhine the lack of joint emergency exercises involving the nuclear reactor parks across the borders in France is not reassuring.

Financially the nuclear power industry is facing a bleak picture in the United States. Not only were there high costs associated with the major accidents, but the growing cost of building the plants without error and to increasingly stringent requirements, coupled with a decline and reversal in the rate of growth of fossil fuel prices, has raised havoc with the economic side of the industry. The Washington Public Power Supply System, which had invested heavily in nuclear power plants, had to mothball two of its reactors and has a multibillion-dollar debt. Electricité de France is the world's largest debtor: $200 billion. In the meantime some incomplete nuclear plants are being converted to fossil fuel operation in the United States.

The nuclear industry and its regulators have not been open with the public. They must have felt that the public cannot be trusted, that it is too easily swayed by "Luddites and scare mongers." In France the protests against nuclear energy have been squelched by strong police measures and secrecy. "You don't tell the frogs when you are draining the march," said the director of EdF (Hawkes et al., 1986, 67). Nevertheless it has been noticed by the public that regulators' figures for "safe" doses of radiation exposure have been lowered again and again over the years. (Further reductions are likely to follow in the wake of revelations that the Hiroshima casualties were produced by less radiation exposure than had been calculated hitherto). It is also no secret that insurance companies are not willing to underwrite policies covering the full potential losses incurred by an accident (which makes measurement of perceived risk by the method of examining insurance policies impossible here). Finally, residents near many nuclear plants know how inadequate emergency evacuation plans are.

The public mistrust which the nuclear industry and its regulators have earned is unfortunate because nuclear power is an alternative we must seriously consider as our fossil fuels become scarcer, rise in price once again, or become otherwise inaccessible. Much more is required in the way of candid, intelligent discourse and action if the public is to be expected to underwrite continued experiments with nuclear power. An unusual undertaking in this respect was the 1-year educational program sponsored by the Swedish government prior to a public referendum on whether or not the nuclear energy program in that country should be terminated and existing plants be phased out. Supporters and opponents were given public funds to broadcast programs in support of their positions.

Safe Exit

In our Chap. 2 discussion we based the engineer's responsibility for safety on considerations of ethics. In this chapter we have described how risks are perceived, assessed, and weighed against benefits; also how engineers ultimately are faced with designing as much safety into their products as feasible under constraints of knowledge, time, cost, and clients' wishes. We stated in Chap. 3 that the tough part of the engineering experiment begins when the product is put to use. Let us pick up the thread at that juncture.

It is almost impossible to build a completely safe product or one that will never fail. The best one can do is to assure that a product—if and when it fails—will fail safely, that the product can be abandoned safely, or that the user can safely escape the product. Let us refer to these three conditions as *safe exit*. It is not obvious who should take the responsibility for providing safe exit. But apart from questions of who will build, install, maintain, and pay for a safe exit system there remains the crucial question of who will think of the need for a safe exit.

It is our position that providing for a safe exit is an integral part of the experimental procedure—in other words, of sound engineering. The experiment is to be carried out without causing bodily or financial harm. If it does, it must be terminated safely. The full responsibility cannot fall on the shoulders of a lone engineer, but one can expect the engineer to issue warnings when a safe exit does not exist or the experiment must be terminated. The only way one can justify continuation of an experiment without safe exit is for all participants (including the subjects of the experiment) to have given valid consent for its continuation.

Let us illustrate by examples what this might involve. Ships need lifeboats with sufficient spaces for all passengers and crew members. Buildings need usable fire escapes. Operation of nuclear power plants calls for realistic means of evacuating nearby communities. The foregoing are examples of safe exits for people. Provisions are needed for safe abandonment of products and materials: altogether too many truck accidents and train derailments have exposed communities to toxic gases, and too many dumps have let toxic wastes get to the groundwater table or into the hands of children. Finally, to avoid business failure may require redundant or alternative means of continuing a process when the original procedure fails. An example would be a computer-based data retrieval system backed up by printed copies of the data, or a water supply backed up by a reservoir.

What we have described is risk management; in other words, how do you go about meeting and minimizing the damage identified in a risk assessment exercise. The last line of defense, and the one which must not be omitted, is the safe exit. A key word in this context is "management." Coordination among producers, users, and local communities are required to provide a realistic safe exit. Engineers are the ideal catalysts to set the process in motion. This is an added burden of responsibility and must be balanced by concomitant rights to openly identify the risks and communicate with other producers and users across organizational barriers.

As we stated in Chap. 2, the responsibility of engineers for safety derives from clients' and the public's right not to be endangered without prior warning in a manner understandable to them. Only with adequate knowledge can persons become willing participants in an engineering project qua experiment, decide not to participate, or decide to oppose it. An ethics based on distributive justice which gives rights to clients (and the public) and to engineers (and their managers) supports this view, but it can also be founded on duty- or goal-based theories of ethical behavior. Engineers need to handle safety issues with great care, but they need not reassess each detail on ethical bases if by habit they have acquired the appropriate "virtues" of responsible engineering.

Study Questions

1 Collect some examples of literature promoting and criticizing use of nuclear power. A good start could be made with Gueron (pro, 1984) and MacKenzie (con, 1984). Are the statements factually complete? Do you find yourself agreeing with what-

ever item you have read last? Discuss the responsibility of experts to reasonably educate the populace about the issues involved, pro and con.

2 It has been said that Three Mile Island showed us the risks of nuclear power and the Arab oil embargo the risk of having no energy. Removing hazardous products or services from the market has been criticized as closing out the options of those with rising aspirations who can now afford them and who may all along have borne more than their share of the risks without any of the benefits. Finally, pioneers have always exposed themselves to risk. Without risk there would be no progress. Discuss this problem of "the risk of no risk." (Compare: Wildavsky, 1980).

3 A number of engineers engaged in nuclear power plant work have expressed their concern over inadequate attention to safety. Some resigned first, then went public with their testimony; some spoke out and were fired; others spoke up but nevertheless retained their positions. Examine the literature about the cases listed below to see if there were any issues of ethical import involved. If so, what were they?

a Carl Houston, 1970, welding superintendent, Stone & Webster (Houston, 1975)

b Peter Faulkner, 1974, systems application engineer, Nuclear Services Corporation (Faulkner, 1981)

c Dale G. Bridenbaugh, 1976, manager of performance evaluation and improvement; Richard B. Hubbard, 1976, manager of quality assurance; Gregory C. Minor, 1976, manager of advanced control and instrumentation; all with Atomic Power Division, General Electric Company, San Jose, California (Kaplan, 1976; Weil, 1977)

d Ronald M. Fluegge, 1976, safety analyst; Demetrios L. Basdekas, 1976, reactor engineer; both with Nuclear Regulatory Commission (*Congressional Record,* 13 Dec 1976)

e Robert D. Pollard, 1976, project manager, Nuclear Regulatory Commission (Friedlander, 1976)

4 Discuss the notion of safe exit, using evacuation plans for communities near nuclear power plants or chemical process plants. Examples (and sample references) you could use include Bhopal's parent plant in Institute, West Virginia (Beck, 1984) and the following nuclear power plants: Diablo Canyon (Gini, 1983), Shoreham (Zorpette, 1987), Seabrook (Larmer, 1987), and early plans for a plant in New York City (Mazuzan, 1986).

5 Search the literature for reports on the Swedish referendum on nuclear power. A popular vote was to be taken after a 1-year public debate on the pros and cons, with proponents and opponents given funds by the government to air their views. Discuss the procedure and its possible applicability in other countries or to other technological issues.

SUMMARY

A *risk* is the potential that something unwanted and harmful may occur. A thing is *safe* for persons to the extent that they judge (or would judge) its risks to be acceptable in the light of full information about the risks and in light of their settled value principles. Thus, in designing for safety, estimates must be made of which risks are acceptable to clients and to the public that will be affected by the projects or products in question.

Many factors influence people's judgments when they decide which risks are acceptable. Most basic is their set of value principles regarding what they care about and to what extent they care about it. Other factors include whether or not a risk is assumed voluntarily, the degree of anxiety connected with any risk, whether or not a risk is immediately evident, whether or not potential victims are identifiable beforehand, and the manner in which statistics about a risk are presented to people.

Thus, for engineers, assessing safety is a complex matter. First, the risks connected to a project or product must be identified. This requires foreseeing both intended and unintended interactions between individuals or groups and machines or systems. Second, the purposes of the project or product must be identified and ranked in importance. Third, the costs of reducing risks must be estimated. Fourth, the costs must be weighed against both organizational goals (e.g., profit, reputation for quality, avoiding lawsuits) and degrees of acceptability of risks to clients and the public. Fifth, the project or product must be tested and then either carried out or manufactured.

Uncertainties in assessing risks arise at all these stages. For example, at the stage of testing a product only one or a few prototypes are typically used, and at that under carefully controlled conditions. Results may not accurately mirror what will happen following mass production and installation under normal operating conditions. Moreover, testing a product to destruction (the most effective way of testing) is sometimes ineffective or inappropriate. It may only be possible to work with simulations, including analytical tools such as fault-tree analysis (tracing possible causes of a systems failure back to the component level).

In spite of the complexities involved, a great many risks can be reduced or eliminated in fairly obvious and routine ways, at least by those engineers and managers who work with an attitude of deep caution. And increasingly the specter of legal liability serves as an incentive toward a preventive, defensive approach to engineering. Conceiving of engineering as a social experiment helps foster such an attitude. Moreover, by emphasizing the notion of informed consent, the experimentation model points out how safety is ultimately a matter of informed judgment about the acceptability of risks.

If the malfunction of a system can lead to serious injuries, death, and other grave consequences, such a system must be equipped with *safe exit* for those who would otherwise be hurt. Safe termination of an experiment in this sense is good experimental procedure and responsible engineering.

PART III

ENGINEERS, MANAGEMENT, AND ORGANIZATIONS

Professionals have to have autonomy. They cannot be controlled, supervised, or directed by the client. Decisions have to be entrusted to their knowledge and judgment. But it is the foundation of their autonomy, and indeed its rationale, that they see themselves as "affected with the client's interest."

Peter F. Drucker

Mankind cannot survive without technology. But unless technology becomes a true servant of man, the survival of mankind is in jeopardy. And if technology is to be the servant, then the engineer's paramount loyalty must be to society.

Victor Paschkis

CHAPTER **5**

PROFESSIONAL RESPONSIBILITY AND EMPLOYER AUTHORITY

The 1970 Clean Air Act requires car manufacturers to conduct 50,000-mile durability tests on new engines using only one tune-up. Test results on emissions must be reported to the Environmental Protection Agency (EPA), which decides whether the engines meet current pollution standards. In May 1972, top managers at Ford Motor Company were eagerly awaiting government approval of the test results they had submitted on the engines for 1973 Ford cars. They had every reason to be confident of the results they had submitted to EPA, which were based upon tests conducted by their own employees; their only concern was about meeting tight production schedules once EPA's approval was received.

Their confidence was shattered, however, when then Ford president Lee Iacocca received a memo from a specialist in the computer division. That computer specialist had been examining the computer tapes from the tests to review the effectiveness of his division in support of engine development. His memo identified numerous irregularities in the test records, showing unauthorized maintenance of which EPA was not notified. The memo also stated that when the specialist sought an explanation of the irregularities from the engine division he was urged to burn the computer tapes and forget the matter.

Intensive research into the matter by management quickly verified the information contained in the memo. Evidently, four "supervisory technical" employees who had conducted the original tests had ordered or engaged in over 300 acts of illegal maintenance on the test engines. Spark plugs and points had been replaced frequently, carburetors cleaned, and ignition timing repeatedly reset. These adjustments lowered the levels of pollutants emitted.

Within 3 days Mr. Iacocca revealed to EPA officials all he had learned about the tests and withdrew Ford's application for certification of four major types of engines. In spite of its full cooperation with EPA investigators, the company was fined $7 million in criminal and civil fines for having conducted improper tests and issued false reports to the government. Because of the record size of the fines, Ford received damaging publicity in front-page newspaper articles (for example, in the *New York Times* and the *Los Angeles Times*, 14 Feb. 1973). It was also hurt by the costs of new tests that had to be conducted on an around-the-clock emergency basis and by having to delay production schedules (*Wall Street Journal*, 25 May and 31 May 1972).

Misguided Loyalty?

Nothing written about the Ford test scandal tells what motivated the Ford supervisors and other engineers and technicians involved. Possibly it was only a self-interested concern—a desire to make themselves look good by ensuring their engines would pass the qualifying tests. But it is equally possible that they were acting as loyal employees. Ford had been late in obtaining some government approvals the previous year, and perhaps the individuals believed—however mistakenly—that they were serving the company's best interests by avoiding such difficulties this year. Perhaps some of them were merely following orders from higher up to tamper with the engines. In any case, management was not particularly punitive: despite the staggering costs incurred, no one who had participated in rigging the tests was fired and the four supervisors were merely transferred to new positions.

This case suggests three points concerning the relationship between professional responsibility and loyalty to companies or employers. First, acting on professional commitments to the public can be a more effective way to serve a company than a mere willingness to do anything one sees as good for the company. Ford would have benefited much more from engineers committed to professional standards than it did by the misguided loyalty shown to it by its employees.

Second, it is clear from the example that loyalty to companies or their current owners should not be equated with merely obeying one's immediate supervisor. It would have shown a greater loyalty to Ford to act in a way consistent with the concerns of higher management, rather than in a manner consistent with the aims of an immediate supervisor.

Third, the case illustrates how an engineer might have professional obligations to both an employer and to the public that reinforce rather than contradict each other. Thus there need be *no general contrast* between the moral status of employees and professionals. In fact, obligations to the public and to one's employer often point in the same direction.

Nevertheless, we have also seen from other cases we have examined that obligations to employers and to the public do not always straightforwardly coincide. Sometimes an engineer seeking to protect the public is overruled by top management for financial reasons. For example, in the DC-10 case, the director of product engineering was told by higher management that it would be too costly for his company to redesign an unsafe cabin floor and cargo door. At other times there are disagreements over technical matters, and engineers are told they must not push their own views further. This we saw illustrated in the case of space shuttle *Challenger*.

The relationship between being a responsible engineer, with obligations to the public, and being a loyal employee is a matter of some complexity. We will explore it first from the direction of professionalism, then from a study of employers' authority, and finally by discussing four topics: conflicts of interest, confidentiality, unionism, and white-collar crime.

PROFESSIONALISM

What is a professional? If we answer that it is someone who is a member of a profession, then what is a profession and how does one become a member of one? Our first concern in dealing with these questions is to understand why there is so much disagreement over how to answer them. A second concern is to sketch a conception of professionalism compatible with viewing employed engineers as professionals having obligations to both employers and the public.

Professions

In one of its senses, the word "profession" is used as a synonym for "job" or "occupation," and to be a professional at some activity means merely to earn one's living through it. Thus we speak of professional football and tennis players, as opposed to amateurs who do not draw an income from these sports. We also speak of professional sanitation workers, taxicab drivers, bartenders, and even mercenaries and killers.

But there is another sense of the word which rules out such examples. "Profession," in this new sense, can be applied only to certain occupations which meet special criteria. Generally the criteria include restrictions of the following sort:

1 The work involves exercising sophisticated skills, judgment, and discretion which is not entirely routine or susceptible to mechanization.

2 Preparation to engage in the work requires extensive formal education, including technical studies in one or more areas of systematic knowledge as well as broader humanistic studies. Generally, continuing education and updating of knowledge are also required.

3 Special societies and organizations controlled by members of the profession are allowed by the public to play a major role in setting standards for admission to the profession, drafting codes of ethics, enforcing standards of conduct, and representing the profession before the public and the government.

4 The occupation serves some important aspect of the public good, as indicated in the codes of ethics. (For example, medicine is directed toward promoting health, law toward protecting the public's legal rights, and engineering toward promoting the public's health, safety, and welfare as they relate to technology.)

There are many debates over just which occupations meet these criteria. The traditional professions of medicine, law, teaching, and the ministry are cited as paradigm or clear-cut examples. So too are professions like engineering and business administration that have emerged more recently. Sanitation work, taxicab driving, and basketball are not counted because of the lack of required advanced education. Disagreements occur over occupations requiring intermediate amounts of formal training: advertising, realty, cosmetology, and some jobs in computer and medical technology.

Membership Criteria

Further disputes arise over how a person does or should become a member of an accepted profession. Such disputes often occur with respect to engineering. Each of the following has been proposed as a criterion for being an engineer or a "professional engineer" in the United States:

1 Earning a bachelor's degree in engineering at a school approved by the Accreditation Board for Engineering and Technology. (If applied in retrospect, this would rule out Leonardo da Vinci and Thomas Edison.)

2 Performing work commonly recognized as what engineers do. (This rules out many engineers who have become full-time managers, and also rules in some people who do not hold engineering degrees.)

3 Being officially registered and licensed as a "Professional Engineer" ("P.E."). Becoming registered typically includes (*a*) passing the Engineer-in-Training Examination or Professional Engineer Associate Examination during the senior year in engineering school, (*b*) working 4 to 5 years at responsible engineering, (*c*) passing a professional examination, (*d*) paying the requisite registration fees. (This rules out a large percentage of unregistered people holding bachelor's, master's, and doctoral degrees in engineering, many of whom work in education or manufacturing industries where they are exempt from registration.)

4 Acting in morally responsible ways while practicing engineering. (This rules out scoundrels, no matter how creative they may be in the practice of engineering.)

The words "profession" and "professional" have acquired positive emotional connotations and suggest a highly desirable status for occupations and individuals. At least part of these connotations derive from the public importance of professional skills and knowledge, and also from the difficulty of acquiring them. Because of these factors, professionals are regarded as deserving high pay, prestige, and other social benefits. Social status is frequently enhanced by a title, such as Doctor or Reverend. In this respect some engineers in the United States, where engineering is often not considered on a par with medicine or the ministry, yearn for more of the open recognition accorded their counterparts in some other countries.

Persuasive Definitions

One could choose any one of the above mentioned criteria for what constitutes an engineer and claim, by assuming a particular value perspective, that it is the only correct definition. The somewhat loose ordinary cognitive meaning (defining criteria) could then be altered by making it more precise and narrow while retaining the ordinary *emotive meaning* (positive connotations). One would then be giving what is called a *persuasive definition* of the term "professional engineer": one used to espouse a particular value perspective (Stevenson, 1938; Cogan, 1955, 105).

As might be expected, such persuasive definitions occur frequently in disagreements over values, and there need be nothing improper about them. But they must be understood for what they are: techniques for altering attitudes, which by themselves do not constitute arguments. They should be critically examined, rather than passively allowed to influence us under the guise of being "truths, by definition." For they are not at all like definitions of triangles as three-sided planar figures or bachelors as adult, unmarried males.

For instance, if a psychologist defines intelligence as simply what certain psychological tests reveal, we should beware of the possible implications of bestowing so much significance on present-day techniques of psychological testing. Again, if medicine is defined as the science of health, and if health is defined as "a state of complete physical, mental and social well-being" (which is how the World Health Organization defines it), we should be wary of how these definitions encourage excessive expectations about what doctors and medical techniques can do.

Similarly, we need to be ready to assess the implications of accepting any given persuasive definition of "professional engineer." The attitudes and value perspectives embodied in such definitions concerning the desirable properties of professional engineers and how best to identify those properties need to be critically examined. For instance, those who seek to restrict the term "professional" to officially registered engineers will view the restriction as a way to ensure that stringent qualifications are met which will maximize benefits to the public. Those who are against this definition, however, may argue that it needlessly increases bureaucracy and is not an effective way of judging engineering qualifications.

Professionalism as Independence

There is one type of persuasive definition of professional engineer which is especially significant for our present purposes. That definition directly ties professionalism to independence and freedom from coercion. One version was given by Robert L. Whitelaw in an essay entitled, "The Professional Status of the American Engineer: A Bill of Rights." Whitelaw sharply contrasts bureaucratic submission to employers with the independence he sees as inherent in professionalism. In fact, he defines professionalism and employee status as logically incompatible: ". . . so long as the individual is looked upon as an employee rather than as a free artisan, to that extent there is no professional status" (Whitelaw, 1975, 37–38).

In Whitelaw's view, only consulting engineers qualify as professionals. The mass of engineers working as employees within corporate or governmental bureaucracies will not become professionals until they are protected by an engineering bill of rights ensuring the freedoms already enjoyed by self-employed engineers. Examples of these rights are "the right to refuse unethical activity without prejudice or loss of contract" and "the right to freedom from surveillance, psychological manipulation, and other job evaluation techniques."

According to Whitelaw's definition, one is not a professional engineer if one acts merely on the basis of an employer's orders in matters where the public good is concerned. Being a professional involves the freedom to act according to one's own judgment about what the correct course of any action should be. It is clear that Whitelaw is reacting sharply to what he views as the excessive domination of engineers by the authority of management. While many of his concerns are legitimate, and while his definition is a potent rhetorical instrument, we must ask whether his definition expresses too extreme a position.

Professionalism as Serving Employers

An opposite type of persuasive definition would treat loyal service to employers (or to clients, in the case of consulting engineers) as the heart of professionalism in engineering. Such a view is implied in Samuel Florman's widely discussed essay, "Moral Blueprints." Florman argues that "it is essential that professionals should serve" (Florman, 1978, 32). Rather than "filtering their everyday work through a sieve of ethical sensitivity," as Florman puts it, professionals have the task of meeting the expectations of their clients and employers. Professional restraints should be laws and government regulations rather than personal conscience.

Florman's essay is devoted to attacking the entry in the code of ethics of the former Engineers' Council for Professional Development which states, "Engineers shall hold paramount the safety, health and welfare of the public in the performance of their professional duties." His response is: "Engineers are obliged to bring integrity and competence to whatever work they undertake. But they should not be counted upon to consider paramount the welfare of the human race" (Florman, 1978, 32).

It is fair to view Florman as expressing the dominant view of management concerning engineering professionalism. And many engineers would concur with the definition. Yet here again we must ask whether this conception of the professional obligations of engineers is one-sided.

An Intermediate Position

We will state, but not attempt to defend in detail, our own "persuasive definition" of professionalism in engineering. Our main concern in this section has been to emphasize that such definitions will generally be an outgrowth of one's perspective on the moral obligations of engineers. Accordingly, in discussions about the subject, attention should be focused on the obligations themselves rather than on how they are reflected in the criteria one espouses in defining the term "professional engineer."

Our view of the obligations of engineers involves a moderate position lying between the extremes represented by Florman and Whitelaw. For us, employed engineers have major moral obligations to both employers and to the public, and we think it a mistake to seize on either obligation as the essence of professionalism. A more useful definition would allow us to speak straightforwardly of "salaried professionals" (contra Whitelaw), and would also enable us to reject the view of professional obligations as essentially service to employers within the limits of law (contra Florman).

Accordingly, we favor viewing professional engineers as meeting two general criteria: (1) Attaining standards of achievement in either education, job performance, or creativity in engineering which distinguish them from engineering technicians and technologists. (We recognize that for legal and educational purposes the nature of those standards will have to be made more clear-cut and explicit.) (2) Accepting as part of their professional obligations at least the most basic moral responsibilities to the public as well as to their employers, clients, colleagues, and subordinates. This latter criterion lends to the term "professionalism" a moral dimension consistent with the fact that "unprofessional conduct" is often used as a synonym for "unethical conduct." Yet it makes no assumptions about which type of obligation is most central to engineering—an issue that should be debated independently of how to define what it means to be a professional engineer.

Obligations to the Public as Paramount

At this point let us set aside the issues that arise when we try to define professionalism and turn directly to the relationship between the two general obligations to the public and to employers. Should we agree that the obligation to protect the public health, safety, and welfare is *paramount,* as recent codes have stated?

In our view, yes, so long as "paramount" is understood in its colloquial sense to mean "chief in importance or deserving primary emphasis." We make this judgment against the background of a conspicuous reality: Most employers have enormous power compared with the engineers they employ. They have the power to fire or take other negative sanctions against individuals who fail to meet their obligations to the employer. And engineers have relatively little recourse at present when an employer does not support their efforts to meet their obligations to the public. Hence if "paramount" means "deserving most emphasis in the minds of engineers, engineering societies, and the wider community," then the obligation to the public deserves to be regarded as paramount.

"Paramount," however, can also be construed in a technical philosophical sense to mean that whenever the obligations to employers and the public come into conflict (creating a moral dilemma), the obligation to the public *always* takes precedence. Thus it can mean that, whenever these two prima facie duties conflict, one's actual duty—what one *ought* to do, all things considered—is always to meet one's public obligations.

We doubt that this technical sense of "paramount" is what drafters of the codes had in mind. In any case, it seems to us to be a dubious view if carried to its extreme. Consider the following case: A design group develops a new electronic circuit to be used in clock radios which would extend their average life from 5 to 7 years at a cost that would raise manufacturing expenses by only 1 percent. After presenting their arguments to top management, however, the latter reject the proposal as not being cost-effective. Does the design group's obligation to the public outweigh its employer's directives to drop further work on the circuit?

In this case it would undermine an employer's legitimate authority to say that engineers must subsume their obligations as employees to their obligations to the public. Of course the obligation to the public should override the obligation to the employer in cases where something of extreme importance is at stake for the public: generally where lives are seriously threatened, serious financial corruption is involved, or serious economic loss might result. Many instances of justified whistle-blowing fall into this category, as we shall argue in Chap. 6. But this does not mean that the first priority is *always* the public good whenever that good conflicts morally with an employer's good.

Engineers, in short, must weigh their obligations to the public, their employers, their colleagues, and others when conflicts between such obligations arise. A simple, exceptionless ordering of priorities is not always possible.

Study Questions

1 Comment on the following definitions, or partial definitions, of professionalism in engineering. In each case, do you agree that the passage presents something es-

sential to an understanding of professionalism? Is the definition a controversial persuasive definition with which you disagree? Why?

a "Professionalism implies a certain set of attitudes. A professional analyzes problems from a base of knowledge in a specific area, in a manner which is objective and independent of self-interest and directed toward the best interests of his client. In fact, the professional's task is to know what is best for his client even if his client does not know himself" (Storch, 1971, 38).

b "A truly professional man will go beyond the call to duty. He will assume his just share of the responsibility to use his special knowledge to make his community, his state, and his nation a better place in which to live. He will give freely of his time, his energy, and his worldly goods to assist his fellow man and promote the welfare of his community. He will assume his full share of civic responsibility" (Simrall, 1963, 39).

c "If they mean to be professionals, engineers themselves will have to take moral responsibility for their work rather than unquestioningly accepting whatever orders come down to them from Government or employers" (Walters, 1973, 42).

d "A profession, in contradistinction to a trade...[is] a body of persons with learned knowledge having an ability to examine itself and its purposes; an ability to link its body of knowledge with other bodies of knowledge to achieve common purposes; the ability to defend dissent, not just within the society but dissent by its members in conflicts with their employer organizations or their government agencies or corporations; and above all the ability to pioneer new policies that are not brought into effect by market incentives" (Nader, 1972, 14).

2 Discuss under what circumstances you think engineers are justified in participating in the design and manufacture of products with built-in obsolescence. Such products wear out rapidly and cannot be repaired.

EMPLOYERS' AUTHORITY

Salaried engineers have obligations to respect their employers' legitimate authority. But what is the nature of this authority? How far should it be recognized by salaried professionals as being morally justified?

In order to address these questions we will begin with a discussion of how and why authority arises within institutions. Then several distinctions will be drawn which make clear why such authority is not automatically the same as moral authority.

Goals of Institutions

Engineers work within virtually all forms of modern organizations. These organizations vary enormously in the specific goals they are created and maintained to serve. Two general types are (1) service organizations, and (2) business, or profit-making, organizations (Drucker, 1973, 131).

Service institutions have as their primary purpose to provide selected services to the public or to other organizations of which they are parts. Universities provide education, hospitals give health care, court systems serve legal needs, professional societies serve professionals, and so on for churches, the military and government, and "natural monopolies" like telephone and utility companies. They operate under the economic restraints of a budget allocated by supervising government agencies or based on their income from the services they provide.

Business, or profit-making, institutions are established primarily to produce income. The criterion for performance is taken by some to be maximum profits (Friedman, 1979, 192), and by others to be a reasonable return on investment. Still other observers will include making social contributions as well. As the necessary means to achieving the primary purpose of producing income, business institutions must of course provide some product or service which customers will purchase. Moreover, businesses must do so within the boundaries delimiting the public good set by the government, which grants the businesses the charters that allow them to operate. Thus a fuller specification of the purpose of profit-making institutions is to make a profit by providing a product or service which the public finds useful.

Both service and business organizations may take on further secondary goals. In order to meet their primary goals, they generally adopt the goals of survival and of maintaining adequate degrees of freedom from outside control (Galbraith, 1971, 170). In practice this latter goal often means resisting extensive government regulation.

Institutional Authority

In order to meet their institutional goals, organizational rules are created. Typically these rules attach specific duties to positions within the organization. The rules may also allow one person to assign duties to others. Thus, an *institutional duty* is any assigned task within an organization, whether the assignment is directly or indirectly rule-specified. Managerial tasks, for example, may be to allocate money or other resources, to make policy decisions or recommendations, or to oversee projects and issue directives to subordinates on particular topics.

The need for authority relationships in meeting organizational goals is clear. Decisions must be made in situations where allowing everyone to exercise unrestrained individual discretion would create chaos. Moreover, clear lines of authority provide a means for identifying areas of personal responsibility and accountability.

In order to enable people holding managerial positions to meet their institutional duties, the rules also assign them the requisite authority. This is *institutional authority,* since it is acquired, exercised, and defined within institutions. It may be defined as the *institutional right* given to a person to exercise power based on the resources of the institution (Pichler, 1974, 428).

Institutional rights (authority) and duties are for the most part two sides of the same coin, and they deal with precisely the same activities and functions. Project engineers, for example, have the institutional duty to ensure that the projects they supervise are successfully completed, and they are given the institutional rights or authority necessary to carry out this duty. Obviously, too, these rights involve a certain amount of freedom or liberty: It would be self-defeating for an institution to assign tasks but to deny the freedom from interference necessary to perform them.

Institutional versus Expert Authority

It clearly benefits institutions to give authority to the individuals best qualified to serve the institution's goals in a given capacity. But in practice there is not always a perfect match between the authority granted and the qualifications needed to exercise it. Incompetence is found in all large institutions, and there is some truth in the cynical remark that in bureaucracies people tend to rise to their own level of incompetence.

Thus institutional authority should not be equated with expert authority. *Expert authority* is the possession of special knowledge, skill, or competence to perform some task or to give sound advice. In this sense, doctors are authorities on health and civil engineers are authorities on structures and transportation. One of the key competencies for management is leadership ability, which has its own kind of expert authority that has been called the "authority of leadership": the expertise to effectively direct others (Barnard, 1968, 173).

It is possible for engineers to have expert authority in matters for which they have little or no institutional authority to make decisions. Their institutional authority may extend no further than the right to provide management with analyses of possible ways to perform some technical task, after which they are restricted to following management's directives about which option to pursue.

Authority versus Power

Institutional authority must also be distinguished from power. Institutional authority typically carries with it an allotment of the resources needed to complete tasks. Yet ineffectual persons may not be able to summon the power which their position allows them to exercise. A manager, for example, who lacks the skills of leadership may be unable to inspire and encourage employees to produce in ways the institution requires, much in the way a conductor may fail with an orchestra.

Conversely, people who are especially effective may acquire great power or influence—power which goes well beyond the authority attached to the positions they hold. Charismatic leaders often have influence outside their domains of authority. And highly respected engineers of proven integrity may have power within an organization exceeding their explicit institutional rights.

Authority and Managerial Strategies

As we noted, institutional authority often gives one the prerogative to issue orders in a given area and to expect them to be complied with. But it would be a serious misconception to equate managing people with issuing orders and then standing on one's authority in demanding unquestioning obedience (McGregor, 1960).

To manage people is to guide and integrate their work, and there are many general approaches or strategies for doing so. One is the direct assertion of authority over a subordinate: "I'm in charge—obey or I'll fire you." But repeated use of this approach would be viewed negatively within business as an authoritarian abuse of authority. Other strategies include a heavy mixture of persuasion and rational argument. Some emphasize mutual decision making, or decision making based on full consultation with subordinates.

A consensus approach may be slower, but it is more effective and prudent in the long run. And in dealing with salaried professionals, it is more than prudent. A strictly authoritarian approach can easily lead to the demise of moral integrity among employees, with a resultant weakening of felt obligations vis-à-vis both employer and the public.

Morally Justified Authority

The preceding distinctions clear the way for making two observations. First, an employer may have the institutional authority to direct engineers to do something which is not morally justified. Second, engineers may have an institutional duty to obey a directive which is morally unjustified and which it is their moral duty, all things considered, to disobey.

To repeat: Institutional authority is the institutional right to exercise certain kinds of power, and this right is merely the liberty which the rules of the institution say a person has. Institutional duties are the duties specified by the rules of the institution, either directly as attached to offices and positions or indirectly as delegated by a superior (who in turn derives such authority from the rules of the institution). These rights and duties may be established as means to the end of meeting institutional goals. But they are not thereby moral rights and duties, or morally justified institutional rights and duties.

Before concluding that a specific act of exercising institutional authority is morally justified, we would need to know (1) whether the institutional goals are themselves morally permissible or morally desirable and (2) whether that act violates basic moral duties.

Engineers do take on some moral obligations to meet their institutional duties when they accept employment—but only so long as meeting those institutional duties is morally permissible. An employment contract can be viewed as a morally conditioned mutual promise. Promises to act immorally are either invalid or automatically overridden by moral considerations.

The relationship between *moral* rights and duties and *institutional* rights and duties is complex. Only a few further observations will be made here.

Recall that in earlier chapters we distinguished between general human moral rights and special moral rights. Obviously human rights and institutional rights cannot be equated. By definition, human rights (such as the rights to life, liberty, and the pursuit of happiness) are possessed by virtue of being a person, not by virtue of being a member of an institution.

However, some institutional rights and duties can be equated with special moral rights and duties—namely, those which are morally justified. For example, through employment agreements employees acquire a special institutional duty to protect proprietary information, and employers have an institutional right to require that employees do so. And to the extent that those duties and rights can be morally justified, either through some argument deriving from the employment contract itself or because of other, independent considerations, they are also moral duties and rights.

Accepting and Obeying Authority

Let us now shift perspective from the authority of employers to the recognition of that authority by their employees. Employees recognize their employer's authority when for the most part they accept the guidance and obey the directives issued by the employer having to do with the areas of activity covered by the employer's institutional authority. There are exceptions, since it is possible in special cases to recognize someone's authority but to disobey an order on moral grounds. But our present concern is to obtain a clearer idea of what accepting authority under normal conditions should and should not involve.

In his classic text, *Administrative Behavior,* Herbert Simon states: "A subordinate is said to accept authority whenever he permits his behavior to be guided by the decision of a superior, without independently examining the merits of that decision" (Simon, 1976, 11). In general, authority relationships are "all situations where suggestions are accepted without any critical review or consideration" (Simon, 1976, 128). Again, "the characteristic which distinguishes authority from other kinds of influence is . . . that a subordinate holds in abeyance his own critical faculties for choosing between alternatives and uses the formal criterion of the receipt of a command or signal as his basis for choice" (Simon, 1976, 127). In Simon's view, the reasoning of subordinates in their role as subordinates is at most aimed at anticipating commands by asking themselves how their superiors would wish them to behave in a given situation.

Simon notes that all employees place limits on the "zone of acceptance" in which they are willing to accept their employer's authority. But within that zone, an "individual, relaxing his own critical faculties, permits the communicated decision of another person to guide his own choice" (Simon, 1976, 151).

Simon provides an influential picture of what obedience involves. But its limitations must be kept in mind. Employees are generally not inclined to make an issue of every incident of questionable morality, sometimes because of moral inertia, at other times out of a reluctance to generate an overload of complaints or a willingness to give their employer a certain amount of leeway within which to operate, or even because of a wish to save the strongest arguments and the possible risk of losing their job for the most serious infractions. While this automatic obedience within the "zone of acceptance" of an employer's authority is understandable, it also carries with it the risk of becoming blind and unthinking in regard to moral matters. The problem which arises then is that the boundaries of tolerance are easily expanded and rationalized when expediency so dictates. Thus the size of any person's "zone of acceptance" could become a measure of the lack of that individual's moral integrity. To avoid this problem, employees must be reflective concerning the justified extent of their "zone of acceptance" of employers' authority. In a sense, then, they should never suspend their critical review of employers' directives in the manner Simon describes.

From a different direction, therefore, we have reached the same conclusion we came to in the last section: As professionals, engineers have obligations to accept their employers' institutional authority. But this is not an obligation to obey blindly. Professional autonomy entails exercising independent judgment, even though it does not mean disregarding legitimate directives. The basic moral task of salaried engineers is to be aware of their obligations to obey employers on the one hand and to protect and serve the public and clients on the other. Most of the time there is no conflict between the two. But when, occasionally, genuine conflict arises, it must be resolved by the exercise of an autonomous moral judgment.

Loyalty

Let us return for a moment to the topic of loyalty to company and employer, a topic mentioned in connection with the Ford case at the beginning of this chapter. The word "loyalty" suggests something more than merely recognizing and accepting the authority of the employer. It implies, at least in ordinary language, doing so from certain kinds of motives. People who detest their employers and companies and who obey grudgingly and spitefully are not considered loyal. A loyal person shows at least some degree of genuine concern in serving the interests of those to whom she or he is loyal.

Actually there are two different concepts of loyalty. According to the first, to be loyal and faithful is to seek to meet one's moral duties to a person or organization, and to do so willingly, with an attitude of devotion and personal attachment and identification (Ladd, 1967, 98). In this sense loyalty is an inherently good thing. Indeed, it is a moral virtue.

According to the second concept, by contrast, loyalty is not automatically a good thing. Here, to be loyal and faithful means to be devoted and obedient to or zealously supportive of a cause, person, or organization, but not necessarily out of (nor in a way restricted by) moral duty. People loyal in this sense try to promote the interests of whatever or whomever they are loyal to and they do so out of genuine concern. But whether it is good or obligatory to be loyal in this way depends upon the specific person, organization, or cause the loyalty is directed toward, and upon the circumstances in which the loyalty is displayed. There is a moral obligation *not* to act loyally in situations where violations of important moral duties could occur (Baron, 1984).

Hence loyalty to one's employer in this second sense can be misguided in two ways: (1) by being based on a mistake about what is good for one's company (as in the Ford Motor case opening this chapter), and (2) by failing to be in accord with duties owed to other people.

When codes of ethics state that engineers ought to be loyal to employers, or that they should act as their employer's or client's "faithful agents or trustees," the word is generally meant in its moral sense, as is suggested by the subheadings under the injunction to be a faithful agent or trustee. Typically those subheadings list specific moral duties: to avoid conflicts of interest, to inform employers of any possible conflicts of interest, to protect confidential information, to be honest in making estimates, to admit one's errors, and so on.

Yet it is important to bear in mind the possible ambiguity in speaking of loyalty. A call for loyalty to a company may be intended as more than a call for meeting one's moral obligations, and may involve the second concept of loyalty. It can be a tacit urging of close emotional identification with, and personal commitment to, the company's good. Urging loyalty to an employer can even mean recommending unquestioning obedience and devotion to the employer.

Loyalty and faithfulness in this second sense can be very valuable in creating a climate of mutual concern and commitment to shared goals among members of an organization. Such loyalty can add a human and personal dimension to the workplace, as well as aid in meeting the organization's goals. Yet it also has the potential in some situations of leading employees to disregard their wider moral obligations. For it can encourage the uncritical attitude that whatever is good for one's company is automatically good for the public.

Study Questions

1 Consider the following series of events:

> An applicant for employment in a number of companies accepted employment with Company X, knowing that he preferred employment in Company Y. He did not get an offer from Company Y until after he had worked for Company X for three months. He then changed to Company Y, and after several months there he discovered that employment conditions were not as good as they were in Company X. He then applied at Company X for re-employment (Alger, Christensen, and Olmsted, 1965, 219).

Did the person in the case fail to act loyally to Company Y? In answering this question, distinguish and discuss both concepts of loyalty mentioned in this section. Touch also upon the element of duration of service as it may relate to loyalty.

2 During a 1973 CBS interview, a chief executive of Phillips Petroleum was asked what Phillips sought in prospective employees (Baron, 1984, 1). The executive stated that loyalty was by far the most important feature sought. He went on to explain that in his view loyalty meant buying Phillips's products rather than those of competitors, voting in local, state, and national elections in favor of policies that would benefit Phillips, and staying to work for Phillips unless moving became unavoidable. (The wives of prospective male employees were screened to see if they had careers which might interfere with their husbands staying at Phillips.) Did the authority of the executives at Phillips morally justify the call for loyalty of this sort? Which of the two senses of "loyalty" do you think the Phillips executive had in mind?

3 The moral complexities related to obeying authority arise in most contexts where authority is needed for meeting specific goals of a group. In this connection, discuss any analogies or dissimilarities you see between the obligations of employed professionals to obey employers and accept their authority and (*a*) professional baseball players obeying umpires, especially in cases where the umpire makes a bad call; (*b*) children respecting their parents' authority; (*c*) soldiers on a battlefield obeying their commanders; (*d*) college students recognizing their professor's authority to direct a class; (*e*) nurses obeying doctors' orders and the directives of hospital administrators; (*f*) musicians obeying a conductor. Which of these contexts has the closest analogies to why employees generally ought, and perhaps occasionally ought not, to obey their employers? In presenting your answer consider some examples where those in authority make an incorrect decision or issue a poor directive.

4 How can the concept of employees' loyalty to employers be upheld in the case of a company which falls into one or more of the following categories: (*a*) rapidly expands its work force—including engineers—when its business is good, but equally rapidly lays off employees when business begins to drop; (*b*) is bought out by a conglomerate with headquarters in a distant city and with more apparent interest in the acquired company's profit-making potential than its products; (*c*) is owned by shareholders who buy or sell shares at a moment's notice, depending on the daily stock market report.

CONFLICTS OF INTEREST

Engineers are expected to avoid conflicts of interest and to protect confidential information. Traditionally these two obligations have been given prominence in engineering codes of ethics, in management policy statements, and in the law. Indeed, next to following legitimate directives, they are probably the most emphasized aspects of loyalty to employers and companies. This section will focus on conflicts of interests, and the next section on confidentiality.

Definition

In a wide sense, conflicts of interest arise whenever people or groups have interests which if pursued could keep them from meeting at least one of their obligations. We are concerned here with the obligations of employees to serve the interests of their employers or companies. Thus we will mean by "employee conflicts of interest" any situation where employees have an interest which if pursued might keep them from meeting their obligations to serve the interests of their employers or companies.

Sometimes such an interest involves serving in some other professional role—say, as a consultant for a competitor's company. Other times it is a more personal interest, such as making substantial private investments in a competitor's company.

These side interests are generally understood to threaten the fulfillment of employer-related obligations in one main way: They have the potential to deflect or distort the judgment of at least some people who find themselves in that type of situation. Thus an alternative definition of employee conflicts of interest is the following: Situations in which employees have side interests substantial enough potentially to affect their independent judgment, or the independent judgment of a typical person in their situation, in serving their company's interests. The qualification concerning "a typical person" is necessary. There might be conclusive evidence that the actual people involved would never allow a side interest to affect their judgment. But they could still be said to be in a conflict of interest situation.

Being in such a situation is not merely being confronted with conflicting interests (Margolis, 1979, 361). A student, for example, may have interests in excelling on four final exams. She believes, however, that there is time to study adequately for only three of them, and so she must choose which interest not to pursue. Or an investor may strongly desire to invest in two stocks but have sufficient funds for investment in only one. In these cases "conflicting interests" means a person has two or more desires which cannot all be satisfied given the circumstances. But there is no suggestion that it is *morally* wrong or problematic to try pursuing them all. By contrast, in conflicts of interest it is often physically or economically possible to pursue all of the conflicting interests, but it is morally problematic whether one should do so.

Conflicts of interest should also be distinguished from moral dilemmas, even though in some situations both are involved. Moral dilemmas occur when two or more moral obligations, rights, or ideals come into conflict and not all of them can be met. By contrast, it is often possible for an employee caught in a conflict of interest to pursue both the obligation to the employer and the side interest.

Examples

A wide variety of circumstances might arise which create conflicts of interest for employees. One type already mentioned is having an interest in a competitor's business. This might involve actually working for the competitor as an employee or consultant. Or it might involve partial ownership or substantial stockholdings in the competitor's business.

A variation on this type of situation is when engineers prepare to leave a corporation to form their own competing businesses. It would be a clear conflict of interest if they sought to lure customers away from their current employers while still working for them.

A second important category involves using "inside" information to gain an advantage or set up a business opportunity for oneself, one's family, or one's friends. Thus, for example, engineers might tell their friends about their corporation's plans for a merger which will greatly improve the worth of another company's stock. In doing so, they give those friends an edge on an investment promising high returns. In general, the use of any company secrets by employees to secure a personal gain is felt to threaten the interests of the company the employees are supposed to serve and thus to constitute a conflict of interest.

A third typical variety arises when employees benefit from personal involvements with suppliers, subcontractors, or customers. An obvious example is accepting bribes directly intended to influence judgment. Bribes may be in the form of cash, gifts, loans, services, trips, or entertainment. Another blatant example is working for a subcontractor or supplier who deals with one's corporation. Conflicts of interest also arise when one holds substantial stock or other investments in a firm with which one's company does business.

Sometimes it is difficult to determine just when conflicts of interest exist. Does holding a few shares of stock in a company one has occasional dealings with constitute a conflict of interest? How about occasional luncheons paid for by vendors giving sales presentations? Or those free pens and bottles of wine from salespersons? What about a gift one believes is based on friendship rather than intended to influence one's judgment?

The guidelines for use with the fundamental canons of ethics of the Accreditation Board for Engineering and Technology (ABET) seems to recommend a hard line on such gratuities: "Engineers shall not solicit nor accept gratuities, directly or indirectly, from contractors, their agents, or other parties dealing with their clients or employers in connection with work for which they are responsible" (Sec. 4-e).

Yet most employers would consider this position excessive. Company policies generally ban any gratuities which have more than "nominal" value, or which have any realistic potential for biasing judgment. In part the specific criteria for "nominal" value will be what is widely and openly accepted as normal business practice. In part it will be assessed by a person's own awareness of what might influence his or her judgment. And in part it must be weighed according to how others might perceive (or misperceive) the gratuities. Companies also typically formulate policies stating what is a nonnominal gift from a salesperson: for example, an item worth more than $10, or items totaling over $30 per year.

Moral Status of Conflicts of Interest

There are many other kinds of conflicts of interest we could mention. For example, there is taking additional outside employment—moonlighting—in situations where it harms on-job performance (Reed, 1970, 19–23). And there are cases involving confidential information, to be discussed in the next section. Conflicts of interest can arise in innumerable ways, and with many degrees of subtlety. But let us now ask the following: What is wrong with employees having conflicts of interest?

Most of the answer is obvious from our definition: Employee conflicts of interest occur when employees have interests which if pursued could keep them from meeting their obligations to serve the interests of the company for which they work. Such conflicts of interest should be avoided because they threaten to prevent one from fully meeting those obligations.

More than this, however, needs to be said. Why should mere threats of possible harm always be condemned? Suppose that substantial good might sometimes result from pursuing a conflict of interest?

In fact it is not always unethical to pursue conflicts of interest! In practice some conflicts are unavoidable, or even acceptable. One illustration of this is how the government allows employees of aircraft manufacturers, like Boeing or McDonnell Douglas, to serve as government inspectors for the Federal Aviation Agency (FAA). The FAA is charged with regulating airplane manufacturers and making objective safety and quality inspections of the airplanes they build. Naturally the two roles of government inspector and employee of the manufacturer being inspected could lead to a conflict of interest and biased judgments. Yet with careful screening of inspectors, the likelihood of such bias is said to be outweighed by the practical necessities of airplane inspection. The options would be to greatly increase the number of nonindustry government workers (at great expense to taxpayers) or to do without government inspection altogether (putting public safety at risk).

Where conflicts of interest are unavoidable or reasonable, employees are still obligated to inform their employers and obtain an approval. This suggests a fuller answer to why conflicts of interest are generally prohibited: The professional obligation to employers is (1) very important in that it overrides in the vast majority of cases any appeal to self-interest on the job and (2) easily threatened by self-interest (given human nature) in a way which warrants especially strong safeguards to ensure that it is fulfilled by employees.

As a final point, we should note that even the appearance of seeking a personal profit at the expense of one's employer is considered unethical since the appearance of wrongdoing can harm a corporation as much as any actual biasing that might result from such practices. For example, using inside information to gain a personal advantage for oneself or one's family may not directly hurt a company—indeed, it directly harms only those who are thereby denied a fair opportunity to compete for the advantage. But if such activities become generally known, the company's public image can be hurt—a state of affairs no employer wants to see.

Study Questions

1 "Facts: Engineer Doe is employed on a full-time basis by a radio broadcast equipment manufacturer as a sales representative. In addition, Doe performs consulting engineering services to organizations in the radio broadcast field, including analysis of their technical problems and, when required, recommendation of certain radio broadcast equipment as may be needed. Doe's engineering reports to his clients are prepared in form for filing with the appropriate governmental body having jurisdiction over radio broadcast facilities. In some cases Doe's engineering reports recommend the use of broadcast equipment manufactured by his employer.

"Question: May Doe ethically provide consulting services as described?" (*NSPE Opinions of the Board of Ethical Review,* Case No. 75.10)

2 "Henry is in a position to influence the selection of suppliers for the large volume of equipment that his firm purchases each year.

"At Christmas time, he usually receives small tokens from several salesmen, ranging from inexpensive ballpoint pens to a bottle of liquor. This year, however, one salesman sends an expensive briefcase stamped with Henry's initials" (Kohn and Hughson, 1980, 104).

Should Henry accept the gift? Should he take any further course of action?

3 "You were an engineer in partnership with Richard Jones. On May 10th, you sold your interest in the partnership to Jones and a day later accepted appointment as county director of public works. A few days later (and quite to your surprise) Jones sold your former firm to Octopus Enterprises, Inc., and became an officer of the corporation. It is now May 20th. You have tentatively decided to award an important engineering contract to Octopus. Would there be anything wrong if you did?" (Wells, Jones, Davis, 1986, 44; based on *NSPE Opinions of the Board of Ethical Review,* Case No. 77.9)

Compare and contrast this case with the case of Spiro Agnew described at the beginning of Chap. 2.

4 Read the case study at the beginning of Chap. 6. Was Brown and Root Corporation caught in a conflict of interest by being both the original designer of the road and the subsequent overseer of construction? If so, was the conflict permissible?

CONFIDENTIALITY

Many instances of failing to protect confidential information qualify as conflicts of interest. Nevertheless, the necessity of protecting such information is a distinct obligation of engineers and important in its own right. Indeed, keeping confidences is one of the most central and widely acknowledged duties of any professional. Defense attorneys must keep information clients tell them confidential, doctors and counselors must keep information on their patients confidential, and so too employed engineers must keep privileged information about their companies and their clients confidential.

Definition

What sort of information are engineers obligated to keep confidential? Actually, two questions are involved here. First, what is meant by the term "confidential information?" Second, exactly how can we identify what data should be kept confidential?

The first question is easier to answer. *Confidential information* is information which prima facie ought to be kept secret. "Kept secret" is a relational expression. It always makes sense to ask, "Secret with respect to whom?" In the case of some government organizations, such as the FBI and CIA, highly elaborate systems for classifying information have been developed that identify which individuals and groups may have access to what information. Within other governmental agencies and private companies, engineers and other employees are usually expected to withhold information labeled "confidential" from unauthorized people both inside and outside the organization.

The second question, which concerns the criteria for identifying what information should be treated as confidential, is somewhat more difficult to answer. One criterion is suggested in the code of ethics of the Accrediting Board for Engineering and Technology: "Engineers shall treat information coming to them in the course of their assignments as confidential" (Sec. 4-i). But this is too broad. Some of the information acquired on assignments is routine and widely known. For example, it may be knowledge about new company facilities or plans which is readily available to anyone. Or while working on a project an engineer may become familiar with technical processes known generally throughout the industry.

A different criterion would identify any information which if it became known would cause harm to the corporation or client. Yet there are always questions about just what information would produce that result. To give a precise answer one would need the talents of a fortune-teller.

Most businesses tacitly adopt yet another criterion: Confidential information is any information which the employer would like to have kept secret in order to compete effectively against business rivals. Often this is understood to be any data concerning the company's business or technical processes which are not already public knowledge. While this criterion is somewhat vague, it clearly points to the employer as the main source of the decision as to what information is to be treated as confidential. It is the criterion we will adopt.

Related Terms

Several related terms need to be distinguished. *Privileged information* is an expression often used as a synonym for "confidential information." Literally it means "available only on the basis of special privilege," such as the privilege accorded an employee working on a special assignment. It covers information which has not yet become public or widely known within an organization.

Proprietary information is information which a company owns or is the proprietor of. This term is used primarily in a legal sense, just as "property" and "ownership" are ideas carefully defined by law. Normally it refers to new knowledge generated within the organization which can be legally protected from use by others.

A rough synonym for "proprietary information" is "trade secrets." A *trade secret* can be virtually any type of information which has not become public and which an employer has taken steps to keep secret. It may be data about designs and technical processes, organization of plant facilities, quality control procedures, customer lists, business plans, and so on (Popper, 1980, 101). Trade secrets are given limited legal protection against employee abuse. They are protected by common law—law generated by previous court rulings—rather than by statutes passed legislatively. An employer can sue employees for divulging trade secrets, or even for planning to do so. To win such a case, the employer must be able to prove the information had been or is being actively protected (for example, by showing it was or is available only to special employees for specific purposes, that contracts require subcontractors to keep the data secret, and so forth).

Patents differ from trade secrets. Patents legally protect specific products from being manufactured and sold by competitors without the express permission of the patent holder. Trade secrets have no such protection. A corporation may learn about a competitor's trade secrets through legal means—for instance, "reverse engineering," in which an unknown design or process can be traced out by analyzing the final product. But patents do have the drawback of being public and thus allowing competitors an easy means of working around them by finding alternative designs. Also, patents can be held for only 17 years, whereas trade secrets, so long as they can be kept secret, are under no time restrictions.

Patents are protected by statute laws passed in order to provide incentives for creativity (Vaughn, 1977, 34). In effect they give the patent holder the reward of a legally protected monopoly. By contrast, the legal protection accorded trade secrets is limited to upholding relationships of confidentiality and trust.

Moral Basis of the Confidentiality Obligation

Upon what moral basis does the confidentiality obligation rest, with its wide scope and obvious importance? Specifically, why are employers allowed to determine what information is to be treated as confidential? And what are the moral limits or restrictions on the confidentiality obligations of employees?

The major ethical theories can be applied to answer these questions. Advocates of every theory would probably agree that employers have some moral and institutional rights to decide what information relating to their organizations can be released publicly. They acquire these rights as part of their charge to protect the interests of their organizations—whether those interests be a company's competitive edge (in the case of profit-making institutions) or the safety and well-being of clients or consumers. In addition, in the case of information like trade secrets developed by the company, there can be a right of ownership of the intellectual property (Schwarze). But different ethical theories will justify the rights differently and will also differ in the limits they place on them.

Briefly, *rights ethicists* will appeal to more basic considerations: for example, the right of stockholders to have management pursue a course consistent with their own best interests and the general rights of property ownership. This right, in turn, might be grounded in the fundamental moral right of the stockholders to pursue their legitimate interests within a socially accepted free enterprise system. However, the right of employers to establish what information should be treated as confidential will be limited by other legitimate moral rights: Minimally, no employer has a right to prevent engineers from blowing the whistle in cases where public knowledge of information would save human lives and thereby protect the rights of people to live.

Duty ethicists will emphasize the basic duties of both employers and employees to maintain the trust placed in them at the time they committed themselves to an employment agreement, a commitment that is understood to extend beyond the time of actual employment. They may also appeal to general duties not to abuse the property of others. Such duties, though, can be overridden by others, such as the duty to protect innocent lives, that might occasionally require whistle-blowing.

Utilitarians will view the authority of employers to determine the rules governing confidentiality as justified to the extent that it produces the most good for the greatest number of people. What this extent might be will depend on the particular theory of goodness subscribed to and the means required in any given situation to produce the most good. *Act-utilitarians* will focus on each instance where an employer decides on what is to count as confidential information. Is that act the most beneficial for everyone affected by it? *Rule-utilitarians,* by contrast, will emphasize the general benefits that result from having rules to protect confidential information. For example, investors' profits benefit from guarding trade secrets from competitors, and all society benefits from a system of limited protection of secrets to the extent that it stimulates creation of alternative products. The limits of the confidentiality obligation for utilitarians will depend on when acts of or rules for keeping information confidential do not produce the best consequences.

Confidentiality and Changing Jobs

The obligation to protect confidential information does not cease when employees change jobs. If it did, it would be impossible to protect such information. Former employees would quickly divulge it to their new employers, or perhaps for a price sell it to competitors of their former employers. Thus the relationship of trust between employer and employee in regard to confidentiality continues beyond the formal period of employment. Unless the employer gives consent, former employees are barred indefinitely from revealing trade secrets. This provides a clear illustration of the way in which the professional integrity of engineers involves much more than mere loyalty to one's present employer.

Yet thorny problems arise in this area. Many engineers value professional advancement more than long-term ties with any one company and so change jobs frequently. Engineers in research and development are especially likely to have high rates of job turnover. They are also the people most likely to be exposed to important new trade secrets. Moreover, when they transfer into new companies they frequently do the same kind of work as before—precisely the type of situation in which trade secrets of their old companies may have relevance.

Donald Wohlgemuth and B.F. Goodrich Consider, for example, the case of Donald Wohlgemuth, a chemical engineer who at one time was manager of B.F. Goodrich's space suit division (Baram, 1968, 208). Technology for space suits was undergoing rapid development, with several companies competing for government contracts. Dissatisfied with his salary and the research facilities at B.F. Goodrich, Wohlgemuth negotiated a new job with International Latex Corporation as manager of engineering for industrial products. International Latex had just received a large government subcontract for developing the Apollo astronauts' space suits, and that was one of the programs Wohlgemuth would manage.

The confidentiality obligation required that Wohlgemuth not reveal any trade secrets of Goodrich to his new employer. But this was easier said than done. Of course it is possible for employees in his situation to refrain from explicitly stating processes, formulas, and material specifications. Yet in exercising their general skills and knowledge, it is virtually inevitable that some unintended "leaks" will occur. An engineer's knowledge base generates an intuitive sense of what designs will or will not work, and trade secrets form part of this knowledge base. To fully protect the secrets of an old employer on a new job would thus virtually require that part of the engineer's brain be destroyed—a solution no one would recommend on moral grounds!

Is it perhaps unethical, then, for employees to make job changes in cases where unintentional revelations of confidential information are a possibility? Some companies have contended that it is. Goodrich, for example, charged Wohlgemuth with being unethical in taking the job with International Latex. Goodrich also went to court seeking a restraining order to prevent him from working for International Latex or any other company which developed space suits. The Ohio Court of Appeals refused to issue such an order, although it did issue an injunction prohibiting Wohlgemuth from revealing any Goodrich trade secrets. Their reasoning was that while Goodrich had a right to have trade secrets kept confidential, it had to be balanced against Wohlgemuth's personal right to seek career advancement. And this would seem to be the correct moral verdict as well.

Management Policies

What might be done to recognize the legitimate personal interests and rights of engineers and other employees while also recognizing the rights of employers in this area? And how can obligations to maintain confidences of former employers be properly balanced against obligations to faithfully serve the interests of new employers? There are no simple answers to these questions. Difficult dilemmas will always arise which call for sensitive and creative moral judgment. But while neither Congress nor the states have found it wise to pass strict legislation in this complicated area, some general management policies are being explored (Baram, 1968, 212–215).

One approach is to use employment contracts that place special restrictions on future employment. Traditionally those restrictions have centered on geographical location of future employers, length of time after leaving the present employer before one can engage in certain kinds of work, and the typeof work it is permissible to do for future employers. Thus Goodrich mighthave required as a condition of employment that Wohlgemuth sign an agreement that if he sought work elsewhere he would not work on space suit projects for a competitor in the United States for 5 years after leaving Goodrich.

Yet such contracts are hardly agreements between equals, and they threaten the right of individuals to pursue their careers freely. For this reason the courts have tended not to recognize such contracts as binding, although they do uphold contractual agreements forbidding disclosure of trade secrets.

A different type of employment contract is perhaps not so threatening to employee rights in that it offers positive benefits in exchange for the restrictions it places on future employment. Consider a company which normally does not have a portable pension plan. It might offer such a plan to an engineer in exchange for an agreement not to work for a competitor on certain kinds of projects for a certain number of years after leaving the company. Or another clause might offer an employee a special postemployment annual consulting fee for several years on the condition that he or she not work for a direct competitor during that period.

Other tactics aside from employment contract provisions have been attempted by various companies. One is to place tighter controls on the internal flow of information by restricting access to trade secrets except where absolutely essential. The drawback to this approach is that it may create an atmosphere of distrust in the workplace. It might also stifle creativity by lessening the knowledge base of engineers involved in research and development.

There have been unwritten agreements among competing corporations not to hire one another's more important employees. But the problem here is that when such practices become widespread in a given industry, some of the best engineers may be turned away to other fields offering more job options.

One potential solution is for employers to help generate a sense of professional responsibility among their staff that reaches beyond merely obeying the directives of current employers. Engineers can then develop a real sensitivity to the moral conflicts they may be exposed to by making certain job changes. They can arrive at a greater appreciation of why trade secrets are important in a competitive system and learn to take the steps necessary to protect them. In this way professional concerns and employee loyalty can become intertwined and reinforce each other.

Study Questions

1 Consider the following example:

> Who Owns Your Knowledge? Ken is a process engineer for Stardust Chemical Corp., and he has signed a secrecy agreement with the firm that prohibits his divulging information that the company considers proprietary.
>
> Stardust has developed an adaptation of a standard piece of equipment that makes it highly efficient for cooling a viscous plastics slurry. (Stardust decides not to patent the idea but to keep it as a trade secret.)
>
> Eventually, Ken leaves Stardust and goes to work for a candy-processing company that is not in any way in competition. He soon realizes that a modification similar to Stardust's trade secret could be applied to a different machine used for cooling fudge, and at once has the change made (Kohn and Hughson, 1980, 102).

Has Ken acted unethically? Defend your view.

2 Answer the following questions asked by Philip L. Alger, N. A. Christensen, and Sterling P. Olmsted in their book *Ethical Problems in Engineering:*

> If an engineer has been unjustly discharged, must he keep confidential in later employment the trade secrets of his original employers? In general, is it wise to follow the doctrine of an "eye for an eye and a tooth for a tooth"? (Alger, Christensen, and Olmsted, 1965, 111)

3 Alger, Christensen, and Olmsted also give the following example:

> Client A solicits competitive quotations of the design and construction of a chemical plant facility. All the bidders are required to furnish as a part of their proposals the processing scheme planned to produce the specified final products. The process generally is one which has been in common use for several years. All of the quotations are generally similar in most respects from the standpoint of technology.
>
> Contractor X submits the highest-price quotation. He includes in his proposals,

however, a unique approach to a portion of the processing scheme. Yields are indicated to be better than current practice, and quality improvement is apparent. A quick laboratory check indicates that the innovation is practicable.

Client A then calls on Contractor Z, the low bidder, and asks him to evaluate and bid on the alternate scheme conceived by Contractor X. Contractor Z is not told the source of alternate design. Client A makes no representation in his quotation request that replies will be held in confidence.

Is Client A justified in his procedure? (Alger, Christensen, and Olmsted, 1965, 177)

4 American Potash and Chemical Corporation advertised for a chemical engineer having industrial experience with titanium oxide. It succeeded in hiring an engineer who had formerly supervised E. I. Du Pont de Nemours and Company's production of titanium oxide. Du Pont went to court and succeeded in obtaining an injunction prohibiting the engineer from working on American Potash's titanium oxide projects. The reason given for the injunction was that it would be inevitable that the engineer would disclose some of du Pont's trade secrets (Carter, 1969, 54). Defend your view as to whether the court injunction was morally warranted or not.

UNIONISM

Is it possible for an engineer to be a professional, dedicated to the highest ethical standards of professional conduct, while simultaneously being a member and supporter of a union? The question, we feel, is too complex to warrant a simple answer. Before answering it we would need to know what kind of union and union activities are at issue. Lacking this information, the answer would seem to be: sometimes yes and sometimes no.

Yet many observers have argued that the ethical aspects of professionalism in engineering are inherently inconsistent with unionism—that is, with union ideology and practice. In *Engineers and their Professions,* for example, John Kemper writes:

> There is little doubt that unionism and professionalism are incompatible. Professionalism holds that the interests of society and of the client (or employer) are paramount. Unions are collective bargaining agents that sometimes place the economic interests of the members ahead of those of the client or employer (Kemper, 1982, 267).

A number of professional societies have also held that loyalty to employers and the public is incompatible with any form of collective bargaining. The National Society of Professional Engineers (NSPE) has fervently led the opposition to union organizing of engineers (Seidman, 1969, 224) and similar activities. Its position is reflected in the NSPE code of ethics: "Engineers shall not actively participate in strikes, picket lines, or other collective coercive action" (Sec. III, 1e). Before discussing two arguments for this view, let us take note of a few historical facts about unions and engineering.

Historical Note

The beginnings of engineering unionism in the United States occurred during World War I (Seidman, 1969, 229). Marine architects and drafters were dissatisfied with their salaries at a time of both rising living costs and rising wages of blue-collar workers. Various groups were organized which later unified to become the American Federation of Technical Engineers, an affiliate of the former American Federation of Labor (AFL).

Most contemporary engineering unions, however, had their origin during the 1940s. These groups usually remained independent of the large national unions like the AFL and the CIO (Congress of Industrial Organizations). World War II and its aftermath brought widespread job insecurity, unhappiness with salaries, and lessened professional recognition (Walton, 1961, 18–45). Yet engineering unions were never able to organize most engineers. In fact, at their peak during the late 1950s, engineering unions had only 10 percent of the total number of engineers as members.

Beginning around 1960, what unionism there was in engineering declined. Now about 25,000 engineers, scientists, and technicians still belong to unions (Asbrand). One major factor for the decline is that engineering salaries have risen favorably in comparison with salaries of graduates of other 4-year programs (although they have not necessarily risen in terms of real income). With many new technologies developing, moreover, engineers have been in great demand.

These observations do not necessarily apply to all industries. In the aerospace industry, for instance, a history of high job turnover has created a highly mobile group of engineers with lessened job security. When engineers at two major aerospace firms were polled in a study by Archie Kleingartner, 30 percent of those eligible to join were found to be members of unions. While this is by no means a majority, a surprising number of engineers working for the two firms disagreed with the statement that "it is impossible for an engineer to belong to a union and at the same time to maintain the standards of his profession." The percentages disagreeing ranged from 68 percent among low-level professionals to 91 percent among high-level professionals (Kleingartner, 1969, 230). As a result of his study, Kleingartner concluded that

> ...the majority of engineers interviewed...do not view unionism as threatening their professionalism, and very likely also they do not see it coming between them and management in any fundamental way. They attribute substantially less importance to the potentially disrupting effects of unions than does management. The engineers view unions as limited institutions performing certain limited functions (Kleingartner, 1969, 235).

When a union is viewed as an external service organization and not as an embodiment of collective will, its size will depend greatly on how well it fulfills its functions (Latta, 1981). Lacking real bargaining power, engineers' unions find it hard to overcome opposition from management and professional societies, even when the quality of worklife is low and attitudes toward management are negative (Manley et al, 1979).

Engineers also show an increasing interest in becoming managers themselves. An engineering degree and several years of experience can open doors in this direction. Employers encourage the trend by making engineers identify with management early on.

Professional societies oppose unionization because of the issue of conflicting loyalties and on the grounds that it is unprofessional. Let us now turn to two arguments in support of this stand as advanced by the NSPE: The first we will call the "faithful agent argument" and the second the "public service argument."

The Faithful Agent Argument

In the current NSPE Code the ban on the use of "collective coercive action" appears as one of the principles of obligation concerning professional integrity (Sec. III 1-e). Yet in earlier versions it was placed prominently in the first section, which dealt with loyalty to employers:

> Section 1—The Engineer will be guided in all his professional relations by the highest standards of integrity, and will act in professional matters for each client or employer as a faithful agent or trustee....
>
> f. He will not actively participate in strikes, picket lines, or other collective coercive action (1979 NSPE Code).

The implication is that being the faithful trustee of one's employer is incompatible with actively supporting collective action aimed against that employer.

In a number of NSPE publications this position has been explicitly endorsed. In 1976, for example, NSPE's Board of Ethical Review reiterated it in discussing a hypothetical example (Case No. 74-3). The case concerned the unionized employees in a state highway department. The employees, most of whom were not engineers, voted to strike when their demands for a pay increase of 60 percent and other benefits were denied. The Board of Ethical Review insisted that it was unethical for the engineers to participate actively, even though not to do so might mean facing union penalties. Passive participation, such as not crossing picket lines, was ruled permissible if it was necessary to avoid physical danger or abuse. The argument given was concise: "the engineers have a higher standard than self-interest; they have the necessary ethical duty to act for their employer as a faithful agent or trustee."

Obviously the Board saw active support of a strike or other collective action used against an employer as a violation of professional *ethics,* which it identified with the duty engineers have to serve as their employer's "faithful agents or trustees." Many people involved in engineering would agree with such a view, and certainly a case can be made for it. The conduct under discussion involves several features, any one of which might seem inconsistent with loyalty to employers: (1) It goes against the desires or interests of theemployer, (2) it uses coercion or force against the employer, and (3) it involves *collective* and organized opposition. Certainly we can all think of behavior along these lines which is unprofessional and disloyal. The difficulty, however, is that not every instance of such conduct is unethical, as the following two examples show.

Consider three supervisory engineers who have good reason to believe they are being underpaid. After individually reasoning with their bosses to no avail they threaten—in a polite way—to seek employment elsewhere. In doing so, they act against the desires and interests of their employer, and they use a type of collective coercion. But they have not acted unethically or violated their duty to their employers. The point, which should by now be familiar, is that the duty to an employer has limits. Loyalty and faithfulness do not always require sacrificing one's own self-interest to an employer's business interests.

Or consider this second case: Management at a mining and refinery operation have consistently kept wages below industry-wide levels. They have also sacrificed worker safety in order to save costs by not installing special structural reinforcements in the mines, and they have made no effort to control excessive pollution of the work environment. As a result the operation has reaped larger than average profits. Management has been approached both by individuals and by representatives of employee groups about raising wages and taking the steps necessary to ensure worker safety, but to no avail. A nonviolent strike is called and the metallurgical engineers support it for reasons of worker safety and public health. Here collective action aimed at coercing an employer is being used—and specifically a strike. But is it unethical or unprofessional?

It will be objected that these are special cases, and of course they are. They were designed expressly to show that it is not always obvious that a strike or other collective, forceful, action on the part of employees is unprofessional, excessively self-interested, or disloyal to employers. One must look at specific unions, specific strikes, specific situations. Even in the case discussed by the Board of Ethical Review we would want to know whether the union demands were reasonable. Were the workers so seriously underpaid that a 60 percent raise was not as fantastic as it sounds? And did the other benefits demanded relate to worker safety, compensation for injury, or possibly even public highway safety?

The examples suggest two generalizations. First, employee duty to employers does not entail unlimited sacrifice of economic self-interest. "Faithful agency" primarily concerns carrying out one's assigned tasks; it does not mean that one should never negotiate salary and other economic benefits from a position of strength.

Second, as the NSPE code itself states, the duty to employers is limited by the more paramount duty to protect public health, safety, and welfare. Moreover, duty to employers is also limited by considerations such as worker safety and the right to refuse to obey illegal or unethical directives. Collective action of a coercive nature might sometimes be the only effective way to pursue these concerns of overriding importance. Professional societies have themselves engaged in a type of collective, coercive, action when they print editorials in official journals exposing companies for abuses they have committed against engineers.

NSPE recommends the use of a sounding board, composed of a mix of employees and managerial engineers, to settle disputes with employees through reasonable dialogue. Certainly where feasible this is preferable to the use of collective force. Yet only a confirmed optimist could think that this procedure will always provide adequate support for salaried engineers.

The Public Service Argument

A second general argument against unions begins by emphasizing that the paramount duty of engineers is to serve the public. It then notes that by definition unions seek to promote the special interests of their members, not the interests of the general public. It is inevitable, so the argument continues, that clashes will occur, posing a threat to the meeting of professional commitments to the public. Strikes, which are the ultimate source of power for unions, may wreak havoc with the public good. Witness what has happened in recent strikes by police officers, firefighters, teachers, and nurses. Then imagine what would happen to the economy if all computer engineers and technicians were to go on strike!

There is force in this argument. Yet once again it points out only the dangers of unions, even using the worst possible scenario of what might happen, and assumes that engineering unions must act irresponsibly. Of course many unions have acted in that way, but not all.

It is at least possible that a collective bargaining group for engineers, whether called a union, a guild, or an association, led by professional engineers, could devote itself to promoting the interests of engineers only within the limits set by professional concern for the public good. It could also devote itself to giving positive support to ethical conduct by engineers—which, after all, is part of the self-interest of morally concerned engineers. As we shall see more fully in Chap. 6, engineers who have sought to protect the public have not always fared well at the hands of management. The collective power of a guild or union might prevent the vindictive firing of responsible whistle-blowers (Shapley, 1972, 620). It might also secure certain economic benefits, such as portable pensions, which would allow engineers a greater measure of freedom to act in the face of possible dismissal for whistle-blowing or for refusing to act unethically.

Conclusion

What we have said is neither a general endorsement of unionism nor a blanket condemnation of it. Our intention was a limited one: to question two of the main arguments used to show that there is an inherent inconsistency between professionalism and unionism. Whether collective bargaining and its tactics are unethical or not depends on the details of any given situation.

We would agree that unions often enough have abused their power and irresponsibly disregarded the public good, so that the formation of any new union carries with it new risks to professionalism. But to conclude, therefore, that the formation of engineering unions is *always* unprofessional is like arguing that because a new technology involves risks it should never be developed.

The moral assessment of unions is complex, and a considerable number of morally relevant facts must be considered before a judgment can be made about any specific case or before a generalization can be formed. Disputes over unionism itself, however, typically involve disagreements over claims like those given in the following two lists (Burton, 1978, 129; Kemper, 1982, 263–270).

Union Critics

1 Unions are a main source of inflation, which can devastate the economy of a country. Unions harm the economy by placing distorting influences on efficient uses of labor.

2 Unions encourage adversary, rather than cooperative, decision making. They also remove person to person negotiations between employers and employees and make the individual worker a pawn of the collective bargaining group.

3 Unions promote mediocrity and discourage initiative by emphasizing job security and by making job promotion and retention rest on seniority. Management is prevented from rewarding individuals by having to negotiate salaries according to job description and length of company service rather than according to personal achievement. A further side effect is the pigeonholing of employees in narrow job classifications to which the salary scales are attached.

4 Unions encourage unrest and strained relations between workers and management.

Union Supporters

1 Unions have been the primary factor in creating healthy salaries and the high standard of living enjoyed by today's workers. Even nonunionized workers have benefited since their employers must pay salaries comparable to those unions win for their workers.

2 Unions give employees a greater sense of participation in company decision making. For example, the European practice of codetermination, in which union representatives serve on boards of directors, has contributed to labor peace.

3 Unions are a healthy balance to the power of employers to fire at will. They give workers greater job security and protection against arbitrary treatment. Employees with union backing are more able to resist orders to perform unethical acts.

4 Unions yield stability by providing an effective grievance procedure for employee complaints. They are also a counterforce to radical political movements which exploit worker dissatisfaction and alienation.

Study Questions

1. Present and defend your view as to when collective action aimed at employers does or does not involve unfaithfulness and disloyalty on the part of the employees. In doing so distinguish between the two senses of "loyalty" given earlier in this chapter. Consider issues like salary, harmful labor practices, and the public good. Also consider the use of collective action by different groups, such as (*a*) unions, (*b*) professional societies, (*c*) nonunion employee groups, and (*d*) manufacturer's associations and trade organizations.
2. Answer the questions asked by Philip M. Kohn and Roy V. Hughson in regard to the following case. Give reasons.

> Reginald's company pays its engineers overtime plus a bonus to work during a strike. The plant is being struck over "unsafe" working conditions, a claim that the company disputes. Reginald, considered by the company to be "management," believes conditions *may* be unsafe, even though no government regulations apply. Should Reginald:
>
> **1** Refuse to work, because he thinks the union's allegations may have merit?
> **2** Refuse to work, because he believes that strike-breaking is unethical?
> **3** Work, because he feels this is an obligation of all members of management?
> **4** Work, because it is a great way to catch up on some of his bills, or earn the down payment on a car, etc.?
> **5** Work, because he believes he may be fired if he doesn't?
> **6** Other? (Please specify) (Kohn and Hughson, 1980, 102 and 105)

WHITE-COLLAR CRIME: CASE STUDIES

White-collar crime is the secretive violation of laws regulating work activities, usually, but not always, committed by white-collar workers. It ranges in severity from pilfering cash registers to bribing public officials. This section presents examples of three types of cases: stealing trade secrets, conspiring to fix prices, and endangering lives. The cases are offered as further contexts for discussion of the central themes in this chapter: professionalism, loyalty, conflicts of interest, and confidentiality.

Espionage in Silicon Valley

Santa Clara Valley in Northern California is a marvel of the high-tech and computer industries. For two decades it has been a major center for development and manufacture of integrated-circuit microprocessors, or "computer chips." The Valley has attracted vast numbers of creative engineers and entrepreneurs. It has also attracted industrial espionage on an unprecedented scale.

Several factors contributed to make the Valley an ideal environment for industrial espionage, that is, for stealing and illegal spying in industry. First, the development of computer chips is intensely competitive and fast-paced. Innovation is so rapid that products are often outdated within 2 years. Fortunes can be made or lost in months, depending on how quickly new products are developed and marketed.

Second, computer chips can be extremely expensive to develop; it may cost hundreds of thousands or millions of dollars to get a chip into production. Enormous savings are possible through legal reverse engineering. This involves literally "dismantling" a competitor's device—either mentally, physically, or by tests. The device is then "reconstructed" to produce an identical or better device which can be offered at a lower price because development costs were less or nonexistent. Even greater savings are possible by illegally acquiring design information from competitors.

Third, computer chips and the tools used to produce them are so small that it is easy to smuggle them out of offices and buildings. Stopping the smuggling would require body searches of the sort used in prisons. As it is, the chances of being caught are low.

Fourth, law enforcement has been ineffective, weakening the role of punishment in deterring crime. Most crimes go unreported to police. Managers often prefer to avoid bad publicity and embarrassment before stockholders. Until recently police lacked the sophistication even to understand the complicated nature of the materials being stolen. And even when tried and convicted, white-collar criminals suffer relatively modest penalties.

Fifth, employees who betray company secrets need not be artful criminals. Criminal "expertise" is provided by go-between criminals who buy trade secrets from one company and sell them to others.

Consider the case of Peter Gopal, who for a decade ran a lucrative trade as a go-between until he was caught in 1978 (Halamka, 1984; Hiltzig, 1982; Samuelson, 1982). Gopal was a semiconductor expert who worked for a number of high-tech companies before establishing his own consulting firm in 1973. He became a familiar figure in the Valley, and he developed numerous contacts which enabled him to buy and sell competitors' secrets.

One contact was James Catanich, a skilled electronics draftsperson who worked for Gopal on a moonlighting basis in addition to his regular job at National Semiconductor Corporation. Gopal loaned Catanich $10,000 for a home loan. Later he urged Catanich to pay off the debt with documents stolen from National Semiconductor. Catanich found this an easy way out of his financial difficulties, especially since his desk was located next to his supervisor's desk, which contained key circuitry documents.

Gopal sold National Semiconductor's secrets to Intel Corporation. He also stole from Intel to sell to National Semiconductor. Intel has one of the tightest security systems in Silicon Valley. Its security includes magnetic switches and alarms over all doors, closed-circuit cameras in offices, passes worn by employees, strict control of access to documents, and armed guards. But Gopal learned that many Intel manufacturing materials were stored at NBK, an Intel subcontractor which lacked comparable security. NBK kept chip "reticles," the palm-sized glass plates which display magnified chip circuitry. It also stored "masks"—prints of a reduced image of the reticle—and data tapes giving design information. Gopal purchased copies of reticles and masks from Lee Yamada, the supervisor at NBK, who had easy access to everything Gopal needed.

Finally, Silicon Valley corporations have high employee turnover rates because of opportunities for advancement with competitors. Gopal found it easy to buy dozens of major trade secrets from former employees.

It required a complicated undercover operation conducted jointly by National Semiconductor, Intel, and the police to capture Gopal. After arresting him, police searched his apartment to find 27 reticles for a recent Intel chip and assorted loot from other companies. Gopal was convicted of domestic crimes involving American corporations, but there was strong evidence that he had also sold to European companies that deal with eastern bloc countries. His tax reports, it might be added, listed his annual income as $30,000 despite the fact that he probably made millions of dollars.

Price Fixing in the Electrical Equipment Industry

In 1890 Congress passed the Sherman Antitrust Act. It forbids companies from jointly setting prices in ways that restrain free competition and trade. The Act has frequently been violated in the electrical equipment industry, where large contracts and few competitors are the norm.

For example, in 1983 six large electrical contractors, together with eight company presidents and vice presidents, were indicted on charges of conspiring to fix bids on four or five public power plants to be built in the state of Washington. The plants were valued at more than $250 million. Company officers were charged with discussing the bids each would submit, sharing pricing information, and agreeing on the low bidder for each project. This ensured lucrative business for each company without having to beat the competition with low bids.

The most famous violation of the Sherman Act in the electric power industry was prosecuted in 1961 (Fuller, 1962; Herling, 1962; Geis, 1977; Bane, 1973). Forty-five individuals from twenty-nine corporations pled guilty or entered pleas of nolo contendere (i.e., "no contest," a plea that allows for some face saving).

Top officials of Westinghouse and General Electric were indicted, although their presidents were evidently kept ignorant of the conspiracy (analogously to how President Reagan is said to have been kept ignorant of the diversion of funds in the Iran-Contra scandal of 1987). Westinghouse and General Electric received fines of several thousand dollars, insignificant sums for companies of their size. But subsequent civil suits by clients for triple damages ran in the hundreds of millions of dollars. Jail sentences of 30 days were imposed on seven defendants: four vice presidents, two division managers, and one sales manager.

The conspirators would allocate bids based on their companies' previous market shares. A company with 20 percent of the market, for example, would be allowed to submit the lowest bid for 20 percent of the new contracts. Occasionally the low bid was not accepted because of another company's better reputation, and then special adjustments would be made, sometimes involving heated negotiations. A few contracts were allocated on a rotating plan code-named "phase of the moon."

The participants were highly respected officials of their companies and members of their communities. Several were deacons in their churches. One was president of the local chamber of commerce. What could motivate such otherwise decent citizens to break the law?

Surprisingly, most of them did not view their activities as criminal or harmful, even though they knew they were "technically" illegal. In fact, many of them defended their conduct as beneficial. A Westinghouse executive offered the following testimony before a Senate subcommittee on antitrust and monopoly.

> *Committee attorney:* Did you know that these meetings with competitors were illegal?
>
> *Witness:* Illegal? Yes, but not criminal. I didn't find that out until I read the indictment. . . . I assumed that criminal action meant damaging someone, and we did not do that. . . . I thought that we were more or less working on a survival basis in order to try to make enough to keep our plant and our employees. (Geis, 1977, 122)

Several conspirators also argued that the price fixing benefited the public by stabilizing prices.

The practice of price-fixing had been so widespread in the industry for so long that it became accepted as proper. A General Electric vice president testified that in 1946 his superior casually introduced him to the practice and presupposed that he would cooperate. At the time, he was a recent graduate in electrical engineering and was rapidly moving up the ranks of management.

This same man, incidentally, expressed indignation at his company for refusing to pay him his regular salary during the month he served in jail. "When I got out of being a guest of the government for thirty days, I had found out that we were not to be paid while we were there [a matter of some $11,000 for the jail term], and I got, frankly, madder than hell" (Geis, 1977, 127).

Killing in Manufacturing

Employers who expose their employees to safety hazards usually escape criminal penalties. Victims will often sue companies for damages under tort (i.e., civil) law, which allows them to gain compensation without having to prove a crime has been committed. This is true even when people die as a result of horrendous corporate negligence.

No example is more shocking than that of the companies in the asbestos industry, especially Manville Corporation (formerly Johns-Manville Corporation), which is the largest producer of asbestos. Manville knew from the 1930s and 1940s onward that asbestos fibers in the lungs cause asbestosis, an incurable form of cancer. For three decades it concealed this information from workers and the public who had a right to give informed consent to the dangers confronting them. In 1949 Manville's company physician defended a policy of not informing employees diagnosed with asbestosis: "As long as the man feels well, is happy at home and at work and his physical condition remains good, nothing should be said" (Brodeur, 1985, 174–175). When Manville was finally brought to trial, company officials claimed that some 1300 of the company's own studies of asbestos had mysteriously disappeared from its files.

One recent study showed that 38 percent of asbestos insulation workers die of cancer, 11 percent from asbestosis. It is predicted that "among the twenty-one million living American men and women who had been occupationally exposed to asbestos between 1940 and 1980 there would be between eight and ten thousand deaths from asbestos-related cancer each year for the next twenty years" (Brodeur, 1985, 6). The actor Steve McQueen is just one individual included among these grim statistics. In his youth he held a summer job handling asbestos insulation and two decades later died of asbestosis.

It seems doubtful that many, if any, of Manville's employees will be prosecuted. Tens of thousands of victims and their families have filed civil suits for damages, seeking monetary compensation rather than criminal justice. In order to postpone settling the flood of lawsuits, Manville filed for bankruptcy in 1982. (Its assets of $2 billion made it the largest American corporation ever to do so.) A court agreement reached in 1985 allows it to continue operating while paying some $2.5 billion in lawsuits over the next 25 years.

The year 1985 also saw a highly unusual court verdict in a different case. For the first time in history, a judge convicted three officials of a company for industrial murder (Frank, 1987). Film Recovery Systems was a small corporation which recycled silver from used photographic and x-ray plates. Used plates were soaked in a cyanide solution to leach out their silver content. Other companies use this process safely by protecting workers against inhaling cyanide gas and making skin contact with the liquid. Standard safety equipment includes rubber gloves, boots, and aprons, as well as respirators and proper ventilation.

None of these precautions were used by Film Recovery Systems. Workers were given useless paper face masks and cloth gloves. Ventilation was terrible, and respirators were not provided. Workers frequently became nauseated and had to go outside to vomit before returning to work at the cyanide vats. This continued until an autopsy on one employee, a Polish immigrant, revealed lethal cyanide poisoning.

Charges were brought against the executives of Film Recovery Systems under an Illinois statute which states that "a person who kills an individual without lawful justification commits murder if, in performing the acts which cause the death...he knows that such acts create a strong probability of death or great bodily harm to that individual or another" (Frank, 1987, 104). During the trial it was proven that the company president, the plant manager, and the plant foreperson all knew of the dangers of cyanide. They also knew about the hazardous conditions at their plant. Each was sentenced to 25 years in jail and fined $10,000.

Study Questions

1 Discuss the cases in this section in light of the concepts of loyalty presented earlier in this chapter. In doing so, evaluate the following claim (taking account of different senses of "loyalty"): The Silicon Valley espionage involved disloyalty by employees and former employees; the cases of Manville and Film Recovery Systems involved lack of loyalty by employers to their employees; and the electrical equipment case involved misguided loyalty to the company.

2 Employers have often been reluctant to prosecute employees who commit crimes against them. It is easier just to fire them, thereby avoiding court hassles and bad publicity. Given that companies need to make profits, is this reluctance to bring criminal charges against employees morally permissible and responsible?

3 Criminal penalties for white-collar crimes have been relatively light, at least until recently. This is due, in part, to the belief that white-collar crimes are usually "victimless crimes," since corporations rather than individuals are harmed. Discuss this belief with respect to the cases of Silicon Valley industrial espionage and the electrical equipment price fixing. Are any individuals hurt in those cases, and how badly? Should those crimes be treated more lightly than crimes involving burglary, violence, or threatened violence? Would your answer be the same with respect to the cases of Manville and Film Recovery Systems?

4 In the Silicon Valley case, was Catanich in an immoral conflict of interest simply by moonlighting for Gopal?

5 The executives of Film Recovery Systems were convicted of murder. Critics have disagreed with this conviction on the grounds that murder involves intentional and purposeful killing. At most, say the critics, the executives committed manslaughter, which is killing due to negligence or indifference (such as when drunk drivers kill). Do you think the executives of Manville should be charged with manslaughter, murder, or no crime at all?

6 *Self-deception* is the intentional avoiding of truths which are painful to recognize (Martin, 1986). One might suspect or have general knowledge about an unpleasant truth and then turn away before learning more about it. Or one might engage in *rationalization:* giving biased explanations of one's motives and actions in order to maintain a flattering view of oneself. Discuss the possible role of self-deception in the electrical equipment case. Consider, for example, the distinction the conspirators drew between "illegal" and "criminal" conduct, and their belief that their actions were beneficial to the public. What personal benefits might have led them to

believe that no one was hurt by the price fixing? How did this belief benefit their self-esteem?

7 Find the names of the main conspirators who were found guilty in the electric power equipment prosecutions of 1961 by consulting one or more of the books cited (Fuller, 1962; Herling, 1962; Geis, 1977; Bane, 1973). Then trace these persons' careers before and after 1961 by referring to *Who's Who in America.* How did the companies treat them? How would you have treated them? Do their civic involvements constitute mitigating circumstances?

8 One way to control white-collar crime is to use polygraph (lie detector) tests. Are companies justified in giving their employees an annual polygraph test in order to ferret out employees who are stealing from them? (Consider this question again after reading Chap. 6.)

9 Plan a role-playing session in which some participants defend and others attack various kinds of white-collar crime. Include typical occurrences not mentioned expressly in this chapter, such as padding pay rolls or falsifying test results. (Further examples appear in Chapter 7, sections 3 and 4.)

SUMMARY

Professions are occupations requiring sophisticated skills, extensive formal education, group commitment to some public good, and a significant degree of self-regulation. *Persuasive definitions* are frequently given for terms like "professionalism," "professional," and "profession." That is, special criteria (cognitive meaning) are applied to the terms with their generally positive connotations (positive emotive meaning). For example, some people think of the "professionalism of salaried engineers" as being centered in loyal service to employers, while others have seen it as freedom from control by employers; both, however, are persuasive definitions in that they link the emotional connotations of a term to special (and in this case controversial) cognitive criteria for applying the term.

In our view, the duty of engineers to the public is paramount in the sense that it deserves special emphasis given the contemporary obstacles to meeting that duty. Yet it is too much to say that obligations to the public always and everywhere should override obligations to employers. Both obligations are important. When they come into conflict it is necessary to examine the specific situation before deciding which ought to take precedence.

The relationship between loyalty to employers and other professional obligations is complex. Loyalty to employers can mean (1) meeting one's moral obligations to employers—in which case loyalty is automatically good; (2) being zealously supportive of the employers' interests—in which case there are limits to how far loyalty is good.

Institutional authority involves the right of employers and managers to exercise power so employees will meet their institutional duties, and the prerogatives it entails are specified by rules designed to further the institution's good; it is not the same as *expert authority* (special knowledge or expertise). Institutional authority is morally justified only where the goals of the institution are morally permissible or desirable and when the way in which it is exercised does not violate other moral duties.

Authority relationships between employers and employees are normally necessary for avoiding the negative effects of unlimited individual discretion. And the employment contract constitutes a promise on the part of employees to recognize legitimate institutional authority. The obligation to obey authoritative directives, however, should not be construed as an obligation to suspend one's critical faculties and blindly follow those directives regardless of their moral content.

Employee conflicts of interest occur when employees have side interests which if pursued could prevent them from meeting their obligation to serve the interests of their employers. Such side interests are generally understood to threaten employer interests in one main way: They have the potential to bias the employee's independent judgment. Examples of conflicts of interest include moonlighting for a competitor, misusing inside confidential information for personal gain, and accepting substantial gifts from clients or suppliers. Some conflicts of interest are permissible, however, subject to the employer's approval.

Confidential information is information which an employer or client judges should be kept secret to serve the company's or client's interests. *Proprietary information* and *trade secrets* are information which is protected by the courts. The confidentiality obligation can be justified in rights-based theories (for example, by reference to the rights of stockholders or the rights to intellectual property of corporations), in duty-based theories (by reference to the mutual promises of the employment contract), and in utilitarian theories (by reference to the benefits derived by companies and the public). Moral dilemmas can arise for engineers when they move to new jobs since they may possess privileged information from their old jobs which they carry with them. The confidentiality obligation extends beyond the old job, however, and places reasonable restraints on engineers in regard to how and when they may work for new employers. The confidentiality obligation is limited by the public's right to be warned of potential hazards.

Unionism and professionalism seem inherently incompatible when the duty of employees to employers is seen as paramount and unlimited. But when that duty is viewed as limited by both a legitimate degree of self-interest on the part of employees and the wider good of the public, the incompatibility becomes less clear. Rather, individual unions and union tactics must be assessed in terms of their positive and negative effects in specific situations.

White-collar crime is the secretive violation of laws regulating work activities, whether or not by white-collar workers. It is motivated by personal greed, corporate ambition, misguided company loyalty, and many other motives. Only recently have penalties begun to toughen sufficiently to provide deterrence for individuals for whom ethical motivation does not suffice.

CHAPTER **6**

RIGHTS OF ENGINEERS

Several years ago Charles Pettis was sent by Brown and Root Overseas, Inc., to serve as resident engineer in Peru. Brown and Root had been hired by Peru's government to protect its interest on a project being undertaken by another firm, Morrison-Knudsen. The project was breathtaking: construction of a 146-mile highway across the Andes Mountains. The highway would open major new trade routes between Peru's coast and its isolated inner cities on the other side of the Andes. At age forty-four, with years of experience as a geological engineer behind him, Pettis was given the key assignment of ensuring that contract agreements between the Peruvian government and Morrison-Knudsen were met. His signature on the payroll certified that the interests of the Peruvian government were being served.

Almost immediately Pettis experienced doubts about the project. The design for the highway, which had originally been done by Brown and Root and was therefore a source of potential conflict of interest, called for cutting deep channels—some of them 300 feet deep—through the mountains with cliffs rising sharply on both sides of the road. Unfortunately, the Andes Mountains are known for their instability, and not enough geological borings had been taken to identify potential slide areas. Pettis's worries about this problem were confirmed when several slides and other construction incidents killed thirty-one workers.

Morrison-Knudsen instructed Pettis to add to the payroll in order to cover the substantial costs of slide removals. Pettis viewed this as padding and as not justified by anything in the contract. At first Brown and Root supported him. But later Morrison-Knudsen had exerted sufficient pressure on Brown and Root management that they ordered Pettis to add the slide-removal costs to the payroll. He continued refusing to do so, insisting it would be a violation of the Peruvian government's interests, which he was charged with protecting. At that point Brown and Root relieved him of responsibility for payroll authorization.

Suspicions, however, had been aroused in the minds of Peru's transportation officials. They sought direct assurances from Pettis that the work was proceeding properly. Brown and Root placed enormous pressure on him to give those assurances, even promising him his pick of jobs if he cooperated. But Pettis refused to lie to his client. As a result Brown and Root fired him.

When Pettis later learned of Senator William Proxmire's investigations into the contract policies of multinational construction companies, he volunteered to testify before officials of the General Accounting Office, and while doing so he blew the whistle on Brown and Root and Morrison-Knudsen. The General Accounting Office was able to confirm Pettis's charges of corporate misconduct (Peters and Branch, 1972, 183–186; Nader, Petkas, and Blackwell, 1972, 135–139).

Issues

Do engineers have a moral right to refuse to carry out what they consider to be unethical activity? How far are employers obligated to respect this right and to forgo the use of coercion and retribution in dealing with those employees who exercise it? Should engineers be recognized as having rights to speak out to clients, government regulators, and others concerning their employers' misconduct?

It may seem that endorsing such rights is incompatible with allowing employers full charge to direct a company. Isn't the position Pettis took inconsistent with recognizing management's rights? And what if management honestly disagrees with an engineer's safety judgments or interpretations of a contract?

Issues concerning the rights of engineers and other professionals working within organizations were usually given little attention. Only recently has the topic of the rights of employees been as seriously discussed as their duties and responsibilities. Indeed, the 1980s promises to be the decade when discussion of the rights of employed professionals reaches full maturity (Westin and Salisbury, 1980, xi).

PROFESSIONAL RIGHTS

Engineers have different types of moral rights, which fall into the sometimes overlapping categories of human, employee, contractual, and professional rights. As human beings, engineers have fundamental rights to live and freely pursue their legitimate interests. For example, a human right we will discuss later in this chapter, in the section titled "Discrimination," is the right not to be unfairly discriminated against in employment on the basis of sex, race, or age. Another example, mentioned in connection with confidentiality in Chap. 5, is the human right to pursue one's career.

As employees, engineers have special rights, some of which will be explored later in this chapter, in the section titled "Employee Rights." Some of those include institutional rights which arise from specific agreements in the employment contract. For example, there is the right to receive one's salary and other company benefits in return for performing one's duties. However, other employee rights are not reducible to purely institutional rights. For example, the right to engage in the nonwork political activities of one's choosing, without reprisal or coercion from employers. Employers ought to respect this right, whether or not it is explicitly recognized in a contract or employment agreement.

Finally, engineers as professionals have special rights which arise from their professional role and the obligations it involves. Those include the right to form and express one's professional judgment freely (without intimidation), the right to refuse to carry out illegal and unethical activity, the right to talk publicly about one's work within bounds set by the confidentiality obligation, the right to engage in the activities of professional societies, the right to protect clients and the public from the dangers or harm that might arise from one's work, and the right to professional recognition (including fair remuneration) for one's services. All these, as we shall see, can be viewed as aspects of one fundamental professional right.

The Basic Right of Professional Conscience

There is one basic or generic professional right of engineers: the moral right to exercise responsible professional judgment in pursuing professional responsibilities. Pursuing those responsibilities involves exercising both technical judgment and reasoned moral convictions. For brevity, this basic right can be referred to as *the right of professional conscience.*

If the duties of engineers were so clear-cut that in regard to every situation it was obvious to every sane person what it was morally acceptable to do, there would be little point in speaking of "conscience" in specifying this basic right. Instead, we could simply say it is the right to do what everyone agrees it is obligatory for the professional engineer to do. But as we have seen throughout this book, engineering calls for as morally complex decisions as any other major profession does. It requires autonomous moral judgment in attempting to uncover the most morally reasonable courses of action, and the correct courses of action are not always obvious.

As with most moral rights, the basic professional right is an entitlement giving one the moral authority to act without interference from others. It is what we earlier called a "liberty," since it places an obligation on others *not* to interfere with its proper exercise.

Yet occasionally special resources may be required by the engineer seeking to exercise it in the course of meeting his or her professional obligations. For example, conducting an adequate safety inspection may require that special equipment be made available by employers. Or, more generally, in order to feel comfortable about making certain kinds of decisions on a project, the engineers involved may need an environment conducive to trust and support which management may be obligated to help create and sustain. In this way the basic right is also in some respects a "positive right," placing on others an obligation to do more than merely not interfere.

Institutional Recognition of Moral Rights

Having a moral right is one thing. Having it respected by others and given recognition within a corporation is quite another. When engineers appeal to the basic right of professional conscience they may be arguing for its institutional recognition by employers.

Consider in this connection the following comments made by two engineers at the 1975 Conference on Engineering Ethics:

> H. B. Koning: I think that one item that should be in the code of ethics is that engineers have the right at all times to exercise the dictates of their own consciences. For example, they need not apply their knowledge, skill and energy to scientific or technical business actions or plans which they feel will violate or lead to the violation of their personal or professional ethical standards (*Conference,* 99).

> N. Balabanian: Few engineers are self-employed. The vast majority work for others. What is desperately needed for engineer employees is to have a right of conscience. It isn't so much a matter of forcing engineers to conduct themselves ethically but to give them room—room for action—to carry out their own personal ethical convictions without threats of retribution (*Conference,* 101).

The first speaker is appealing to the moral right of professional conscience which engineers *do* have, even though it is not formally recognized in codes. He is arguing that this right should be stated formally and given official recognition. The second speaker seems to mean by a "right to conscience" an institutionally recognized right, one which engineers will have only after employers acknowledge and respect it. Both speakers are arguing for similar points, but using different language.

Specific Professional Rights

The right of professional conscience is the most basic—but also the most abstract—generic, professional right. It encompasses many other more particular rights. As with professional duties, specific professional rights can be stated in different ways involving different levels of generality.

For example, engineers have a general *obligation* to protect the safety and well-being of the public. Correspondingly, they have a general *right* to protect the safety and well-being of the public. As we will discuss in the next section, that obligation to the public might in special situations require whistle-blowing. Thus engineers have a (limited) right to whistle-blow. In turn, the whistle-blowing right becomes more precisely specified by listing conditions under which whistle-blowing is permissible. In general, as a particular professional obligation is more narrowly delineated, the corresponding professional right is also more precisely specified.

Realizing that professional rights can be stated with different degrees of abstraction helps us avoid two mistakes. First, just because some talk about professional rights is couched in abstract and general terms should not lead us to dismiss its significance. The same potential difficulty surrounds other rights. Consider the right to live. In the abstract, it sounds like it entails a right never to be killed. But it does not: for example, in situations where the only way to prevent a murder from occurring is to kill the murderer first. Such tacit limits on the right to live do not lead us to reject that right as nonsense. Similarly, sensitivity to the necessary limits on the rights of professional conscience within organizations should not lead us to dismiss lightly the importance of those rights.

A related second danger is that talk about rights may be used too loosely and not made specific with respect to given contexts. It will not do, for example, to object to every negative action by an employer as violating the rights of engineers. Even such vitally important rights as protection of public safety may in some situations be limited by the legitimate rights of employers—at least this possibility has to be explored. Neither the rights of engineers nor those of employers are unrestricted moral "passes," and there will always be difficult moral dilemmas involving conflicts between them. Such dilemmas can be resolved by developing cogent arguments for why one right should be limited in a specific context by another right.

Both the importance and the difficulty of applying professional rights in specific circumstances can be illustrated by the examples of the right of conscientious refusal and the right to professional recognition.

Right of Conscientious Refusal The right of conscientious refusal is the right to refuse to engage in what one believes and has reason to believe is unethical behavior, and to refuse to do so solely because one views it as unethical. This is a kind of second-order right. It arises because other rights to pursue moral obligations within the authority-based relationships of employment sometimes come into conflict.

There are two situations to be considered: (1) where there is widely shared agreement in the profession as to whether or not an act is unethical and (2) where there is room for disagreement among reasonable people over whether an act is unethical.

It seems clear enough that engineers and other professionals have a moral right to refuse to participate in activities which are straightforwardly and uncontroversially unethical (e.g., forging documents, altering test results, lying, giving or taking bribes, or padding payrolls). And to coerce them into doing so by means of threats (e.g., to their jobs) plainly constitutes a violation of this right.

The troublesome cases concern situations where there is no shared agreement about whether or not a project or procedure is unethical. Possibly the Charles Pettis case involved different assessments of whether or not slide-removal charges could ethically be charged to Peru's government under the contract agreement. Do engineers have any rights to exercise their personal consciences in these more cloudy areas?

Let us approach this question with a rough analogy from medical ethics. There is no shared agreement over whether abortions are morally permissible or not. Yet, as is widely acknowledged, nurses who believe them to be immoral have a right to refuse to participate in abortion procedures. This is so even though nurses function under the institutional authority of doctors, clinics, and hospitals in ways analogous to how engineers work under the authority of management. Nevertheless, nurses' rights do not extend so far as to give them the right to work in an abortion clinic while refusing to play their assigned role in performing abortions.

Likewise, we believe engineers should be recognized as having a *limited* right to turn down assignments which violate their personal consciences in matters of great importance, such as threats to human life, even where there is room for moral disagreement among reasonable people about the situation in question. We emphasize the word "limited" because the right is contingent on the organization's ability to reassign them to alternative projects without serious economic hardship to itself.

For example, consider an engineer who requests not to work on a South African project because she views such work as supporting a racist regime. Her corporation should be willing to try to find an alternative assignment for her, without any implication that she is being disloyal to the company. Yet if the bulk of the work for which she is needed is on South African projects, she must be willing to seek employment elsewhere. The right of professional conscience does not extend to the right to be paid for not working.

Right to Recognition Engineers have a right to professional recognition for their work and accomplishments. Part of this has to do with fair monetary remuneration, and part has to do with nonmonetary forms of recognition.

The right to reasonable remuneration is sufficiently clear that it can serve as a moral basis for arguments against corporations which make excessive profits while engineers are paid below pay scales of blue-collar workers. It can also serve as the basis for criticizing the unfairness of patent arrangements which fail to give more than nominal rewards to the creative engineers who make the discoveries leading to the patents. If a patent leads to millions of dollars of revenue for a company, it is unfair to give the discoverer a nominal bonus and a thank you letter.

But the right to professional recognition is not sufficiently precise to pinpoint just what a reasonable salary is or what a fair remuneration for patent discoveries is. Such detailed matters must be worked out cooperatively between employers and employees, for they depend upon both the resources of a company and the bargaining position of engineers.

It may seem, incidentally, that the right to fair remuneration is related merely to the engineer's self-interest, and as such does not properly fall under the basic right of professional conscience. Of course it does centrally involve self-interest. But there are also reasons why it is related to the basic right of conscience. For one thing, without a fair remuneration engineers cannot concentrate their energies where they properly belong—on carrying out the immediate duties of their jobs and on maintaining up-to-date skills through formal and informal continuing education. Their time will be taken up by money worries, or even by moonlighting in order to maintain a decent standard of living. Or consider the seemingly "purely" economic issue of portable pensions. If a company's retirement plan is tied to ongoing employment with that company, engineers will feel considerable pressure not to leave their jobs. This pressure can deflect them from vigorously pursuing their obligations in situations where employers' directives are not in line with the legitimate needs or safety of clients and the public.

Nonmonetary forms of recognition are also important. Consider the following report by a 40-year-old chemical engineer:

> I have had to write papers and sections of books which appeared under the authorship of my supervisor three levels up, on matters he can hardly understand, much less contribute to except by proof reading for grammatical errors. . . . The four key people whose work he became a world-recognized success by are disposed of as follows:
>
> **(1)** Dead, heart attack, age 53, Ph.D. Chemical Engineering
> **(2)** Dead, heart attack, age 42, M.S. Chemistry
> **(3)** Dismissed from his job, age 49, Ph.D. Chemistry
> **(4)** Mental breakdown, 2 months in psychiatric hospital, age 36, Ph.D. Chemical Engineering, currently seeking other employment (Bailyn, 1980, 73).

The point of this medical and obituary report is presumably to underscore how unhealthy it is to work hard at one's job without proper recognition. Unrecognized work is also demeaning. But just how far employers are morally required to go in providing fair recognition for their engineers is again a matter that must be regularly discussed and mutually agreed upon by management and engineers.

Moral Foundation of Professional Rights

Thus far we have said that engineers' professional rights, by definition, are those possessed by virtue of being engineers. More fully, they arise because of the special moral duties engineers acquire in the course of serving the public, clients, and employers. We have given several examples of those rights and illustrated the complexities that arise when we begin to apply them. Next we must inquire into the moral basis or justification for asserting that such rights do indeed exist.

Professional rights and duties are not identical with the rights and duties of nonprofessionals, but neither are they unrelated to them. Professional rights and duties are justified in terms of more basic moral principles which also apply outside the professional job context. One's view of those more basic moral principles will depend, of course, on the particular ethical theory one endorses: rights ethics, duty ethics, or utilitarianism.

There are two general ways to apply ethical theories to justify the basic right of professional conscience. One is to proceed piecemeal by reiterating the justifications given for the specific professional duties. Whatever justification there is for the specific duties will also provide justification for allowing engineers the right to pursue those duties. Fulfilling duties, in turn, requires the exercise of moral reflection and conscience, rather than rote application of simplistic rules. Hence the justification of each duty ultimately yields a justification of the right of conscience with respect to that duty. But throughout this book we have illustrated how to justify various specific duties of engineers by means of more general ethical theories and there is no need to repeat that process here. Instead we shall pursue a second way to justify the right of professional conscience, which involves grounding it more directly in the ethical theories. Here, as elsewhere, we invoke the ethical theories to serve as general models for organizing moral reflections and to provide frameworks for approaching practical problems.

A Rights Model Rights theories, it will be recalled, emphasize human moral rights as at least one ultimate ground of morality. "Ultimate" means that human rights do not themselves need to be justified by referring to other, more fundamental moral principles. Thus a rights-based ethicist will seek to justify professional rights—in particular the basic right of professional conscience—by reference to human rights.

Let us follow A. I. Melden in viewing the most basic human right as the right to pursue one's legitimate interests. "Legitimate interests" will be those which do not violate others' rights. Hence the rights of any one individual must be understood within the context of a community of people, each of whom has rights which limit the extent of others' rights. Melden emphasizes that this community is a moral community, based upon ties of mutual understanding and concern (Melden, 1977, 140–145).

Although Melden does not himself apply his theory to professional rights, we would apply it as follows. "Legitimate interests" surely include moral concerns, especially concerns about meeting one's obligations. Thus the right to pursue legitimate interests implies a right to pursue moral obligations. This may be viewed as a human right of conscience directly derived from the most basic human right.

Now as engineers and other professionals take on special professional obligations, this general right of conscience acquires a further extension: It gives rise to a right of professional conscience which relates to specific professional obligations. In this way, the right of professional conscience is justified by reference to human rights as applied to the context of professional activity.

A Duty Model In duty ethics, rights are not the ultimate moral appeal. Instead they are mirror-image correlates of more basic duties. If I have a right to do something it is only because others have duties or obligations to allow me to do so. Within this context, the basic professional right is justified by reference to the duties others have to support or not to interfere with the work-related exercise of conscience by professionals. But who are these others, and what specifically do their duties entail? In regard to professionals, the "others" are their employers. And most importantly in regard to professional engineering, employers have a duty not to harm the public by placing handicaps in the way of the engineers they employ as those engineers seek to meet their obligations to the public. In addition, employers are directly obligated to professionals not to use coercion (i.e., not to threaten negative sanctions) which would encourage any compromise of personal moral integrity. To return to an earlier example, no hospital administrator has the right to pressure a Catholic nurse to participate in an abortion by threatening to fire her; to do so would show an utter disregard for her dignity as a moral agent (to use Kant's language). Similarly no employer has the right to threaten engineers with the loss of their jobs for refusing to work on projects they see as likely to lead to the death or injury of unsuspecting victims.

A Utilitarian Model Utilitarians will justify the right of professional conscience by reference to the basic goal of producing the most good for the greatest number of people. And no matter how "goodness" is defined, the public good is certain to be served by allowing professionals to meet their obligations to the public. For those obligations arise in the first place because of the role they play in promoting the public good.

Rule-utilitarians will seek to establish the best rule or policy in regard to employee rights for promoting the public good. Act-utilitarians will look at each situation to see whether and how far professionals should be allowed to exercise their consciences in pursuing their duties to the public.

Study Questions

1 Consider the following example by Philip M. Kohn and Roy V. Hughson:

> Jay's boss is an acknowledged expert in the field of catalysis. Jay is the leader of a group that has been charged with developing a new catalyst system, and the search has narrowed to two possibilities, Catalyst A and Catalyst B.
>
> The boss is certain that the best choice is A, but he directs that tests be run on both, "just for the record." Owing to inexperienced help, the tests take longer than expected, and the results show that B is the preferred material. The engi-

> neers question the validity of the tests, but because of the project's timetable, there is no time to repeat the series. So the boss directs Jay to work the math backwards and come up with phony data to substantiate the choice of Catalyst A, a choice that all the engineers in the group, including Jay, fully agree with (Kohn and Hughson, 1980, 103).

What should Jay do, and does he have a moral right to not do as he is directed?

2 Comment on the following passage, making any suggestions about how engineers might be protected against such situations:

> Older engineers, in particular, find job security in competition with ethical instinct. With considerable sympathy, I recall the dilemma of an older PE, in the shadow of a comfortable retirement, who was confronted by a new general manager of the plant in which he was employed as a facilities engineer. In consideration of plans for a plant expansion, the general manager insisted that the PE reduce footings and structural steel specifications below standards of good practice. The PE was told to choose between his job and his seal on the plans. Did he really have a choice? (Howard, 1966, 48)

3 In 1971 Louis V. McIntire and Marion McIntire published a novel entitled *Scientists and Engineers: The Professionals Who Are Not*. The story was about the problems encountered by J. Marmaduke Glumm, a chemist working for the Logan Chemical Company. It portrayed the disillusionment of scientists and engineers pressured into becoming managers and thus forced to move away from their original areas of expertise. It also described the tactics management used to cheat employees out of bonuses, to show unjustified favoritism, to take unfair advantage of employees in employment contracts, and to coerce professionals into going along with management's views on safety and health hazards. The novel recommended that engineers form a national federation to seek laws protecting and favoring engineers working as employees.

The novel was a thinly disguised satire of the company Louis McIntire had worked for during the past 17 years: Du Pont. When McIntire's employers learned about the novel in 1972 they fired him. In 1974 McIntire sued Du Pont, but his claim that the First Amendment protected him from being fired was rejected by the courts.

Present and defend your view as to whether McIntire had a moral right not to be fired for writing the novel. Do you think the courts should have recognized such a right legally?

4 Leonardo da Vinci reported in his journal that he had discovered how to make what today we would call a submarine. He also noted that he refused to reveal the idea to anyone because of what he viewed as its likely misuse. He wrote:

> ...now by an appliance many are able to remain for some time under water. How and why I do not describe my method of remaining under water for as long a time as I can remain without food; and this I do not publish or divulge on account of the evil nature of men who would practice assassinations at the bottom of the seas, by breaking the ships in their lowest parts and sinking them together with the crews who are in them. (da Vinci, 850)

Suppose that da Vinci discovered this idea while he was employed as a military engineer for Cesare Borgia or other military leaders, as he was at times in his career. Would he have had a moral right to refuse to reveal the idea to his employer? Would he be disloyal to the employer if he did refuse to reveal it? Why draw a line now? Defend your view by means of one of the ethical theories outlined earlier.

WHISTLE-BLOWING

We have seen how obligations to the public and to clients may come into conflict with obligations to employers. And considering the importance of obligations to the public, especially the obligation to inform those members of the public affected by "social experimentation" through engineering, we have suggested that sometimes, though not always, obligations to the public override obligations to an employer. In seeking to meet those obligations to the public, engineers and others have sometimes engaged in what is known as "whistle-blowing."

A variety of normative moral issues arise in connection with blowing the whistle on organizations: Is it ever morally permissible to do so? When? Is whistle-blowing ever morally obligatory? Is it always an act of disloyalty to an organization, or could it sometimes be consistent with company loyalty—even an expression of it? Should it sometimes be viewed as an act of moral heroism which goes beyond the call of duty? What procedures ought to be followed in blowing the whistle? And to what extent do engineers have a right to "whistle-blow?"

Before considering some of these questions, though, we need to define whistle-blowing.

Definition of Whistle-Blowing

Whistle-blowing is sometimes defined as making public accusations concerning misconduct by one's organization (Bok, 1980, 277; James, 1980, 99; Bowie, 1982, 142). This definition, however, is too narrow. On the one hand, an individual need not be a member of an organization in order to blow the whistle on it publicly. Journalists, politicians, and consumer groups may learn of corruption in organizations they do not work for and blow the whistle on them by publishing articles or informing regulatory agencies. Our main interest in this section, however, will be in whistle-blowing by employees (both present and former employees), especially where disobedience of an employer's directives or company policies is involved.

On the other hand, not all whistle-blowing involves going outside the organization. Recall the Ford engine-test case discussed at the beginning of Chap. 5. There the whistle was blown within the organization when the computer specialist wrote a memo to the company president informing him of misconduct in the engine and foundry division.

We shall not attempt to define all types of whistle-blowing in all situations. Instead, we shall list four main features that characterize most cases of whistle-blowing by employees of organizations, whether the whistle is being blown on individuals or problems within the organizations:

1 Information is conveyed outside approved organizational channels or in situations where the person conveying it is usually under pressure from supervisors or others not to do so.

2 The information being revealed is new or not fully known to the person or group it is being sent to.

3 The information concerns what the whistle-blower believes is a significant moral problem concerning the organization. Examples of significant problems are criminal behavior, unethical policies, injustices to workers within the organization, and threats to public safety.

4 The information is conveyed intentionally with the aim of drawing attention to the problem.

Using these four features as our definition, we will speak of *external whistle-blowing* when the information is passed outside the organization. *Internal whistle-blowing* occurs when the information is conveyed to someone within the organization.

The definition also allows us to distinguish between open and anonymous whistle-blowing. In *open whistle-blowing* individuals openly reveal their identity as they convey the information. *Anonymous whistle-blowing,* by contrast, involves concealing one's identity. But there are also overlapping cases, such as when individuals acknowledge their identities to a journalist but insist their names be withheld from anyone else.

Persuasive Definitions of Whistle-Blowing

Notice that the above definition leaves open the question of whether whistle-blowing is justified or not. As we shall suggest in a moment, sometimes it is and sometimes it is not. By contrast, some writers have packed into their definitions of whistle-blowing much of their own particular value perspectives concerning it. In doing so they have created persuasive or prescriptive definitions which sometimes blur the issues. Consider, for example, the following two proposals:

> "Whistle-blowing"—the act of a man or woman who, believing that the public interest overrides the interest of the organization he [sic] serves, publicly "blows the whistle" if the organization is involved in corrupt, illegal, fraudulent, or harmful activity (Nader, Petkas, and Blackwell, 1972, vii).

> Some of the enemies of business now encourage an employee to be disloyal to the enterprise. They want to create suspicion and disharmony and pry into the proprietary interests of the business. However this is labelled—industrial espionage, whistle-blowing or professional responsibility—it is another tactic for spreading disunity and creating conflict (Roche, 1971, 445).

The first definition was set forth by Ralph Nader at the beginning of a book which evaluates whistle-blowing positively. Notice that the definition assumes that whistle-blowing springs from an admirable motive: the belief that one is acting on behalf of the higher of two duties. It also assumes that whistle-blowers hold accurate views about corporate wrongdoing. Thus in two ways it automatically implies a favorable general attitude toward whistle-blowers.

The second passage was written by James M. Roche while he was chairman of the board of General Motors Corporation. It virtually identifies whistle-blowing with motives like disloyalty and a malicious desire to harm the organization.

For the sake of clarity it is preferable to adopt a more value-neutral definition of whistle-blowing, as we have done. Then the evaluative issues can be dealt with on their own merits.

Ernest Fitzgerald and the C-5A

One of the most publicized instances of open, external, whistle-blowing occurred on November 13, 1968. On that day Ernest Fitzgerald was one of several witnesses called to testify before Senator William Proxmire's Subcommittee on Economy in Government concerning the C-5A, a giant cargo plane being built by Lockheed Aircraft Corporation for the Air Force. Fitzgerald, who had previously been an industrial engineer and management consultant, was then a deputy for management systems under the Assistant Secretary of the Air Force. During the preceding 2 years he had reported huge cost overruns in the C-5A project to his superiors, overruns which by 1968 had hit $2 billion. He had argued forcefully against similar overruns relating to other projects, so forcefully that he had become unpopular with his superiors. They pressured him not to discuss the extent of the C-5A overruns before Senator Proxmire's committee. Yet when Fitzgerald was directly asked to confirm Proxmire's own estimates of the overruns on that November 13, he told the truth.

Doing so turned his career into a costly nightmare for himself, his wife, and his three children (A. E. Fitzgerald, 1972; Peters, 1972, 200). He was immediately stripped of his duties and assigned trivial projects, such as examining cost overruns on a bowling alley in Thailand. He was shunned by his colleagues. Within 12 days he was notified that his promised civil service tenure was a computer error. And within 4 months the bureaucracy was restructured so as to abolish his job. It took 4 years of extensive court battles before federal courts ruled that he had been wrongfully fired and ordered the Air Force to rehire him. And years of further litigation, involving fees of around $900,000, were required before, in 1981, he was reinstated in his former position.

Fitzgerald displayed remarkable courage at considerable sacrifice to himself. Was he obligated to do what he did? The Code of Ethics for the United States Government Service says that employees should "put loyalty to the highest moral principles and to country above loyalty to persons, party, or government department" and that they should expose "corruption wherever discovered." A coverup of a $2 billion expenditure of taxpayers' money in contract overruns would seem to qualify as corruption. Is the principle in the code a morally valid one? We believe it is. The alternative would be to endorse a kind of "organizational egoism" where only the good of one's particular in-group is emphasized.

If we feel any hesitation in saying Fitzgerald was obligated to whistleblow, it concerns whether it might be asking too much of someone in his position to do what he did. Perhaps it is beyond the call of duty to require such an incredible degree of personal sacrifice in performing one's job. In any case, his acts seem to us admirable to the point of heroism.

Not all whistle-blowing, of course, is admirable, obligatory, or even permissible. Obligations to an organization are significant. As we have suggested, they are not automatically canceled or outweighed by the obligation to the public in all situations. But Fitzgerald's case seems to us clear-cut because (1) he had made every effort to first seek a remedy to the abuses he uncovered by working within accepted organizational channels, (2) his views were well founded on hard evidence, and (3) the harm done to the Air Force by his disclosures was both a just treatment for its mismanagement of the C-5A project and far outweighed by the benefits that accrued to the public. In addition, (4) Fitzgerald was a public servant with especially strong obligations to the public which his organization, the Air Force, is committed to serve, and (5) to have withheld the information from Senator Proxmire would have involved lying and participating in a coverup. In Fitzgerald's case, as is often true, failure to blow the whistle would have amounted to complicity in wrongdoing.

Carl Houston and Welding in Nuclear Plants

In 1970 Carl Houston was working for Stone and Webster, the contractor for a nuclear power facility being constructed in Surry, Virginia. Houston was assigned as a welding supervisor at the facility and immediately saw that improper welding procedures were being used. Wrong materials were being utilized and the welders had not been properly trained. The situation was especially dangerous since some of the defective welds were appearing on the water pipes carrying coolant to the reactor core. Rupture of the pipes could cause disaster if safety backups failed simultaneously.

Houston reported his observations to Stone and Webster's local manager, who disregarded them. When he threatened to write to Stone and Webster's headquarters, he was told he would be fired. He sought to alert the reactor suppliers to the danger, and shortly thereafter he was fired on the trumped-up grounds that he was not qualified for welding. Afterwards he wrote letters to the governor of Virginia and to the Atomic Energy Commission, which were never answered. Finally, two further letters which he wrote to Senators Howard Baker and Albert Gore (from his home state of Tennessee) had an effect. The senators prompted the Atomic Energy Commission to make investigations which confirmed his allegations (Houston, 1975, 25).

Was Houston justified in going outside Stone and Webster with his warnings? In retrospect, perhaps it would have been better for him to try working further within the organization first, for example, by writing to division headquarters. Yet there seems little doubt that his actions in pursuing the matter were highly praiseworthy. Because there was so much at stake—i.e., since the possible consequences in the event of a nuclear plant accident were so disastrous—the obligation to protect the public had a clear priority in this case.

Moral Guidelines to Whistle-Blowing

Under what conditions are engineers justified in going outside their organizations when safety is involved? This really involves two questions: When are they morally *permitted,* and when are they morally *obligated,* to do so?

Richard T. De George has suggested that it is morally *permissible* for engineers to engage in external whistle-blowing concerning safety when three conditions are met (De George, 6):

> **1** If the harm that will be done by the product to the public is serious and considerable;
>
> **2** If they make their concerns known to their superiors; and
>
> **3** If getting no satisfaction from their immediate superiors, they exhaust the channels available within the corporation, including going to the board of directors.

In order for the whistle-blowing to be morally *obligatory,* however, De George gives two further conditions (De George, 6):

> **4** He [or she] must have documented evidence that would convince a reasonable, impartial observer that his [or her] view of the situation is correct and the company policy wrong.
>
> **5** There must be strong evidence that making the information public will in fact prevent the threatened serious harm.

De George sets forth these conditions as rough general rules, something like moral rules of thumb. Exceptions and additions can be made. His account allows for the possibility of instances of permissible whistle-blowing which do not meet all of conditions 1 through 3. For example, situations of extreme urgency may arise in which there is insufficient time to work through all the normal organizational channels. Also, the first condition should be expanded to include violations of rights and fraud.

We should also add that there may be personal obligations to family and others which militate against whistle-blowing. And where blowing the whistle openly could result not only in the loss of one's job but also in being blacklisted within the profession, the sacrifice may in some cases be too much to demand, or may become supererogatory—more than one's basic moral obligations require. Or, what is more likely, anonymous whistle-blowing may be the only morally mandatory action.

Nevertheless, conditions 1 through 5 give strong support to the importance of engineers' obligations to the public in areas of safety while at the same time allowing for the importance of their obligations to employers by underscoring the need to work first within organizational channels when trying to correct problematic or dangerous situations. And we agree with De George that when all five conditions are met there arises a very strong prima facie obligation to whistle-blow.

George B. Geary and U.S. Steel

We have been focusing upon external whistle-blowing, where information is passed outside the organization. Let us now consider a case of internal whistle-blowing where the information was conveyed within the organization, although outside regular organizational channels.

George Geary had worked 14 years for the U.S. Steel Corporation. In 1967 he was a sales executive with the company's oil and gas industry supply division in Houston. U.S. Steel was about to market a new type of pipe which Geary believed had been insufficiently tested and might be defective. If the pipe should burst or break while in use, not only would property be damaged, but there might be serious injuries to customers and the public.

Geary expressed his strong objections to midlevel management, which decided to go ahead with marketing the new pipe anyway. So, while obeying directives to sell the pipe, he sent his objections to U.S. Steel's higher management. Largely because of his good reputation within the company, top officials took the assertions seriously. They ordered a major reevaluation of the pipe and withdrew it from sales until the tests were completed. Yet shortly thereafter, Geary was fired on the ground of insubordination; the charge was that he went over his manager's head in a matter beyond his area of expertise.

U.S. Steel then attempted to block his unemployment compensation by arguing that he was guilty of willful misconduct. However, the Unemployment Compensation Board of Review, upon hearing the case, reached the following conclusion:

> No company places a man in the position held by the claimant and pays him the salary received by the claimant simply to have him quietly agree to all proposals. The claimant did not refuse to follow orders, but, in fact, agreed to do as instructed despite his opposition to the program proposed. Although he may have been vigorous in his opposition and offended some superiors by going to a vice president, it is clear that at all times the claimant was working in the best interest of the company and that the welfare of the company was primary in his mind. Under these circumstances, giving due regard to the claimant's position with the company, his conduct cannot be deemed willful misconduct (quoted in Nader, 1972, 155–156).

It is possible that Geary's actions prevented injuries, reduced consumer costs, and even saved significant costs to U.S. Steel from premature marketing of a defective product. By all the evidence, he acted as a loyal employee, concerned at once for the good of U.S. Steel and of the public. Yet as a result of communicating information against his immediate superior's wishes, he suffered the same fate as the external whistle-blowers discussed above.

Protecting Whistle-Blowers

The three cases we have examined may seem to present a one-sided, negative picture of what happens to whistle-blowers. Whistle-blowing does not always have such unfortunate results. Yet most whistle-blowers have suffered unhappy, even tragic, fates. In the words of one lawyer who defended a number of them:

> Whistle-blowing is lonely, unrewarded, and fraught with peril. It entails a substantial risk of retaliation which is difficult and expensive to challenge. Furthermore, "success" may mean no more than retirement to a job where the bridges are already burned, or monetary compensation that cannot undo damage to a reputation, career and personal relationships (Raven-Hansen, 1980, 44).

Yet the vital service to the public provided by many whistle-blowers has led increasingly to public awareness of a need to protect them against retaliation by employers. Government employees have won important protections. Various federal laws related to environmental protection and safety and the Civil Service Reform Act of 1978 protect them against reprisals for lawful disclosures of information believed to show "a violation of any law, rule, or regulation, mismanagement, a gross waste of funds, an abuse of authority, or a substantial and specific danger to public health and safety" (Raven-Hansen, 1980, 42; Unger, 1982, 94). The fact that few disclosures are made appears to be due mostly to a sense of futility—the feeling that no corrective action will be undertaken. In the private sector, employees are covered by statutes forbidding firing or harassing of whistle-blowers who report to government regulatory agencies the violations of some twenty federal laws, including those covering coal mine safety, control of water and air pollution, disposal of toxic substances, and occupational safety and health. In a few instances unions provide further protection.

Aside from these exceptions, however, most states still allow employers to fire employees they consider "disloyal" at will. There has yet to be full legal recognition of the right of salaried engineers to adhere to professional codes of ethics. But the laws concerning whistle-blowing are in transition, and a number of observers believe they are moving in directions favorable to responsible whistle-blowing (Walters, 1975, 34; Ewing, 1977, 113; Westin, 1981, 163–164; Petersen and Farrell, 1986, 20). Protection of whistle-blowers against unjust firing is being added to many specific laws. It is reasonable to hope that more systematic national legislation will be forthcoming to support responsible whistle-blowers.

We should also add that beyond the protection afforded by law in cases of whistle-blowing, there is an important potential role for professional societies to play. Until recently those societies were reluctant to become involved in supporting engineers who followed the entries in their codes of ethics calling for members to "notify proper authorities" when overruled by their superiors in their professional judgments about dangers to the public. But this is changing. In the BART case, as we will see in the next section, the Institute of Electrical and Electronics Engineers (IEEE) was willing to write a friend of the court brief seeking to establish legal recognition of the right of engineers to act in accordance with professional codes of ethics. The IEEE has also established awards and other forms of honorary recognition for whistle-blowers who act according to its ethical code, and furthermore has helped locate new jobs for discharged engineers. Another avenue of protection for engineers being explored by professional societies is the publication in their journals of the names of companies who take unjust reprisals against whistle-blowers.

Commonsense Procedures in Whistle-Blowing

It is clear that a decision to whistle-blow, whether within or outside an organization, is a serious matter that deserves careful reflection. And there are several rules of practical advice and common sense which should be heeded before taking this action (Unger, 1979, 56–57; Westin, 1981, 160–163; Elliston et al, 1985, 2 books):

1 Except for extremely rare emergencies, always try working first through normal organizational channels. Get to know both the formal and informal (unwritten) rules for making appeals within the organization.

2 Be prompt in expressing objections. Waiting too long may create the appearance of plotting for your advantage and seeking to embarrass a supervisor.

3 Proceed in a tactful, low-key manner. Be considerate of the feelings of others involved. Always keep focused on the issues themselves, avoiding any personal criticisms that might create antagonism and deflect attention from solving those issues.

4 As much as possible, keep supervisors informed of your actions, both through informal discussion and formal memorandums.

5 Be accurate in your observations and claims, and keep formal records documenting relevant events.

6 Consult colleagues for advice—avoid isolation.

7 Before going outside the organization, consult the ethics committee of your professional society.

8 Consult a lawyer concerning potential legal liabilities.

The Right to Whistle-Blow

Whistle-blowers who proceed responsibly and take special care to document their views are fulfilling their obligations to protect and serve the public. To this extent they have a professional moral right to whistle-blow. This important right is a restricted one, however, and its appropriate extent can vary depending on a number of factors.

Engineers working for the government have as public servants an especially strong charge to protect the public. It is appropriate that they have correspondingly strong rights to whistle-blow in the public interest. That right is limited by legitimate needs to keep some information confidential. But not all information stamped "classified" is legitimately confidential. There have been many instances where government corruption has been hidden under claims of confidentiality. The electronic surveillance involved in Watergate is but one example. Moreover, even where legitimate confidentiality is involved, the public interest served by whistle-blowing may be of even greater importance.

Engineers working in the private sector also have obligations to the public, especially those based on the right of the public to make informed decisions concerning the use of technological products. Thus they also have a professional right to whistle-blow when such action is justified by the appropriate conditions. And we along with the observers noted above, hope and believe that the courts and professional organizations will continue to expand the legal and institutional recognition and protection of this right during the coming decade.

Beyond Whistle-Blowing

Sometimes whistle-blowing is a practical moral necessity. But generally it holds little promise as the best possible method for remedying problems and should be viewed as a last resort.

The obvious way to remove the need for *internal* whistle-blowing is to allow greater freedom and openness of communication within the organization. That is, the need to violate the often rigid channels of communication within organizations would be removed by making those channels more flexible and convenient. But this means more than merely announcing formal "open-door" policies and appeals procedures which give direct access to higher levels of management. Those would be good first steps, and a further step would be the creation of an ombudsperson or an ethics review committee with genuine freedom to investigate complaints and make independent recommendations to top management. The crucial factor which must be involved in any structural change, however, is the creation of an atmosphere of tolerance. There must be a positive affirmation of engineers' efforts to assert and defend their professional judgments in matters involving ethical considerations. Any formal policy can be subverted by supervisors who are preoccupied with their own authority or who create a climate of intimidation. It can also be subverted by those engineers who are insensitive to the legitimate needs of management.

Creating such an atmosphere, then, requires the efforts of management and engineers alike. But it falls on the shoulders of top management to give this aspect of the organization equal priority with other organizational needs and goals. Management's tools include the formal ones of classes and workshops for employees. At Fluor Corporation, for example, a course in engineering ethics was an integral part of an in-house masters degree program in engineering which the company financed for its employees. Ultimately, however, it is management's example and style that are decisive in communicating ethical concern to those lower down in the hierarchy. Such concern can be stifled by just one act of inflicting a drastic penalty on an engineer for failing to follow the letter of organizational procedures while zealously pursuing ethical concerns, and it is management's responsibility not to produce such an intimidating atmosphere.

What about *external* whistle-blowing? Much of it can also be avoided by the same sorts of intraorganizational modifications. Yet there will always remain troublesome cases where top management and engineers differ in their assessments of a situation even though both sides may be equally concerned to meet their professional obligations to safety.

To date, the assumption has been that management has the final say in any such dispute. But our view is that engineers have a right to some further recourse in seeking to have their views heard.

It is impossible to generalize concerning what this recourse should be within all contexts. Minimally we think it essential that engineers be allowed to discuss—in confidence—their moral concerns with the ethics committees of their professional societies. And it is highly desirable that representatives from those committees, or perhaps professional arbitrators of some sort, be allowed to enter into discussions between engineers and management which have reached a deadlock—again in confidence, as far as the public is concerned. Such was the purpose of Ralph Nader's short-lived Clearinghouse for Professional Responsibility, which sought to serve as a first-step arbiter to resolve employer-employee conflicts in-house.

Beyond this, ongoing piecemeal changes in the law, within regulatory bodies, and within corporations themselves must be explored. Some will argue for strong legislation favorable to whistle-blowing. But this would allow greater public control over private corporate goals, and management could be expected to resist such outside threats to its autonomy. How far we, as a society, should support laws favorable to whistle-blowing will ultimately be as difficult a decision as any other concerning public regulation of private enterprise.

Study Questions

1 Consider the following example:

> Harry works as a designer for a component supplier and often sits in on meetings with clients to keep abreast of their needs. He attends a meeting of corporate executives who decide to phase out a particular component—an encapsulated assembly on which the company is losing money at current production levels. The company will have a new component on the market in six months that not only performs the same functions, but also does additional peripheral functions. It will not be a plug-in replacement for the older component. The client is obviously making a long-term commitment in his design of a system, and the component apparently is key to that system. The client assumes that, in time, when mainte-

nance is required on his system, he will be able to order replacements, if needed. No other company makes a plug-in equivalent of this component. Harry quietly asks the nearest salesman if he is aware that the component is being discontinued, and finds out that the salesmen know its impending fate, but are remaining silent, since the newer component won't be available for six months, and the client is looking to buy something now. If the company doesn't have something to sell him, he'll look elsewhere. What should Harry do? (Perry, 1981, 56–57)

In answering this question assume that Harry's immediate supervisor tells him not to do anything at all.

2 Present and defend your view as to whether or not, in the case described below, the actions of Ms. Edgerton and her supervisor were morally permissible, obligatory, or admirable. Did Ms. Edgerton have a professional moral right to act as she did? Was hers a case of legitimate whistle-blowing?

In 1977 Virginia Edgerton was senior information scientist on a project for New York City's Criminal Justice Coordinating Council. The project was to develop a computer system for use by New York district attorneys in keeping track of data concerning court cases. It was to be added on to another computer system, already in operation, which dispatched police cars in response to emergency calls. Ms. Edgerton, who had 13 years of data processing experience, judged that adding on the new system might result in overloading the existing system in such a way that the response time for dispatching emergency vehicles might be increased. Because it might risk lives to test the system in operation, she recommended that a study be conducted ahead of time to estimate the likelihood of such overload.

She made this recommendation to her immediate supervisor, the project director, who refused to follow it. She then sought advice from the Institute of Electrical and Electronics Engineers, of which she was a member. The Institute's Working Group on Ethics and Employment Practices referred her to the manager of systems programming at Columbia University's computer center, who verified that she was raising a legitimate issue.

Next she wrote a formal memo to her supervisor, again requesting the study. When her request was rejected, she sent a revised version of the memo to New York's Criminal Justice Steering Committee, a part of the organization for which she worked. In doing so she violated the project director's orders that all communications to the Steering Committee be approved by him in advance. The project director promptly fired her for insubordination. Later he stated: "It is . . . imperative that an employee who is in a highly professional capacity, and has the exposure that accompanies a position dealing with top level policy makers, follow expressly given orders and adhere to established policy" (Edgerton Case, 1978).

3 According to De George's first criterion for justified whistle-blowing, the product involved must actually be seriously harmful. Critics of this view insist that employees need only have very strong evidence that the product is harmful (James, 1984). What is your view, and why is this issue important?

Also, critics have disagreed with De George's fifth criterion, which says there must be good reason to think the whistle-blowing will bring about necessary changes in order for the whistle-blowing to be obligatory (James, 1984). These critics charge that engineers have obligations to warn the public of dangers quite independently of guessing how the public will choose to react to that information. What is your view?

4 A controversial area of recent legislation allows whistle-blowers to collect money. Federal tax legislation, for example, pays informers a percentage of the money recovered from tax violators. And the 1986 False Claims Amendment Act allows 15 to 25 percent of the recovered money to go to whistle-blowers who report overcharging in federal government contracts to corporations. These sums can be substantial because lawsuits can involve double and triple damages as well as fines. Discuss the possible benefits and drawbacks of using this approach in engineering and specifically concerning safety matters. Is the added incentive to whistle-blow worth the risk of encouraging self-interested motives in whistle-blowing?

5 It has been suggested that cases of whistle-blowing which involve organizational disobedience are similar to instances of civil disobedience (Otten, 1980, 182–186; Elliston, "Civil Disobedience and Whistle-blowing", 1982). Civil disobedience was a major social tactic used, for example, in the civil rights movement in the 1960s. It may be defined as having the following features: It involves the intentional breaking of a law or government policy; it is nonviolent; it is conducted publicly (rather than secretively); it is performed by generally loyal citizens (as opposed to anarchists and revolutionaries) seeking to change what they believe to be seriously immoral laws or government actions; and participants do not attempt to evade the legal penalties attached to such activity.

Discuss the similarities and differences you see between civil disobedience and (*a*) the open, external, whistle-blowing of Ernest Fitzgerald, (*b*) the open, internal, whistle-blowing of George Geary, and (*c*) anonymous whistle-blowing.

6 Do you see any special moral issues raised by *anonymous* whistle-blowing? For a helpful discussion consult Frederick Elliston's essay, "Anonymous Whistle-blowing" (Elliston, 1982).

7 June Price Tangney, a psychologist at Bryn Mawr College, published a study which showed that "one out of three scientists at a major university suspect a colleague of falsifying scientific data, and half of them have done nothing to verify or report their suspicions. . . . The scientists' unwillingness to act is particularly disturbing because most cases of scientific fraud are uncovered through whistle-blowing" (Associated Press report in *San Francisco Chronicle,* 30 Aug. 1987). Comment on this and related ethical problems in academe. (See also Study Question 8.)

8 As member of a committee hearing a student's disciplinary case, a professor finds out that the student has committed fraud outside the campus in another case. Realizing that the university administration will not take any action to notify local authorities regarding this latter case, he notifies the district attorney on his own. The university administration censures the professor for breach of confidentiality. Discuss the ethical implications of the professor's and the university's actions.

THE BART CASE

The Bay Area Rapid Transit System (BART) is a suburban rail system that links San Francisco with the cities across its bay. It was constructed during the late 1960s and early 1970s, and its construction led to a now classic case of whistle-blowing. The case is important because it remains controversial, because it involved a precedent-setting intervention by an engineering professional society, and because it became the subject of the first book-length scholarly study of an instance of whistle-blowing (*Divided Loyalties* by Robert M. Anderson et al., 1980).

Background

The example of the pioneering years of railroading indicates that technological experimentation is usually highly fruitful. For example, early fears about the effects of high-speed travel—of sparks showering the countryside, of animals being frightened by the noise and fast movement—were proven to be unfounded. The benefits to agriculture, industry, and commerce, moreover, were immense. And society learned that to secure those benefits it could live with the loss of forests to railroad ties and fuel, or with the cycle of settlement building and abandonment entailed by the construction of new railroads.

As technological innovation in railroading accelerated, however, the trend to do the fashionable thing for its own sake increasingly predominated. For example, railroads took over in instances where common sense would have dictated the continued use of barges on canals. To some extent BART is a recent example of that trend. Developed to incorporate the latest "space age" technology in its design, it ended up as more expensive and less reliable than its traditional counterparts.

The BART system was built with tax funds, and its construction was characterized by tremendous cost overruns and numerous delays. Much of this can be ascribed to the introduction of innovative methods of communicating with individual trains and of controlling them automatically. In addition, plain fail-safe operation was replaced by complex redundancy schemes. (Fail-safe features simply cause a train to stop if something breaks down; redundancy features try to keep trains running by switching the faulted components to alternate ones.) The rationale given for this approach was that the system could be sold to the public only if it involved glamorous and exciting gadgetry.

Responsibility and Experimentation

The opportunity to build a rail system from scratch, unfettered by old technology, was a challenge that excited many engineers and engineering firms. Indeed, altogether the project was an interesting experiment. Yet among the engineers who worked on it were some who came to feel that too much experimentation was going on without proper safeguards. Safety features were given insufficient attention and quality control was poor, they thought.

Three engineers in particular—Holger Hjortsvang, Robert Bruder, and Max Blankenzee—identified dangers that were to be recognized by management only much later. They saw that the automatic train control was unsafely designed. Moreover, schedules for testing it and providing operator training prior to its public use were inadequate. Computer software problems continued to plague the system. Finally, there was insufficient monitoring of the work of the various contractors hired to design and construct the railroad. These inadequacies were to become the main causes of several early accidents (Friedlander, March 1973 and April 1973).

The three engineers wrote a number of memos and voiced their concerns to their employers and colleagues. Their initial efforts were directed through organizational channels to both their immediate supervisors and the two next higher levels of management, but to no avail. Yet they refused to wait passively for accidents to occur, and resolved to do more.

Up to this point Hjortsvang, Bruder, and Blankenzee clearly displayed the kind of moral responsibility described in Chap. 3. They were conscientious in refusing to lose sight of their primary obligation to the public—that is, their obligation to what was, in effect, the "subject" of this particular engineering "experiment." They were imaginative in foreseeing dangers. They were personally and autonomously involved. And they were willing to accept moral accountability for their participation in the project.

Of special interest in the case is that for the most part the three engineers were not specifically assigned or authorized by the BART organization to check into the safety of the automatic control system. Hjortsvang, for example, first identified the dangers when he was sent to Westinghouse (a BART subcontractor) primarily to observe, not supervise, the development of the control system. Similarly, Robert Bruder worked for the construction department, not the operations department which had responsibility for the train control. Thus, both engineers looked to the wider implications of the specific tasks assigned them within the organization. They refused to have their moral responsibility confined within a narrow organizational bailiwick.

Controversy

The controversial events that followed as the engineers sought to pursue their concerns further are described and interpreted from the opposing viewpoints of the engineers and management (and others) in the book *Divided Loyalties* by Robert M. Anderson et al. (cited in the Bibliography). Here is an account of five of those events.

First, Hjortsvang wrote an anonymous memo summarizing the problems, and distributed copies of it to nearly all levels of management, including the project's general manager. The memo argued that a new systems engineering department was needed, a department that Hjortsvang had also requested in an earlier signed memo. Distribution of such an unsigned memo was regarded by management as suspicious and unprofessional since it was done outside the normal channels of accountability within the organization. Later, when its author was identified, management decided Hjortsvang was motivated by self-interest and a desire for power since it could be assumed that he wished to become the head of such a department.

Second, the three engineers contacted several members of BART's board of directors when their concerns were not taken seriously by lower levels of management. By doing so, they departed from approved organizational channels, since BART's general manager allowed only himself and his designates to deal directly with the board. Since BART was a publicly funded organization governed by the public board of directors, it could be argued that this was an instance of internal whistle-blowing.

Third, in order to obtain an independent view, the engineers contacted a private engineering consultant who on his own wrote an evaluation of the automatic train control.

Fourth, one of the directors, Dan Helix, listened sympathetically and agreed to contact top management while keeping the engineer's names confidential. But to the shock of the three engineers, Helix released copies of their unsigned memos and the consultant's report to the local newspapers. It would be the engineers, not Helix, who would be penalized for this act of external whistle-blowing.

Fifth, management immediately sought to locate the source of Helix's information. Fearing reprisals, the engineers at first lied to their supervisors and denied their involvement.

Aftermath

At Helix's request the engineers later agreed to reveal themselves by going before the full board of directors in order to seek a remedy for the safety problems. On that occasion they were unable to convince the board of those problems. One week later they were given the option of resigning or being fired. The grounds given for the dismissal were insubordination, incompetence, lying to their superiors, causing staff disruptions, and failing to follow understood organizational procedures.

These dismissals were damaging to the engineers. Robert Bruder could not find engineering work for 8 months. He had to sell his house, go on welfare, and receive food stamps. Max Blankenzee was unable to find work for nearly 5 months, lost his house, and was separated from his wife for 1½ months. Holger Hjortsvang could not obtain full-time employment for 14 months, during which time he suffered from extreme nervousness and insomnia.

The impact on BART, by comparison, was minor. Subsequent studies proved that the safety judgments of the engineers were sound. Changes in the design of the automatic train control were made, but it is unclear whether those changes would have been made in any case. During its decade of development BART was plagued by many technical problems of the type the engineers drew attention to. And the inability of BART management to deal effectively with the engineers' concerns was typical of many other instances of poor management.

Two years later the engineers sued BART for damages in the sum of $875,000 on the grounds of breach of contract, harming their future work prospects, and depriving them of their constitutional rights under the First and Fourteenth Amendments. A few days before the trial began, however, they were advised by their attorney that they could not win the case because they had lied to their employers during the episode. They settled out of court for $75,000 minus 40 percent for lawyers' fees.

In the development of their case the engineers were assisted in their court case by an amicus curiae ("friend of the court") brief filed by the Institute of Electrical and Electronics Engineers (IEEE). This legal brief noted in their defense that it is part of each engineer's professional duty to promote the public welfare, as stated in IEEE's code of ethics. In 1978 IEEE presented each of them with its Award for Outstanding Service in the Public Interest for "courageously adhering to the letter and spirit of the IEEE code of ethics."

Comments

The study questions below ask you to assess the extent to which the three engineers and BART's management acted responsibly. The complexities revealed in *Divided Loyalties* show the case is hardly a simple one. Here we wish to comment upon two attitudes held by the authors of that book, attitudes germane to the topic of moral responsibility and deserving of mention because of the frequency with which similar arguments are heard in other contexts.

The authors' final verdict is that the BART case "can be viewed as not really involving safety or ethics to any marked degree" (Anderson et al., 1980, 353). We disagree. The main basis for that verdict seems to be the claim that BART's complex organizational structure alone was to blame for the conflicts which helped precipitate the incidents. For example, the engineers were given considerable freedom to determine for themselves the specific tasks they were to pursue, but granted little authority to implement changes they felt were needed. Frustration on their part was therefore to be expected.

This argument, however, fails to show that ethical issues were not involved. On the contrary, it shows how ethical issues can arise out of problems associated with organizational structure. Indeed, the conflicts engendered by the social, political, and economic settings of an organization quite frequently form the background for the ethical problems engineers confront when concerned about how best to ensure the safety of their projects.

The authors' verdict may also have resulted from a lack of clarity about what an ethical problem is. For they emphasize that there were *no villains* in the BART episode. Those involved were basically good people trying in the main to do their jobs responsibly even if they were influenced to some degree by self-interest. This seems to imply that ethical situations must always involve bad people who are opposed by good people—a melodramatic view of morality. Yet surely the question of how best to assure safety in any engineering project is a moral issue, whatever the ultimate personal motivations of the people involved in it. Ethics can involve a decision between good and better just as much as a conflict between good and bad.

Study Questions

1 Present and defend your view as to whether, and in what respects, the BART engineers and BART management acted responsibly. In doing so, discuss alternative courses of action that either or both groups might have pursued. Discuss and apply De George's criteria (from the previous section) for when whistle-blowing is morally permissible and obligatory. Focus especially on (*a*) Hjortsvang's anonymous memo distributed within BART, (*b*) the act of contacting BART's board of directors, and (*c*) lying to the supervisors when questioned about their involvement.

2 The authors of *Divided Loyalties* suggest that "management shares with the three engineers responsibility for the political naiveté which permitted them to carry their grievance as far as they did. It is clear that the engineers took a narrow and technical view of the issues which disturbed them, and failed to place them in the context of the whole BART development. At the same time, management fostered this naiveté by failing adequately to sensitize its professional employees to the political and economic climate surrounding and influencing the activities of the organization" (Anderson et al., 1980, 351). Presumably this is a criticism of the act of contacting the board of directors of a public project for which a positive public image is needed to sustain support and continued funding. Do you agree with these authors that political considerations should have entered into the decisions of the three engineers? Or do you agree with IEEE that the engineers acted in a courageous way in trying to protect public safety?

3 The following lines are from the play *Sarcophagus* by Vladimir Gubaryev (1987). Based on the playwright's imagination of how Chernobyl's director may have been questioned at the time of the reactor accident, they convey the milieu which is found in many bureaucracies. (The real director and two aides have since been sentenced to 10 years at hard labor; others received shorter terms.)

> *Investigator*: ...And do you know why your predecessor in the job was sacked?
>
> *Director*: Everybody knows why. He was a troublemaker. Plus four reprimands for failing to reach his output targets.
>
> *Investigator*: Yet at the station everybody speaks of him with respect. Even with fondness, one might say.
>
> *Director*: All I know is that the authorities found him difficult to get on with.
>
> *Investigator*: Of course. Because he didn't always do as he was told. He used to argue decisions, in fact. Incidentally, on the question of putting No. 4 Reactor on-line ahead of schedule—he was dead against it.
>
> *Director*: That was a matter for decision by higher authority. They're not stupid, the people in the ministry. They know the overall situation and the state of affairs at our station too.

What kind of management style and reporting procedures would you recommend for critical operations which depend on engineers and other technically skilled personnel?

EMPLOYEE RIGHTS

Employee rights are any rights (moral or legal) which have to do with the status of being an employee. They include some professional rights which apply to the employer-employee relationship: for example, the right to disobey unethical directives and to express dissent from company policies without employer retaliation. Thus the professional rights discussed in the previous section are also employee rights insofar as they relate to the condition of being a salaried professional.

Then too, employee rights include fundamental human rights relevant to the employment situation. In the next section we will discuss one of those: the right not to be discriminated against because of one's race, sex, age, or national origin.

And one group of employee rights are institutional rights created by organizational policies or contracts. For example, an engineer whose negotiated salary is $40,000 has a contractual right to that amount of money. He or she may also have contractual rights to various company benefits, such as periodic pay raises and profit sharing. These rights are based solely on employment contracts.

However, a different group of employee rights will be the topic of this section. In contrast with purely contractual rights, these exist even if unrecognized by specific contract arrangements or company policies. Companies and employers *ought* to recognize them, whether or not they actually do. For they are more than mere privileges which employers are permitted to disregard.

Ewing's Employee Bill of Rights

In *Freedom Inside the Organization,* David Ewing, editor of *The Harvard Business Review,* refers to employee rights as the "black hole in American rights." The Bill of Rights in the Constitution was written to apply to government, not to business. But when the Constitution was written no one envisaged the giant corporations which have emerged in our century. Ewing demonstrates compelling parallels between the kinds of threats to liberty posed by large and powerful governments (which the authors of the Constitution sought to protect citizens against) and the kinds of threats to individual freedom posed by present-day business organizations. Corporations wield enormous power politically and socially, and especially over their employees. They operate much as minigovernments, and are often comparable in size to those governments the authors of the Constitution had in mind. For instance, American Telephone & Telegraph in the 1970s employed twice the number of people inhabiting the largest of the original thirteen colonies when the Constitution was written.

Ewing proposes that large corporations ought to recognize a basic set of employee rights. He gives the following concise statement of what those rights should involve:

> No public or private organization shall discriminate against an employee for criticizing the ethical, moral, or legal policies and practices of the organization; nor shall any organization discriminate against an employee for engaging in outside

> activities of his or her choice, or for objecting to a directive that violates common norms of morality.
>
> No organization shall deprive an employee of the enjoyment of reasonable privacy in his or her place of work, and no personal information about employees shall be collected or kept other than that necessary to manage the organization efficiently and to meet legal requirements.
>
> No employee of a public or private organization who alleges in good faith that his or her rights have been violated shall be discharged or penalized without a fair hearing in the employer organization (Ewing, 1977, 234–235).

In previous sections we discussed some of these rights, such as the rights to free speech and dissent and the right of conscientious refusal to obey unethical directives. Here we will examine several others—in particular, those relating to the choice of outside activities, to privacy, and to due process.

Freedom to Choose Outside Activities

All employees have the right to pursue nonwork activities of their own choosing without coercion or retribution from employers. This is part of their basic human right to pursue legitimate interests without interference. But because this right has generally not been protected by state or federal laws, there have been some flagrant violations of it.

For example, a worker in a Ford Motor Company service department was fired because his supervisor learned he had bought a new American Motors Rambler instead of a Ford automobile. Because he happened to be a union member, he was able to regain his job. Others have not been as lucky and have had to buckle to pressures from employers. Or there is the case of an executive for Phillips Petroleum who stated in a national interview that he did not want to see Phillips employees at competitors' gas stations (Ewing, 1977, 120–121).

Such abuses are perhaps becoming rarer, especially in states like California and Florida which have passed laws prohibiting them. But why, we might ask, would employers make such demands and intrusions into the personal purchasing habits of their employees?

No doubt part of the answer lies in an exaggerated concern for company loyalty. Loyalty comes to be viewed as extending beyond the fulfilling of job functions into areas of personal decision making. A more important part of the answer, however, is the extreme concern companies have to present a unified and untarnished image to the public. Even the slightest or most indirect damage to that image, and to the employers' ability to control the image, is perceived as a threat. One such threat lies in the negative attitudes toward the company that could potentially arise when it is learned by outsiders that employees are not purchasing their own company's products. Yet an employer's rightful concern with the company image should not extend to control over employees' personal buying habits.

Consider a different example. In 1971 IBM fired Lawrence Tate, an engineer employed by that company for over 18 years. Tate had a shining record with IBM and was fired solely because of a certain article that appeared in a local newspaper. The article told of Tate's efforts to reform the police department in his area. It also mentioned his subsequent arrest and conviction on two misdemeanor charges and his daughter's arrest for possession of one marijuana cigarette. The convictions were being appealed, and he was countercharging police harassment. Unfortunately, the article also mentioned Tate's place of employment, and it was this which led to his being fired (D. Fitzgerald, 1980, 197–198).

Perhaps some indirect damage to IBM's reputation resulted from the article. But if this sort of thing can be a legitimate reason for firing an employee, none of us has much genuine freedom. The single act of a careless journalist would be enough to undermine our careers.

What about a corporation's right to protect its public image? Surely this places some limits on the rights of employees to pursue outside activities, in spite of what has been said so far. No simple line can be drawn here, but a few generalizations are possible:

First, the rights of employees to pursue outside activities become limited at the point where those activities lead to violations of the duties connected with their jobs. Here what is actually at stake are not the outside activities per se, but their effects on job-related activities. For example, an individual has the right to abuse alcohol without interference in the privacy of his or her own home even though such conduct may be foolhardy and some people upon learning of it may lower their estimate of the company the person works for. But the employer has a right to take action against the employee if the alcohol abuse begins to damage work performance.

Second, employers have the right to take action when outside activities constitute a conflict of interest. Here there may be no actual harm done, as we saw in Chap. 5. But the potential exists for failure to fulfill duties to the organization and for fostering a public image of tenuous employee loyalty to the company. Employers are plainly within their rights, for example, in requiring a person to stop moonlighting for a competitor's business.

Third, employees have no right to consistently sabotage their employers' interests during off-hours. During labor disputes, a mutual recognition of each other's legitimate interests must be maintained.

In every case, however, we should add, the burden of proof should always fall on the employer to establish that the corporation's interests are so compelling as to take precedence over the rights of its employees.

Right to Privacy

The right to pursue outside activities can be thought of as a right to personal privacy in the sense that it means the right to have a private life off the job. In speaking of the *right to privacy* here, however, we mean the right to control access to and use of information about oneself. (As with the right to outside activities, this right is limited in certain instances by employers' rights.)

Consider a few examples of situations in which the functions of employers come into conflict with the right employees have to privacy:

1 Before being hired at a computer center which handles large banking transactions, applicants are required to answer questions about their past criminal records while taking a polygraph (lie detector) test.

2 Job applicants at the sales division of an electronics firm are required to take personality tests which include personal questions about alcohol use and sexual conduct. The rationale given for asking those questions is a sociological study showing correlations between sales ability and certain data obtained from answers to the questionnaire. (That study has been criticized by other sociologists.)

3 A supervisor unlocks and searches the desk of an engineer who is away on vacation without the permission of that engineer. The supervisor suspects the engineer of having leaked information about company plans to a competitor and is searching for evidence to prove those suspicions.

4 A sociologist has been hired as a consultant to a large construction firm which has been having personnel conflicts in one division. Without checking with its employees, management gives the sociologist full access to its personnel files.

5 A large manufacturer of expensive pocket computers has suffered substantial losses from employee theft. It is believed that more than one employee is involved. Without notifying employees, hidden surveillance cameras are installed.

6 A rubber products firm has successfully resisted various attempts by a union to organize its workers. It is always one step ahead of the union's strategies, in part because it monitors the phone calls of employees who are union sympathizers. It also pays selected employees bonuses in exchange for their attending union meetings and reporting on information gathered. It considered, but rejected as imprudent, the possibility of bugging the rest areas where employees were likely to discuss proposals made by union organizers.

Some of these examples involve abuse of employer prerogatives. Most of them involve a clash between the right to privacy of employees and the right of employers to effectively manage a corporation. We may differ in our opinions about some of them. But surely such intrusions are morally problematic and stand in need of special justification. Why is it that privacy is so important, even at work, and what is the basis of the right to it?

In order to answer that question it will be useful to postulate an extreme, hypothetical case (Fried, 1970, 138). Imagine a scenario akin to that described in George Orwell's *1984*, a scenario which is feasible with current technology: Unknown to them, all employees in a firm—engineers, skilled laborers, secretaries—have a tiny electronic device implanted in their plastic identification badges. The device emits steady signals indicating the location of the employee carrying it and can be used by a distant monitoring center to record conversations taking place within a radius of 20 feet around it. Why would this situation upset us if we were the employees?

One answer is that we would object to the deception involved. So to focus matters directly on privacy, let us imagine that we are fully informed of the use and nature of the bugging device. We are not asked to give our consent, however, but are told that allowing the device to be used is a condition of our employment. Why would we still object?

A utilitarian philosopher might answer that such a situation would make us unhappy for various reasons. It would lead to a general apprehensiveness about how our every word might be interpreted by those who could hear our conversations. Thus our conversational spontaneity would be inhibited. Certainly the use of such a device would destroy any sense of being trusted by our employer. And it would open the door to innumerable abuses, such as harassment, by those who might have access to what the device recorded or picked up.

A duty ethicist might argue that use of the device would violate the duty to respect people. Respect for people entails allowing them some degree of control over who has knowledge about their personal conversations (Benn, 1971, 8–9; Reiman, 1979, 387). It also means the duty to allow others the pursuit of intimate relationships—friendships, trust-relationships, etc.—would be harmed by the use of such devices (Fried, 1970, 137–152). Intimacy involves selectively revealing information about ourselves which we otherwise keep secret, and revealing it in the belief that it will not be used to harm us. We would not reveal as much if we knew others were listening. And in the work environment, a climate of fear would prevent even the normal jokes between colleagues about bosses which help contribute to comradeship.

Finally, a rights ethicist would appeal directly to the human right of personal freedom: People should be free to maintain some control over what personal information about themselves is revealed to others. Denying anyone this freedom destroys a rich dimension of choice in expressing oneself and developing personal relationships.

The right to privacy is limited by the legitimate exercise of employers' rights to obtain and use information necessary for the effective managing of an organization. These rights make it legitimate to employ aptitude and skill tests related to job functions. But they do not make it legitimate to use general personality and intelligence tests which have not been established as essential for measuring job performance.

Once gathered, information about employees should be reserved solely for legitimate employer use. It is not permissible to give it to outsiders, or even to members of the corporation who do not need to know it. The personnel division, for example, needs medical and life insurance information about employees, but immediate supervisors usually do not. Only the most routine information about job position and years with the company should be revealed to inquirers from outside the corporation, such as collection agencies, insurance companies, apartment landlords, and private investigators. Such protective procedures can be simply and effectively implemented. IBM led the way years ago in voluntarily initiating such a program, and that program is serving as a model for other corporations (Ewing, 1976, 82).

Employers should be viewed as having the same fiduciary or trust relationship to their employees concerning confidentiality that doctors have to their patients and lawyers have to their clients (Mironi, 1974, 289). In all of these cases personal information is given in trust on the basis of a special professional relationship. Moreover, with rare exceptions involving identification of other parties, employers owe employees the right to examine their dossiers so as to correct outdated or erroneous information.

Due Process

Rights of conscience, free speech, outside activities, and privacy would be of little help to employees were they not given institutional recognition. Institutional recognition includes formal endorsement of the rights, in company policy statements or employment contracts, for example. But it also requires creation of an institutional procedure for protecting those who exercise the rights. Thus, the substantial rights discussed so far imply a right to due process—that is, a right to fair procedures safeguarding the exercise of other rights. The right to due process extends to fair procedures in firing, demotion, and disciplinary actions.

Implementing the right to due process involves two general procedures: First, written explanations are owed to employees who are discharged, demoted, transferred to less enriching work, or in other ways penalized.

Second, an appeals procedure should be established which is available to all employees who believe their rights have been violated. The procedure should be a stable part of the organization, effective, equitable, and efficient. For the sake of both management and employees, it must be easy to use and work quickly, generally yielding a verdict within days after a grievance is filed (Ewing, 1977, 155–174).

Government employees and union members generally have some such procedures available to them, however flawed they may be in practice. Private companies have recently developed a variety of promising procedures, some of which are still largely experimental. Polaroid, for example, has set up a grievance committee composed of members elected by employees. Xerox has a formal employee advisory board. General Electric uses an impartial referee. Some corporations have ombudspersons who hear and investigate complaints.

The power of these various appeals groups and people is limited, but it can be substantial. It is limited because it typically involves making recommendations to top management, without issuing final verdicts binding on management. Binding verdicts, by definition, would remove substantial decision-making authority from management. The ability to issue them would in effect create a new line of management authority. Yet the influence of appeals bodies can be significant: Where employees trust them, an overriding of their decisions by management without very compelling reasons can cause serious personnel problems.

Study Questions

1 Early in 1970 a proposal was made to prohibit all non-American-made cars from parking inside the gates of a major American steel production facility. The proposal grew out of the negative impact imported steel products were having on the domestic steel industry. The rule was to apply to all employees, including management (Barry, 1986, 164–165).

Present and defend your view as to (*a*) whether this proposal should have been approved and (*b*) whether such issues should be decided by a majority vote among employees or in some other way.

2 In 1974 Combustion Engineering required its officers and employees to take lie detector tests. Someone had given confidential information concerning the terms of its nuclear power contracts to the *Wall Street Journal*, and management felt justified in finding out who it was (Ewing, 1977, 131). Explain your view as to whether or not this tactic was justified, or whether it might have been justified depending on further details of the case.

3 Explain and defend your views concerning examples 1 through 6 in the above section on the right to privacy. Was management in each case justified in doing what it did? Can you formulate any general guidelines on the basis of the examples?

4 Some observers have argued that employees have a right to choose the type of clothing and hairstyles they wear to work. The courts, however, have ruled that employers have the right to set reasonable standards, specifically in instances where corporate image and job function are affected. Discuss this issue, defending your own view by reference to utilitarian, duty-based, or rights-based ethical theories.

5 The chairman of the board of directors of a company whose main business comes from government military contracts sends a letter to all employees. The letter, written on company stationery, outlines why it is clearly in the interests of the company that a certain promilitary senator be elected. The letter concludes by stating that while no employee is required to make a campaign donation to the senator, those wishing to do so can have the donation deducted from their next paycheck. Is there anything objectionable about this? Are any rights infringed?

6 Consider the following example:

> Several top executives of a company in a large city are disturbed by community activist organizations that are protesting the treatment of minority groups. The executives feel that the activists, while abstaining from violence, are doing more harm than good to the schools, urban renewal programs, public transportation, and retail business. The chief executive himself has articulated his fears about the activists at local business meetings. However, a young... [engineer working for

> the company] thinks that the activists are on the right track. He goes to work for the leading activist organization, spending many evening and weekend hours doing unpaid volunteer tasks. Occasionally he is quoted in the newspaper and identified with the employer company (Ewing, "What Business Thinks about Employee Rights," 31).

Does the engineer have a right to continue his activities without interference from the employer? Or do the executives have a right to tell him to stop the activities or else be fired?

DISCRIMINATION

Perhaps nothing is more demeaning than to be downgraded for one's sex, race, skin color, age, or religious outlook. These aspects of biological makeup and basic conviction lie at the heart of self-identity and self-respect. Such downgrading is especially pernicious within the work environment, for work is itself fundamental to a person's self-image. Accordingly, human rights to fair and decent treatment at the workplace and in job training are vitally important.

Yet there are challenging moral issues concerning those rights. One concerns the role of government. Should government and the law be used within private enterprise to oppose discrimination? Or do corporations have the right to hire whomever they please in the search for profits and economic efficiency?

A second set of issues concerns the appropriate extent of the right to nondiscrimination. For example, do women and minorities who in the past have been discriminated against have the right to be given preferential treatment in entrance to educational programs, in hiring, and in job retention? Or does such preference violate the right to equal opportunity enjoyed by members of majority groups and thus itself constitute discrimination "in reverse"?

These issues are of significance to engineers. Traditionally, engineering in the United States has been one of the more open avenues for upward mobility of capable but otherwise disadvantaged persons. Partly this has to do with the willingness of engineers to recognize talent where they see it, partly it is the result of government-mandated fair employment practices which must be observed by employers engaged in government contracts. But in times of economic downturns the influx of women, racial minorities, and foreign nationals into the engineering workforce can produce unfavorable reactions. Should there be any conflicts, engineers must carefully examine the ethical bases of their and their colleagues' actions.

Examples

Consider the following examples:

1 An opening arises for a chemical plant manager. Normally such positions are filled by promotions from within the plant. The best qualified person in terms of training and years of experience is a black engineer. Management believes, however, that the majority of workers in the plant would be disgruntled by the appointment of a nonwhite manager. They fear lessened employee cooperation and efficiency. It is decided to promote and transfer a white engineer from another plant to fill the position.

2 Several women engineers work in the sales division of an electronics company. The company prides itself on the practice of hiring proportional numbers of women. Yet the pay scale for these women is systematically lower than that for men having comparable experience and engaging in comparable work. In the absence of objections, management assumes that the women recognize the necessity to pay males more because they are the primary breadwinners for their families.

3 A farm equipment manufacturer has been hit hard by lowered sales caused by a flagging produce economy. Layoffs are inevitable. During several clandestine management meetings, it is decided to use the occasion to "weed out" some of the engineers within 10 years of retirement in order to avoid payments of unvested pension funds.

Definition

The word "discrimination" is used in several senses. Sometimes it means preference on the grounds of sex, race, etc., whether or not such preference is viewed as justified. In everyday speech, however, it has come to mean *morally unjustified treatment of people on arbitrary or irrelevant grounds*. We will use the word in this latter sense. Thus to call something "discrimination" is to condemn it. Where the question of justification is left open for discussion, we will speak of *preferential treatment*.

Most of us would agree that the preceding examples involve discrimination. They also involve violation of antidiscrimination laws. Let us review some of these laws before inquiring into whether they are morally warranted.

Antidiscrimination Laws

The forerunner of antidiscrimination laws was the principle of equality which the Fourteenth Amendment embedded in the Constitution following the Civil War (1868):

> No State shall make or enforce any law which shall abridge the privileges or immunities of citizens of the United States; nor shall any State deprive any person of life, liberty, or property, without due process of law; nor deny to any person within its jurisdiction the equal protection of the laws.

But it was only as recently as 1964 that discrimination by public or private employers was explicitly prohibited legally in the Civil Rights Act:

> It shall be an unlawful employment practice for an employer to fail or refuse to hire or to discharge any individual, or otherwise to discriminate against any individual

> with respect to his compensation, terms, conditions, or privileges of employment, because of such individual's race, color, religion, sex, or national origin (Title VII—Equal Employment Opportunity).

The Equal Employment Opportunity Act of 1972 amended and strengthened the Civil Rights Act by giving greater powers of enforcement to the Equal Employment Opportunity Commission.

Several supporting Executive Orders, having the force of law, were issued under Presidents Kennedy and Johnson. They required that businesses receiving government contracts develop affirmative action programs to remedy underrepresentation of women and minorities on their staffs. "Underrepresentation" usually meant that the percentage of women and minority workers hired for a given type of job did not roughly parallel the percentage of those available locally to be hired.

As originally mandated, affirmative action programs had a dual emphasis. On the one hand, they sought to increase the number of women and minority applicants for jobs and professional education. This was to be accomplished largely through wide advertising of the positions and opportunities involved, assuring applicants that they would be considered on a nondiscriminatory basis. On the other hand, the programs called for setting concrete goals for hiring, including timetables for achieving racial or sexual job parity. Explicit numerical quotas, however, were not endorsed. Reaching those goals usually meant hiring women and minorities over equally qualified white males. In practice, government pressure also sometimes led to giving them preference over better-qualified white males, although this was not justified by the law.

Finally, age discrimination was prohibited by the Age Discrimination in Employment Act of 1967, amended in 1974. That act states:

> Sec. 4. (a) It shall be unlawful for an employer—(1) to fail or refuse to hire or to discharge any individual or otherwise discriminate against any individual with respect to his compensation, terms, conditions, or privileges of employment, because of such individual's age;
>
> (2) to limit, segregate, or classify his employees in any way which would deprive or tend to deprive any individual of employment opportunities or otherwise adversely affect his status as an employee, because of such individual's age (quoted in Sethi, 1977, 334).

Moral Justification of Nondiscrimination Laws

The equal opportunity laws forbid the kinds of discimination involved in the examples given earlier. But are they morally justified? Should the government be allowed to use law to force private corporations to treat people equally?

The libertarian (or conservative free market) position answers these questions in the negative. It argues that government should not meddle in private economic dealings except where necessary to protect contracts and free competition. According to this view, all fair employment legislation is unjust. As Milton Friedman wrote in *Capitalism and Freedom:* "Such legislation clearly involves interference with the freedom of individuals to enter into voluntary contracts with one another" (Friedman, 1962, 111). Forcing a grocer to hire a black clerk in a neighborhood which hates blacks is, he says, unfairly hurting the grocer's business. In general, fair employment legislation lessens economic efficiency and violates the right of businesses to hire whomever they please.

Friedman adds that he personally finds racism and sexism deplorable. But he regards this merely as his personal preference, and insists that neither he nor the majority has a right to force their preferences on others. "It is hard," he writes, "to see that discrimination can have any meaning other than a 'taste' of others that one does not share" (Friedman, 1962, 110).

Friedman denies there are moral rights bearing on employment which are not reducible to contracts and to the conditions which make contracts possible. In our view this is a serious mistake. It is not a mere personal preference to condemn racist and sexist practices. Laws forbidding prejudicial employment practices are not a matter of a majority imposing its tastes on others. For those laws protect the basic moral rights everyone possesses to have a fair opportunity of working to obtain social benefits.

Those rights are not absolute in the sense that nothing could ever override them. Presumably if utter economic disaster resulted from recognizing them, they might have to be somewhat restricted for a time. But it is simply not true that fair employment legislation has drastically disrupted the economy. The laws have placed some new restrictions on the exercise of the economic right to pursue profits. But the vital importance of being able to pursue work and careers without crippling interference from discriminatory practices seems a sufficient reason to justify such restrictions.

Let us now turn to the issues surrounding reverse preferential treatment. These are more troublesome because at least on the surface preferential treatment is a violation of the very concept of equality used to condemn past racist and sexist practices.

Preferential Treatment

Hiring a woman or a member of a minority over an equally qualified white male is only one form of reverse preferential treatment. Let us call it the *weak* form. The *strong* form, by contrast, consists in giving preference to women or minorities over better-qualified white males. The strong version, of course, is the more highly controversial one. It is one thing to give preference by tipping an equally balanced scale. It is quite another to load the scale from the outset. The following discussion is focused upon strong preferential treatment, as are most current debates over reverse discrimination.

Reverse discrimination, as we shall define it, occurs when preference is given to a member of a group which in the past has been the object of discrimination. Typically this occurs when preference is given to a woman or a member of a previously downgraded minority at the expense of a white male. The question is whether or not it is unfair or otherwise unjust to give preference to some less-qualified women or minorities.

The Bakke Decision Reverse discrimination in regard to education received national attention at the time of the Supreme Court's 1978 Bakke decision. Allan Bakke, a white engineer, was twice denied admission to the medical school at the University of California, Davis. His grades and scores on entrance exams were significantly higher than those of most of the minority students admitted under a special admissions program. Of one hundred openings, the program reserved sixteen for Blacks, Chicanos, Asians, and American Indians. Minority applicants for those sixteen positions were judged only against each other, and not by comparison with applicants for the other, unreserved openings. Bakke went to court, charging that he had been unfairly discriminated against in violation of the Civil Rights Act.

The University of California at Davis Davis attempted to defend its special admissions program on four grounds: (1) it increased the number of minorities in the medical profession; (2) it countered the bad effects of past racial discrimination; (3) it increased the number of physicians likely to serve in inner cities and other areas underserved medically; and (4) it created a mixed student body, which in turn widened the educational benefits available to students thus exposed to a greater intellectual and social diversity.

In a split vote (5 to 4), the Supreme Court ruled that the first three grounds did not justify treating candidates differently because of race. It also ruled, however, that the last ground—the educational importance of a mixed student body—did justify counting race as one relevant consideration in screening candidates.

Nevertheless, it agreed that the U.C. Davis quota system unfairly discriminated against Bakke and other whites. For by reserving certain positions for minorities, the Court argued, it prevented complete comparisons being made among all applicants. Thus Bakke won his case and was allowed to enter the medical school (and has since then been graduated).

The Weber Decision One year after the Bakke case, the Supreme Court made a ruling which to some people has seemed incompatible with both the Civil Rights Act and the Bakke condemnation of quota systems. The second case concerned a job-training program at the Gramercy, Louisiana, plant of Kaiser Aluminum and Chemical Corporation. The area surrounding Gramercy has a 40 percent black population. Yet only a small number of skilled crafts workers at the Kaiser plant were black. This was due to the unavailability of skilled black workers, not to any policy of discrimination at the Kaiser plant.

In 1974 Kaiser and the United Steelworkers of America entered into a collective bargaining agreement to give preference to black applicants for job-training programs. Half the positions in these programs were to be reserved for black workers. In addition, a previous eligibility requirement of work experience was abolished. Seniority counted for admission to the programs, but two separate seniority lists were formed, one for black and one for white workers. This policy of preferential treatment was to continue until the percentage of skilled black workers employed by the plant was roughly equal to 40 percent of the skilled work force.

That same year Brian Weber, a white worker, was denied entrance to one of the programs. Weber had worked at the plant since 1969 and had more seniority than some of the black workers admitted. He filed a class action suit against Kaiser, alleging a violation of Title VII of the Civil Rights Act.

A literal reading of the Civil Rights Act would condemn the Kaiser job training program:

> It shall be an unlawful employment practice for any employer, labor organization, or joint labor-management committee controlling apprenticeship or other training or retraining, *including on-the-job training programs,* to discriminate against any individual because of his race, color, religion, sex, or national origin in admission to, or employment in, any program established to provide apprenticeship or other training [42 U.S. Codes 2000e-2 (d) (1970), emphasis added; quoted in C. Cohen, 1981, 375].

Yet the Supreme Court ruled that the literal reading was not the correct one. The justices wrote, "It is a 'familiar rule, that a thing may be within the letter of the statute and yet not within the statute, because not within its spirit, nor within the intention of its makers.'" They contended that the intent of both the Civil Rights Act and the Kaiser program was to eliminate traditional patterns of racial segregation and inequality, and on that basis ruled against Weber (443 U.S. 193 [1979]).

Let us now sketch a few of the main arguments used to justify strong preferential treatment of women and minorities, then some used to prove that it is unjustified and should be considered in effect "reverse discrimination."

Arguments For A rights-ethics argument favoring strong preferential treatment emphasizes the principle of compensatory justice: Past violations of rights must be compensated, where possible "in kind." Taking property from others minimally requires returning it. Similarly, members of groups who have suffered job discrimination in the past are owed special advantages in obtaining jobs today. Ideally such compensation should be given to *individuals* who in the past were denied jobs. But the costs and practical difficulties of determining such discrimination on a case by case basis through the job-interviewing process and the legal system force a more global approach. Preference, therefore, is to be given on the basis of membership in a group which has been disadvantaged in the past.

Those utilitarians who favor preferential treatment have additional arguments. They point to the importance of integrating women and minorities into the economic and social mainstream. Only thus can the benefits of harmony between the races and sexes be achieved. And only thus can our society benefit from the resources of individuals whose potential has been negated in the past. Given the central role work plays in building self-identity and self-esteem, in allowing people to develop their resources to the greatest extent possible, and in providing role models, therefore, preferential treatment in job hiring is seen as the best way to attain these goals.

Arguments Against There are also forceful arguments against strong preferential treatment. Such preference, it can be argued, is a straightforward violation of other people's rights to equal opportunity. Just as the rights of minorities and women to such equality were regularly violated prior to enactment of the Civil Rights Act, the rights of white males are now being violated whenever their job qualifications are disregarded because of their race and sex. In general, the argument goes, whatever made discrimination against disadvantaged groups unfair in the past also makes such "reverse discrimination" unfair in the present.

Opponents of preferential treatment for minorities often grant that past violations of rights may call for compensation. But they view general policies of hiring the disadvantaged at the expense of equally or even better qualified white males as an improper way of providing that compensation. Most white males have not themselves participated in discriminatory job actions, and to ignore their rights to fair employment opportunity amounts to compensating victims by punishing the innocent. Two wrongs cannot make a right.

Thus blanket compensation to all members of a group, it is argued, should be given at most in the form of special early education and social programs designed to provide the necessary training to yield a fair opportunity later in applying for jobs or professional schools. Even here, however, economically deprived whites should not be excluded from participation.

Utilitarians who are against strong preferential treatment have additional arguments. They say that the harm such policies do goes beyond that involved in violating the job rights of white males. For example, there is the intense resentment generated among white males and their families. Those feelings only intensify racial tensions and ultimately work against the goals of integration. Moreover, preferential treatment subtly but insidiously encourages traditional stereotypes: A sense of inferiority may arise in women and minorities who come to feel they cannot make it on their own without special help. Finally, there is the economic harm that results from a policy of not consistently hiring the best qualified person.

Intermediate Positions Recently various attempts have been made to develop intermediate positions sensitive to all the above arguments for and against strong reverse preferential treatment. Two of these will be mentioned here.

The first rejects all blanket preferential treatment of special groups as inherently unjust and a violation of the right to equal treatment of other groups not so treated. Hiring by competence is seen as vital to the well-being of society. Yet the principle of compensatory justice as applied to individuals is also vitally important. If it can be shown that a given company has discriminated against specific individuals, then that company is obligated—and should be required—to give the next available jobs to those individuals. It is true that this might involve discrimination against other, perhaps better-qualified, applicants for those jobs. But the principle of compensatory justice, so the argument goes, in effect automatically closes new jobs to any further applicants beyond the individuals previously discriminated against (Goldman, 1979, 120–127).

In contrast, the second view seeks to justify preferential treatment of special groups. It contends that racist and sexist attitudes are still widespread, at least at a visceral or gut level. Mere affirmative action programs are not sufficient to counterbalance the subtle impact of these attitudes on employment practices. The only adequate way to provide such a counterbalance is to allow and encourage strong preferential treatment. Admittedly some violations of other people's rights to equal employment opportunity will occur in the process, at times with tragic results. But that is the necessary evil we must live with to remedy the deeper tragedy resulting from racism and sexism (Beauchamp, 1983, 625–635).

These two intermediate views, together with the preceding pro and con arguments, make it clear that the issues surrounding strong preferential treatment are subtle and complex. Compelling arguments exist on both sides. In resolving the issue, rights ethicists must identify a reasoned perspective from which to weigh rights to equal treatment against rights to compensation. Duty ethicists must balance duties to treat people equally with duties to provide compensation for past wrongs. Utilitarians must struggle to find a proper way to sum up the positive and negative consequences that can result from such policies. And since nearly all of us will at some time or other be involved in a situation involving preferential treatment, each of us individually needs to find a way to balance these considerations in a carefully reasoned manner.

Sexual Harassment

Sexual harassment is a particularly invidious form of sex discrimination, involving as it does not only the abuse of gender roles and work-related power relationships, but the abuse of sexual intimacy itself. The following discussion focuses on the most widespread type of sexual harassment: male harassment of females. And in regard to the field of engineering, the female may be an engineer, a technician, or a secretary, and the male may be, for example, an engineer-manager or an engineer-colleague.

The term "sexual harassment" is currently applied to a wide variety of physical and psychological attacks, coercion, and sexual practices. One definition of it as applied to women is: "any sexual oriented practice that endangers a woman's job—that undermines her job performance and threatens her economic livelihood" (quoted by Backhouse, 1981, 32). Another definition is: "the unwanted imposition of sexual requirements in the context of a relationship of unequal power" (MacKinnon, 1978, 1).

Sexual harassment may come in many forms: (1) Following an interview for a job as a secretary, a woman is told that the job is hers if she is willing to grant sexual favors to the interviewer. (2) A woman is told by her superior that she will have first priority for receiving a promotion if she is "nice" to him, and talk of a motel makes it clear what is meant by the term "nice." When she refuses to be that "nice," she is not given the promotion and thereafter is assigned less challenging work. (3) Against her will, a woman is grabbed and kissed by her employer, who had asked her to stay after hours at work. She resists and is fired the following day. (4) A woman turns down her boss's request for a date. She makes it clear she is not interested in going out with him ever, but to her chagrin he continues repeatedly to ask her out during the following weeks. (5) The male colleagues of a woman continually leer at her and make sexually suggestive comments about her clothing and body. (6) A male engineer enjoys telling his secretary about his sex life, disregarding her protests against hearing about it.

Although there is evidence that sexual harassment is a widespread and serious problem, the courts have only recently begun to take action. Since 1976 there have been a series of rulings which recognize at least two forms of sexual harassment as instances of sex discrimination prohibited by the Civil Rights Act: first, where a supervisor requires sexual favors as a condition for some employment benefit (a job, promotion, raise); second, where there is job-related retaliation by an employer or supervisor when a sexual request is refused. However, a third, more prevalent, form of sexual harassment has yet to be given serious attention in court rulings: that is, where the harassment functions as part of the everyday work environment among coworkers—for example, in situations where women must put up with repeated, unwanted sexual proposals or lewd comments (MacKinnon, 1978, 57–82).

What is morally objectionable about sexual harassment at the workplace? Any answer must take into account the generally inferior economic status of women. Such harassment takes advantage of that condition, and it is no surprise that it is directed mostly toward secretaries, clerical workers, and other low-paid female workers. As with rape, sexual harassment is a display of power and aggression through sexual means. Accordingly it has appropriately been called "dominance eroticized" (MacKinnon, 1978, 162).

Insofar as it involves coercion, sexual harassment constitutes an infringement of one's autonomy to make free decisions concerning one's body. But whether or not coercion and manipulation are used, it is an assault on the victim's dignity. In abusing sexuality, such harassment degrades people on the basis of a biological and social trait central to their sense of personhood.

Thus a duty ethicist like Kant would condemn it as violating the duty to treat people with respect, to treat them as having dignity and not merely as means to personal aggrandizement and gratification of one's sexual and power interests. A rights ethicist would see it as a serious violation of the human right to pursue one's work free from the pressures, fears, penalties, and insults that typically accompany sexual harassment. And a utilitarian would emphasize the impact it has on the victim's happiness and self-fulfillment, and on women in general.

Study Questions

1 Present and defend your view as to (*a*) whether or not weak preferential treatment of minorities is ever justified; (*b*) whether or not strong preferential treatment of minorities is ever justified; (*c*) whether or not government intervention in the form of enforcing laws like the Civil Rights Act is morally justified.

2 In the early 1960s Motorola screened some 20,000 job applicants each year. In order to increase the efficiency of the hiring process and decrease the costs of screening, the company administered a 5-minute technical test. The Illinois Fair Employment Practices Commission charged that the test was discriminatory against black applicants. Consider the following facts known about the test: (*a*) The same test was used for both whites and blacks, but the score considered passing was based upon an original test-standardization group which was predominantly white. (*b*) There was a proven correlation between general technical ability and high scores on the test. (*c*) While the test was felt to be a generally reliable indicator for technical trainability, it was known that at least some qualified applicants had been ruled out on the basis of their scores after taking it. (*d*) There was no evidence either way concerning whether job performance of blacks was more erroneously predicted by results of the test than job performance of whites (Garrett, 1968, 47–50).

Would any of these facts show the test to be unfairly discriminatory? What further facts would be relevant in determining the answer to this question?

3 While engineering remains one of the most male-dominated professions, strong efforts have been made to encourage women to enter it. About 14 percent of entering engineering classes are composed of women. The salaries of beginning women engineers are on the average slightly higher than those of men (Vetter, 1980, 29–30). Yet women engineers are often subjected to greater pressures than males, both in school and on the job (Davies, 1981, 32). Discuss why you think that is so and what might be done about it.

4 Imagine two applicants for a construction supervisor's position. One is a 55-year-old white male engineer. The second is a 30-year-old white male engineer. Both have sufficient professional credentials for the job, but the younger man has fewer years of work experience. The 30-year-old man is hired.

Redescribe the example more fully (without contradicting the given information) in a way that would make it evident that unfair age discrimination was involved in the hiring decision. Then embellish on the example once more (again without contradicting the facts as given) in a way which would indicate that the correct choice was made in a nondiscriminatory way.

5 A company advertises for an engineer to fill a management position. Among the employees the new manager is to supervise is a woman engineer, Ms. X, who was told by her former boss that she would soon be assigned tasks with increased responsibility. The prime candidate for the manager's position is Mr. Y, a recent immigrant from a country known for its confining roles for women. Ms. X was alerted by other women engineers to expect unchallenging, trivial assignments from a supervisor with Mr. Y's background. Is there anything she can and should do? Would it be ethical for her to try to forestall the appointment of Mr. Y?

SUMMARY

Salaried engineers have several overlapping types of moral rights:

1 Human rights—possessed by virtue of being people or moral agents
Examples:
Fundamental right to pursue legitimate interests
Right to make a living

2 Professional rights—possessed by virtue of being professionals having special moral responsibilities
Examples:
Basic right of professional conscience (the right to exercise professional judgment in pursuing professional obligations)
Right to refuse to engage in unethical activity
Right to express one's professional judgment, including the right to dissent
Right to warn the public of dangers
Right to fair recognition and remuneration for professional services

3 Employee rights—rights which apply or refer to the status of employees
a Contractual—arising solely out of an employee contract
Example: Right to receive a salary of a certain amount
b Noncontractual—existing even if not formally recognized in a contract or company policy
Examples:
Right to choose outside activities
Right to privacy and employer confidentiality
Right to due process from employer
Right to nondiscrimination and absence of sexual harassment at the workplace

Professional and employee rights can be justified by reference to ethical theories. For example, a rights theory would derive the right of professional conscience from a fundamental human right to pursue legitimate interests, where such interests include moral obligations. A duty theory might appeal to the fundamental human duty employers have not to harm others (e.g., the public) by handicapping engineers seeking to meet their professional obligations. A utilitarian theory would argue that the greatest good is promoted by allowing engineers to pursue their obligations. In general, the importance of professional *duties* means that the importance of the *right* to meet those duties must be recognized.

Whistle-blowing most often means intentionally conveying new information outside approved organizational channels or against a supervisor's orders with the aim of drawing attention to a significant moral problem. In *external whistle-blowing* the information is passed outside the organization, while in *internal whistle-blowing* it is conveyed to someone within the organization. In *open* whistle-blowing the whistle-blowers identify themselves, while in *anonymous* whistle-blowing they do not.

Whistle-blowing concerning safety is morally *permissible* in those situations where the problem involved is serious and the whistle-blowers have first made reasonable attempts to warn others about it through regular organizational channels. It is (prima facie) morally *obligatory* where, in addition, those blowing the whistle are certain they could convince reasonable observers that their views are right and company policy wrong and where there is strong evidence that going public with their information will lead to positive remedies. While in the past the fate of whistle-blowers has usually been unhappy, current laws, government policy, and responsible management are moving in the direction of an increased recognition of a limited whistle-blowing right.

Both employee rights and professional rights must be subordinate in some respects to the rights of employers to promote company interests. For example, the right to privacy is limited by the need employers have to acquire relevant information about employee skills. But not just any company interest can override employee rights. This is especially clear in regard to the right not to be discriminated against because of sex, race, age, or religion.

Contemporary disagreements over how to deal with discrimination center on the issue of reverse preferential treatment. *Weak preferential treatment* involves giving an advantage to members of traditionally discriminated-against groups over equally qualified applicants who are members of other groups. *Strong preferential treatment* involves giving preference to minority applicants or women over better-qualified applicants from other groups. Arguments for and against such treatment focus on the right to equal employment opportunity, the right to receive and duty to give compensation for past wrongs, and the best way to achieve social integration.

PART IV

GLOBAL AWARENESS AND CAREER CHOICE

We have not been seeing our Spaceship Earth as an integrally-designed machine which to be persistently successful must be comprehended and serviced in total.

Buckminster Fuller

We must hold a man amenable to reason for the choice of his daily craft or profession. It is not an excuse any longer for his deeds, that they are the custom of his trade. What business has he with an evil trade?

Ralph Waldo Emerson

The rate of progress is such that an individual human being, of ordinary length of life, will be called upon to face novel situations which find no parallel in his past. The fixed person for the fixed duties, who in older societies was such a godsend, in the future will be a public danger.

Alfred North Whitehead

CHAPTER **7**

GLOBAL ISSUES

On December 3, 1984, the operators of Union Carbide's plant in Bhopal, India, became alarmed by a leak and overheating in a storage tank. The tank contained methyl isocyanate, a toxic ingredient used in pesticides . Within an hour the leak exploded in a gush that sent 40 tons of deadly gas into the atmosphere. The result was the worst industrial accident in history: nearly 3,000 people killed, 10,000 permanently disabled, and 100,000 others injured (Shrivastava, 1987).

The disaster was caused by a combination of extremely lax safety procedures and gross judgment errors by plant operators, as we shall see in more detail later. Union Carbide had transferred legal responsibility for safety inspections to the overseas operators, even though it retained general management and technological control. Was this legal transfer morally responsible? Should Carbide have done more to ensure safety standards in dealing with such dangerous material? In general, what are the responsibilities of engineering corporations and engineers doing business in foreign countries?

The word "global" in the title of this chapter refers both to the international context of engineering and to the increasingly wide social dimensions of engineers' work. As responsible social experimenters, engineers need to take these dimensions into account in making engineering decisions and career choices. In this chapter, we will explore these dimensions by discussing four topics: multinational business, the environment, computer ethics, and weapons development.

MULTINATIONAL CORPORATIONS

Multinational corporations do extensive business in more than one country. For example, Union Carbide in 1984 operated in 38 countries, and it is only a medium-sized "giant" corporation, ranking thirty-fifth in size among U.S. corporations. Generally multinationals establish foreign subsidiaries, such as Union Carbide of India, retaining 51 percent of the stock and allowing foreign investors to own the remainder.

The benefits to U.S. companies of doing business in less-developed countries are clear: inexpensive labor, availability of natural resources, favorable tax arrangements, and fresh markets for products. The benefits to the participants in developing countries are equally clear: new jobs, jobs with higher pay and greater challenge, transfer of advanced technology, and an array of social benefits from sharing wealth.

Yet moral difficulties arise, along with business and social complications. Who loses jobs at home when manufacturing is taken "offshore"? What does the host country lose in resources, control over its own trade, and political independence? And what are the moral responsibilities of corporations and individuals operating in less-developed countries? This last is the question we shall focus on here.

Three Senses of "Relative"

According to one popular cliché, deciding how corporations and individuals ought to act in foreign countries is not too difficult. "Values are relative, so when in Rome do as the Romans do." However, the claim that values are relative is ambiguous, and it may be true or false depending on what is meant.

There are many versions of relativism, depending upon how, and with respect to what, values are supposed to be relative (Taylor, 1975 13–30). Here are three versions.

Ethical conventionalism: Actions are morally right when (and because) they are approved by law, custom, or other conventions.

Descriptive relativism: Value beliefs and attitudes differ from culture to culture.

Moral relationalism: Moral judgments should be made in relation to factors that vary from case to case, usually making it impossible to formulate rules which are both simple and absolute (i.e., exceptionless).

The first version of relativism was discussed in Chap. 2. As we said there, that view is clearly false since it implies absurdities. It would justify government-supported cruelty, such as practiced in Nazi Germany, Stalinist Russia, and nineteenth-century U.S. slave states. Laws and conventions are not morally self-certifying. Instead, they are always open to criticism in light of moral reasons concerning human rights, the public good, duties to respect people, and virtues.

The second version, Descriptive Relativism, is obviously true. It merely says there are differences between the moral beliefs and attitudes of various cultures. Descriptive Relativism does not entail Ethical Conventionalism, as is sometimes thought. Mere diversity in what people believe hardly shows there is no objective truth—truth which holds across cultural boundaries. The fact that nearly everyone once believed the earth to be flat is irrelevant to determining the earth's actual shape. Likewise, the mere fact that other cultures accept moral beliefs differing from our own does not mean there is no objective truth which can be used in evaluating customs and laws.

Early cultural anthropologists tended to overemphasize the extent of moral differences between cultures. Preoccupied with exotic practices like head-hunting, human sacrifice, and cannabalism, they moved too quickly from "Moral views differ greatly" to "Morality is simply what a culture says it is." More recent anthropologists have drawn attention to underlying similarities between cultural perspectives. They have noted that virtually all cultures show some commitment to promoting social cooperation and to protecting their members against needless death and suffering. And beneath moral differences often lie differences in circumstances and in beliefs about facts, rather than differences in moral attitude.

For example, the Eskimos and some American Indians used to leave their elderly to die. At first glance this seems radically inconsistent with our notions of respect for human life. Yet the Eskimos and Indians lived under harsh conditions requiring migration and rationing if the group was to survive. In such circumstances the elderly would voluntarily agree to abandonment out of respect for the group and a wish not to be a burden to their families. Or again, the Aztecs' practice of human sacrifice seems a sign of cruelty and lack of concern for life. But a deeper examination reveals that they believed their gods required such sacrifice to ensure the survival of their people as a whole, and it was considered an honor for the victims.

Most of us will also readily endorse the third version of relativism, Moral Relationalism. It is a reminder that moral judgments are made *in relation to* a wide variety of variable factors, and that as a result it is usually impossible to formulate simple and absolute moral rules. A rule like "Lying is immoral," for example, has many exceptions. Special circumstances can arise where people have to lie in order to save a life or to protect their privacy from mischievous intruders.

Virtually all philosophers have accepted Moral Relationalism (with the exception of Kant and some divine command ethicists). Contemporary duty and rights ethicists, for example, emphasize how everyday rules such as "Don't lie" and "Keep promises" can have valid exceptions when they conflict with other rules (creating moral dilemmas). To respect people requires being sensitive to special circumstances. Utilitarians emphasize that in assessing good and bad consequences attention must be paid to the context of conduct. And virtue ethicists stress the role of practical wisdom in identifying facts relevant to assessment of virtuous conduct.

Relationalism readily allows that the customs of cultures may be morally pertinent considerations that require us to adjust moral judgments and conduct. For example, we should remove our shoes before entering the home of a traditional Japanese family since in that context that is a sign of respect, even though it is not in other situations.

Relationalism is important in multinational engineering contexts involving different conventions encountered around the world. However, it is also important to realize that Relationalism only says that foreign customs may be morally *relevant*. It does not say they are always *decisive* in determining what should be done. This crucial difference sets it apart from Ethical Conventionalism.

"When in Rome"—or South Africa

Which standards should guide engineers' conduct when working in foreign countries? Ethical Conventionalism supports the maxim, "When in Rome do as the Romans do." That is, it would have us believe there is no real problem: One should just drift with the conventions dominant in the local area. An equally extreme opposite view would have one retain precisely the same practices endorsed at home, never making any adjustments to a new culture. Both of these choices, however, seem to us unacceptable.

Consider the choice faced by engineering firms and other companies doing business in Italy, where they are confronted with a tax system which differs strikingly from the American one. According to one account, corporations routinely submit a tax return showing profits of only 30 to 70 percent of their actual levels (Kelly, 1983, 37–39). Government officials generally estimate any corporation's taxes at several times over what that corporation claims it owes. Then the two sides negotiate the final figure as a compromise. The Italian revenue agents take as their "fee" an unreported cash payment directly from the representative for the corporation. The size of this payment influences how much or how little the revenue agent will demand in taxes for the Italian government.

Anything like such a practice, of course, would be considered corrupt in the United States. It involves misrepresenting actual profits, secret payments, and an open door to biased tax assessments. Some would describe it in terms of deception, dishonesty, bribes, and conflicts of interest. Yet companies doing business in Italy that accurately report their profits would be assessed taxes much higher than they deserve. It would simply be unworkable to try doing business while refusing to participate in the practice.

Of course such "pragmatic" considerations concerning workability do not automatically justify participation in a practice. But several features should be noted.

First, the practice is both officially sanctioned by the government and also tolerated by the Italian people. A shared understanding exists as to its legitimacy within Italy. Second, in some respects the practice is analogous to bargaining in the market square and bluffing in poker. It could at least be argued that sometimes deception is permissible when all participants accept it and no serious moral harm comes from it. Third, and perhaps most important, no basic human rights are violated by the practice. To be sure, the custom is in many ways undesirable because it lends itself to abuses our own system tries to discourage. And doubtless it encourages laxity about truthfulness in public life and may lessen the public trust needed for effective government. Yet it is not obvious that foreign companies participating in it are actually doing any harm.

Next consider the very different problem posed by the political situation in the Republic of South Africa, a country whose political and economic institutions are based on the system of racial separation called *apartheid* (Callaghan, 1978, 325–359). The economic aspects of the system include differential wage scales, with black workers paid considerably less than the wages paid to whites for similar work. Advancement to high managerial positions within companies has either been limited to whites or made much more difficult for black employees. Segregation has extended to rest rooms, lounge areas, and assembly lines.

These economic aspects of apartheid are backed by political repression. Black people constitute over 70 percent of the population, but they are allowed to live permanently on only 13 percent of the land, generally the least desirable, barren areas. Male workers must leave their families to seek work far away. No black person has the right to vote, to free expression and association, or to participate fully in government. Attempts to gain political rights have resulted in brutal suppression. In 1976, for example, police killed over 700 demonstrators in Soweto, over two dozen after they were arrested and placed in detention.

A number of international engineering firms do business in South Africa. Most of them, although not all, have in the past adopted at least some of the South African policies. Was this morally permissible? Were they perhaps even obligated to conform to the dominant foreign customs? Was profit a sufficient justification, especially given that American subsidiaries in South Africa often made twice the profits of their American counterparts?

We take the view that it was neither obligatory nor, in most cases, permissible to do so. Racial discrimination was wrong in the United States even during the time it was accepted by the majority of citizens. It is equally wrong now, and seriously so. Few practices so clearly violate fundamental moral principles of human equality, dignity, and rights.

The major ethical theories can be invoked to support this judgment. From Kant's duty perspective, the practice constitutes treating the black Africans as a mere means to promoting the advantages of the white minority and foreign corporations seeking inexpensive labor. From Locke's rights perspective, it amounts to a blatant infringement of human rights. From a utilitarian perspective, the immense suffering of the black majority in South Africa outweighs the financial benefits of the already advantaged groups, so that the practice does not produce the most good for the most people. From a virtue ethics perspective, the virtue of justice prohibits supporting grossly unjust systems.

Is our harsh judgment applicable in all instances? Might there be mitigating circumstances? Our earlier comments on relationalism would require us to look more closely at the practices of particular American companies before making unqualified condemnations. Are they taking steps to increase wages for black workers—steps that follow a definite timetable so that equal work will eventually be rewarded by equal pay? Are they working toward bringing more blacks into management positions by offering them the required training now? To the extent that such efforts are underway and indeed constitute more than tokenism, any judgment of blame would have to be qualified.

More than 100 of the American companies doing business in South Africa did attempt to bring about positive change by adopting the Sullivan Principles. These principles were set forth in 1977 by Reverend Leon Sullivan, a black minister in Philadelphia and a member of General Motors' board of directors. The principles called for nonsegregation at the workplace, equal pay for the same job, and equal promotion opportunities. When adhered to, these principles were more than token gestures: They went far beyond the South African conventions and violated South African law, although the government decided to indulge them.

In 1987, however, Reverend Sullivan disavowed his principles. He noted that many companies who professed the principles found ways around them. For example, different job titles would be given to blacks and whites doing essentially the same work, thereby permitting the company to use differential pay scales. Again, companies took down racial segregation signs, but maintained segregation by assigning separate work areas to salaried workers (mostly white) and hourly workers (mostly black). Even when the principles were adhered to in good faith, Reverend Sullivan concluded, the apartheid system was being supported, not improved, by American business. American tax dollars, technology, jobs, and prestige all worked to the advantage of the South African government.

The examples of Italian tax laws and South African apartheid suggest it is neither always right nor always wrong to work within parameters set by the undesirable customs of other countries. In general, foreign customs provide morally relevant data which should enter into our ethical deliberations. But neither they nor current U.S. practice automatically dictates how engineers and corporations ought to act in unusual social settings. Each case, or at least each type of case, must be examined on its own merits.

Technology Transfer and Appropriate Technology

Before returning to the Bhopal disaster, it will be helpful to introduce the concepts of technology transfer and appropriate technology. *Technology trans-*

fer is the process of moving technology to a novel setting and implementing it there (Heller, 1985). Technology includes both hardware (machines and installations) and technique (technical, organizational, and managerial skills and procedures). A *novel setting* is any situation containing at least one new variable relevant to the success or failure of a given technology. The setting may be within a country where the technology is already used elsewhere, or a foreign country, which is our present interest. A variety of agents may conduct the transfer of technology: governments, universities, private volunteer organizations, consulting firms, and multinational corporations.

In most instances, the transfer of technology from a familiar to a new environment is a complex process. The technology being transferred may be one that originally evolved over a period of time and is now being introduced as a ready-made, completely new entity into a different setting. Discerning how the new setting differs from familiar contexts requires the imaginative and cautious vision of "cross-cultural social experimenters" (Heller's phrase).

The expression *appropriate technology* is widely used, but with a variety of meanings. We use it in a generic sense to refer to identification, implementation, and transfer of the most suitable technology for a new set of conditions. Typically the conditions include social factors that go beyond routine economic and technical engineering constraints. Identifying them requires attention to an array of human values and needs that may influence how a technology affects the novel situation. Thus,

> appropriateness may be scrutinized in terms of scale, technical and managerial skills, materials/energy (assured availability of supply at reasonable cost), physical environment (temperature, humidity, atmosphere, salinity, water availability, etc.), capital opportunity costs (to be commensurate with benefits), but especially human values (acceptability of the end-product by the intended users in light of their institutions, traditions, beliefs, taboos, and what they consider the good life) (Heller, 1985, 119).

As examples, we may cite the introduction of agricultural machines and long-distance telephones. A country with many poor farmers can make better immediate use of small, single- or two-wheel tractors which can serve as motorized ploughs, pull wagons, or drive pumps than it can of huge diesel tractors which require collectivized or agribusiness-style farming. On the other hand, the same country may benefit more from the latest in microwave technology to spread its telephone service over long distances than it can from old-fashioned transmission by wire.

Appropriate technology overlaps with, but is not reducible to, *intermediate technology*, which lies between the most advanced forms available in industrialized countries and comparatively primitive forms in less-developed countries (Schumacher, 1973). The British economist E. F. Schumacher argued that intermediate technologies are preferable because the most advanced technologies usually have harmful side-effects, such as causing mass migrations from rural areas to cities where corporations tend to locate. These migrations cause overcrowding, and with it poverty, crime, and disease. Far more appropriate, he argued, are smaller-scale technologies replicated throughout a less-developed country, using low capital investment, labor intensiveness to provide needed jobs, local resources where possible, and simpler techniques manageable by the local population given its education facilities.

We mention intermediate technology, and the movement inspired by Schumacher, not in order to offer a general endorsement (often it has been dramatically beneficial, at other times not particularly effective), but to emphasize that it is only one conception of appropriate technology. "Appropriate technology" is a generic concept which applies to all attempts to emphasize wider social factors when transferring technologies. As such, it reinforces and amplifies our view of engineering as social experimentation.

Bhopal

In retrospect it is clear that greater sensitivity to social factors was needed in transferring chemical technology to India (Shrivastava, 1987; Everest, 1985). By the late 1970s, Union Carbide had transformed its pesticide plant in Bhopal from a formulation plant (mixing chemicals to make pesticides) to a production plant (manufacturing chemical ingredients). It was fully aware of the hazards of the new technology it transferred. For years its West Virginia plant had made methyl isocyanate, the main toxin in two popular pesticides used in India and elsewhere. As a concentrated gas, methyl isocyanate burns any moist part of bodies with which it comes in contact, scalding throats and nasal passages, blinding eyes, and destroying lungs.

Yet, in designing the Bhopal plant Union Carbide did not transfer all the safety mechanisms available. For example, whereas computerized instruments controlled the safety systems and detected leaks at the West Virginia plant, Bhopal's safety controls were all manual and workers were asked to detect leaks with their eyes and noses.

The government of India required the Bhopal plant to be operated entirely by Indian workers. Hence Union Carbide at first took admirable care in training plant personnel, flying them to the West Virginia plant for intensive training. It also had teams of U.S. engineers make regular on-site safety inspections.

But in 1982 financial pressures led Union Carbide to relinquish its supervision of safety at the plant, even though it retained general financial and technical control. The last inspection by a team of U.S. engineers occurred that year, despite the fact that the team warned of many of the hazards that contributed to the disaster.

During the following 2 years safety practices eroded. One source of the erosion had to do with personnel: high turnover of employees, failure to properly train new employees, and low technical preparedness of the Indian labor pool. Workers handling pesticides, for example, learned more from personal experience than from study of safety manuals about the dangers of the pesticides. But even after suffering chest pains, vomiting, and other symptoms they would sometimes fail to wear safety gloves and masks because of high temperatures in the plant—the result of lack of air-conditioning.

The other source of eroding safety practices was the move away from U.S. standards (contrary to Carbide's written policies) toward lower Indian standards. By December of 1984 several extreme hazards, in addition to many smaller ones, were present.

First, the tanks storing the methyl isocyanate gas were overloaded. Carbide's manuals specified they were never to be filled to more than 60 percent of capacity; this was so that in emergencies the extra space could be used to dilute the gas. The tank which was to cause the problem was in fact more than 75 percent full.

Second, a stand-by tank that was supposed to be kept empty for use as an emergency dump tank already contained a large amount of the chemical.

Third, the tanks were supposed to be refrigerated to make the chemical less reactive if trouble should arise. But the refrigeration unit had been shut down 5 months before the accident as a cost-cutting measure, making tank temperatures 3 to 4 times what they should have been.

Six weeks before the catastrophe, production of methyl isocyanate had been suspended because of an oversupply of the pesticides it was used to make. Workers were engaged in routine plant maintenance. A relatively new worker had been instructed by a new supervisor to flush out some pipes and filters connected to the chemical storage tanks. Apparently the worker properly closed valves to isolate the tanks from the pipes and filters being washed, but he failed to insert the required safety disks to back up the valves in case they leaked. (He knew that valves leaked, but he did not check for leaks: "It was not my job." The safety disks were the responsibility of the maintenance department, and the position of second-shift supervisor had been eliminated.)

Two of four valves that should have been open to allow water flow were clogged. The resulting extra pressure was enough to force water to leak into a tank. For nearly 3 hours chemical reactions occurred, generating enormous pressure and heat in the tank.

By the time the workers noticed a gauge showing the mounting pressure and began to feel the sting of leaking gas, they found their main emergency procedures unavailable. The primary defense against gas leaks was a vent-gas scrubber designed to neutralize the gas. It was shut down (and was turned on too late to help), because it was assumed to be unnecessary during times when production was suspended.

The second line of defense was a flare tower that would burn off escaping gas missed by the scrubber. It was inoperable because a section of the pipe connecting it to the tank was being repaired. Finally, workers tried to minimize damage by spraying water 100 feet into the air. The gas, however, was escaping from a stack 120 feet high.

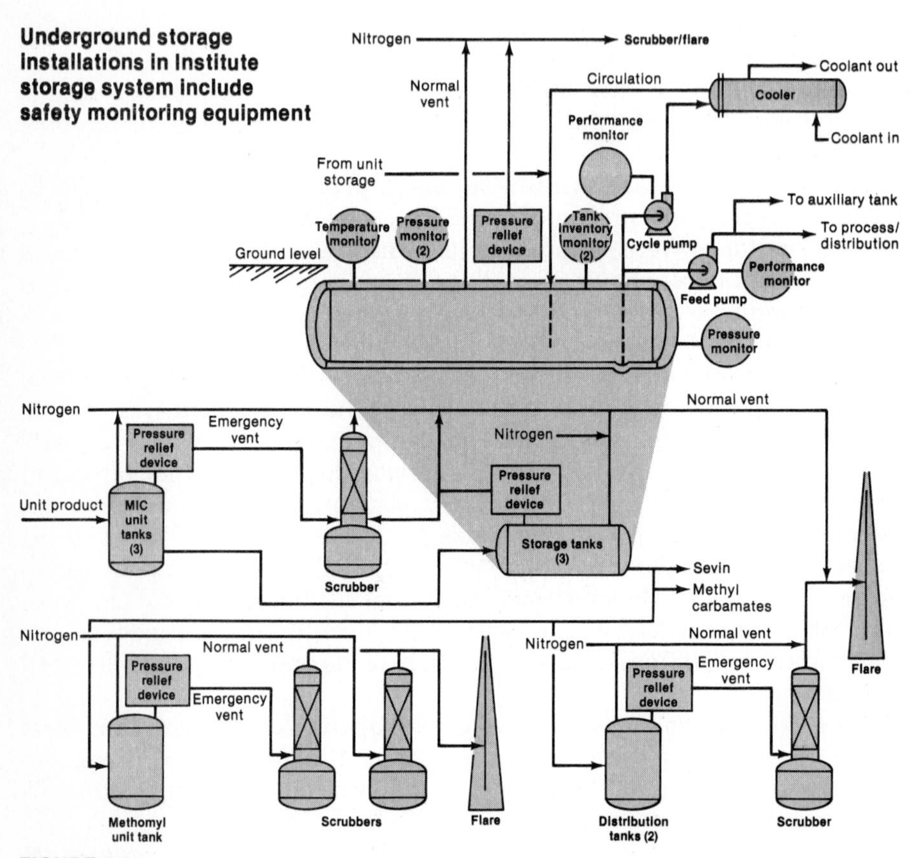

FIGURE 7-1
Diagram of Bhopal system. (*From Ward Worthy, "Methyl Isocyanate: The Chemistry of a Hazard,"* C & EN *February 11, 1985.*)

Within 2 hours most of the chemicals in the tank had escaped to form a deadly cloud over hundreds of thousands of people in Bhopal. As was common in India, desperately poor migrant laborers had become squatters—by the tens of thousands—in the vacant areas surrounding the plant. They had come with hopes of finding any form of employment, as well as to take advantage of whatever water and electricity was available.

Virtually none of the squatters had been officially informed by Union Carbide or the Indian government of the danger posed by the chemicals being produced next door to them. (The only voice of caution was that of a concerned journalist, Rajukman Keswani, who had written articles on the dangers of the plant and had posted warnings: "Poison Gas. Thousands of Workers and Millions of Citizens are in Danger.") There had been no emergency drills, and there were no evacuation plans: the scope of the disaster was greatly increased because of total unpreparedness.

Study Questions

1 Following the disaster at Bhopal, Union Carbide argued that officials at its U.S. corporate headquarters had no knowledge of the violations of Carbide's official safety procedures and standards. This has been challenged as documents were uncovered showing they knew enough to have warranted inquiry on their part, but let us assume they were genuinely ignorant. Would ignorance free them of responsibility for all aspects of the disaster? (In answering this question distinguish between the various senses of "responsibility" set forth in Chap. 2. If possible consult one or more of the many articles and books written about the disaster.)

2 Export of hazardous technologies to less-developed countries is motivated in part by cheaper labor costs, but another factor is that workers are willing to take greater risks. It has been argued that taking advantage of this willingness need not be unjust exploitation if several conditions are met: (1) Workers are informed of the risks. (2) They are paid more for taking the risks. (3) The company takes some steps to lower the risks, even if not to the level acceptable for U.S. workers (DeGeorge, 1986, 368–369). Do you agree with this view? How would you assess Union Carbide's handling of worker safety? Take into account the remarks of an Indian worker interviewed after the disaster. The worker was able to stand only a few hours each day because of permanent damage to his lungs. During that time he begged in the streets while he awaited his share of the legal compensation from Union Carbide. (The Indian government asked Carbide for $3 billion in compensation.) When asked what he would do if offered work again in the plant knowing what he knew now, he replied: "If it opened again tomorrow I'd be happy to take any job they offered me. I wouldn't hesitate for a minute. I want to work in a factory, any factory. Before 'the gas' [disaster] the Union Carbide plant was the best place in all Bhopal to work" (Bordewich, 1987).

3 During 1972 and 1973 the President of Lockheed, A. Carl Kotchian, authorized secret payments totaling around $12 million beyond a contract to representatives of Japan's Prime Minister Tanaka. Later revelations of the bribes helped lead to the resignation of Tanaka and also to new laws in this country forbidding such payments. Mr. Kotchian believed at that time it was the only way to assure sales of Lockheed's TriStar airplanes in a much needed market.

In explaining his actions, Mr. Kotchian cited the following facts (Kotchian, 1977, 67–75): (1) There was no doubt in his mind that the only way to make the sales was to make the payments. (2) No U.S. law at the time forbade the payments. (3) The payments were financially worthwhile, for they totalled only 3 percent of an expected $430 million income for Lockheed. (4) The sales would prevent Lockheed layoffs, provide new jobs, and thereby benefit workers' families and their communities as well as the stockholders. (5) He himself did not personally initiate any of the payments, which were all requested by Japanese negotiators. (6) In order to give the TriStar a chance to prove itself in Japan, he felt he had to "follow the functioning system" of Japan. That is, he viewed the secret payments as the accepted practice in Japan's government circles for this type of sale.

a Drawing on the distinctions made in this section, explain the several senses in which someone might claim that how Mr. Kotchian ought to have acted is a "relative" matter. Which of these senses, in your view, would yield true claims and which false?

b Develop the strongest moral argument you can think of in favor of Mr. Kotchian's actions, then the strongest argument you can against his actions. Draw upon the ethical theories presented in the previous section: either act-utilitarianism, rule-utilitarianism, Kant's duty-ethics, or Locke's rights-ethics. Then indicate which argument you find most compelling and why you disagree with the opposing view.

4 In 1977 the Foreign Corrupt Practices Act was signed into law. It makes it a crime for American corporations to accept payments from or to offer payments to foreign governments for the purpose of obtaining or retaining business, although it does not forbid "grease" payments to low-level employees of foreign governments (such as clerks), which are part of routine business dealings. Critics are urging repeal of the act because there is no question that it has adversely affected American corporations trying to compete with countries that do not forbid paying business extortion. Is damage to profits a sufficient justification for repealing the act?

5 Some U.S. companies have refused to promote women to positions of high authority in their international operations in Asia, the Middle East, and South America. Their rationale is that business will be hurt because some foreign customers do not wish to deal with women. It might be contended that this practice is justified out of respect for the customs of countries which discourage women from entering business and the professions.

Circuit Judge Warren J. Ferguson argued, however, that such practices are wrong. He ruled that sex stereotypes are not to be used in formulating job qualifications, and that customer preferences do not justify sex discrimination. He added that while our legal system cannot be used to force other countries to stop sex discrimination, other countries cannot dictate sex discrimination for citizens of our country (Mayer, 1981, 2).

Present and defend your view as to whether Judge Ferguson's ruling is morally justified.

ENVIRONMENTAL ETHICS

As human beings we share a common environment, a common ecosphere. Concern for that environment must increasingly become an urgent and united commitment that cuts across national boundaries. It is thus appropriate that thinkers from many disciplines have begun to explore a new branch of applied ethics called *environmental* or *ecological ethics.* This field overlaps with engineering ethics at many points, only a few of which can be mentioned here.*

Many observers warn us that we are misusing our scarce resources, fouling our environment, and in general practicing growth in consumption and population which will eventually make "spaceship earth" too small for us. Others think the situation may not be so bleak. Nature itself, they point out, is continually changing the face of the earth and human activity creates at best a small perturbation. But it is generally agreed that industrial activity does affect the biosphere: It denudes the land, pollutes the waters and the atmosphere, and threatens fragile species. But how damaging such effects are in the long run, or to what extent we can develop countermeasures in time, is difficult to assess. As Herman Kahn, William Brown, and Leon Martel point out:

> It may yet turn out that future man will marvel at the paradoxical combination of hubris and modesty of 20th century man, who, at the same time, so exaggerated his ability to do damage and so underestimated his own ability to adapt to or solve such problems. Or it could be that future man, if he exists, will wonder at the recklessness and callousness of 20th century scientists and governments (Kahn, Brown, and Martel, 1970, 180).

While we do not share the pessimists' views of inevitable doom, we also do not subscribe to the optimists' views that new technologies will invariably spring up to deliver us from the traps set by the older technologies. We do recognize that we cannot turn back the clock, that we have to live with some unalterable changes and have already learned to do so. Nevertheless, we observe that technology and its practitioners have engaged us in an ongoing global experiment that needs to be monitored and guided lest we permit it to cause many unacceptable changes of an irrevocable nature.

Engineers need to be aware of their role as agents of change, as experimenters. There are questions every engineer should ask, whether at the time of choosing a career or when entering a new sphere of activity. To what extent does a particular industry affect the environment? To what extent can those effects be controlled physically or regulated politically? Have all reasonable abatement measures been implemented? Can I be effective as an engineer in helping to assure a decent environment for the generations that will follow? What are my responsibilities in all this?

We begin by sampling a few kinds of the myriad environmental issues of concern to engineers. The first two examples have to do with releasing harmful substances into air and water, the third with using toxic substances in

*The following books provide useful starting points for studying environmental ethics: *Environmental Ethics for Engineers,* edited by Alastair S. Gunn and P. Aarne Vesilind; *Earthbound,* edited by Tom Regan; *Ethics and the Environment,* edited by Donald Scherer and Thomas Attig; *Respect for Nature,* by Paul W. Taylor; and *People, Penguins, and Plastic Trees: Basic Issues in Environmental Ethics,* edited by Donald VanDeVeer and Christine Pierce.

food processing, and the last two with disrupting land and water balances. Following these illustrative cases we will touch upon some foundational issues in environmental ethics and some economic and political means of controlling harm to the environment.

Acid Rain

The disaster at Bhopal occurred with numbing terror during a few hours and days. Other disasters, such as the Chernobyl nuclear plant explosion, have involved longer-term environmental effects, but the patterns of damage are still relatively clear. Some ecological tragedies, however, are far more subtle in their impact. Few may turn out to be comparable to the devastation currently being caused by acid rain and acid deposition.

Normal rain has a pH of 5.6, but the typical rain in the northeastern areas of North America is now 3.9 to 4.3. This is 10 to 100 times more acidic than it should be—about as acidic as lemon juice. In addition, the snowmelt each spring releases huge amounts of acid which were in frozen storage during the winter months. Soil which contains natural buffering agents counteracts the acids. But large parts of the northeastern United States and eastern Canada lack natural buffers.

The results? "Acid shock" from snowmelt is thought to cause annual mass killings of fish. Longer-term effects of the acid harm fish eggs and food sources. Deadly quantities of aluminum, zinc, and many other metals leached from the soil by the acid rain also take a toll as they wash into streams and lakes. In the higher elevations of the Adirondack Mountains over half the lakes that were once pristine can no longer support fish. Hundreds of other lakes are dying in the United States and Canada. Forests have also been steadily killed, larger animals have suffered dramatic decreases in population, and some farmlands and drinking-water sources are damaged.

These results have occurred during only a few decades. The next decades will multiply them many times over. It is believed that North America is just slightly behind Scandinavia, where thousands of lakes have been "killed" by acid rain. In both locations the cause is now clear: the burning of fossil fuels which release large amounts of sulfur dioxide (SO_2)—the primary culprit—and nitrogen oxides (NO_X). In both instances major sources of the pollutants are located hundreds and even thousands of miles away, with winds supplying a deadly transportation system to the damaged ecosystems. Much of Sweden's problem, for example, is traceable to the industrial plants of England and northern Europe. Acid rain problems in Canada and the northeastern United States derive in large measure from the utilities of the Ohio Valley, the largest source of sulfur dioxide pollution in this country.

Much remains to be learned about the mechanisms involved in the processes pictured in Figure 7.2. It is still impossible to link specific sources with

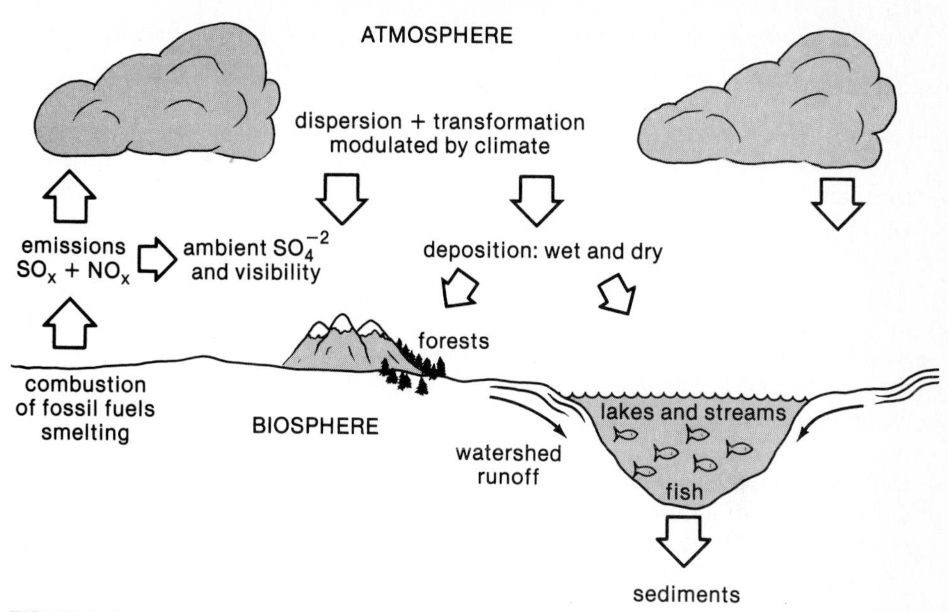

FIGURE 7-2
Acid deposition: diagram of sources and affected ecosystems. (*National Research Council, 1986, p. 11.*)

specific damage. More research into shifting wind patterns and the air transport of acids is needed. Nor is there a reliable estimate of current damage. For example, many believe that microorganisms in soil are being affected in ways that are potentially devastating, but no one knows for sure. Groundwater is undoubtedly being polluted, but it is unclear what that means for human health. Much underground water currently being used was deposited by rainfall over a hundred years ago, and current acid rain may have its main effects on underground water a century from now. Effects on human food sources are also largely unknown (Adams, 1985; National Research Council, 1986; Schmandt, 1985).

Other examples can be given of amorphous patterns of ecological damage, like those of acid rain. Worldwide use of fossil fuels by industrial nations is causing a buildup of carbon dioxide in the atmosphere which could result in a greenhouse effect damaging the entire earth. Similarly, damage to the protective ozone layer of the earth's atmosphere resulting from the release of freon is related to technological products used by the populations of those same nations. And rivers amass pollutants as they wind their way through several states or countries, eventually to dump their toxic contents into an ocean. The Rhine is such a river, and the North Sea, now a "special protection area," is such an ocean.

If all environmental damage were on such a massive scale, the work of engineers would be a continual nightmare. Fortunately, most environmental impacts remain relatively local, as we shall now illustrate.

PCBs and Kanemi's Rice-Oil—Japan, 1968 A decade of rapid industrial growth in Japan had taken its toll of the environment. City dwellers fell ill from air pollution. Some rivers were covered with dead fish floating on the surface. The dreadful Minamata disease, traceable to mercury pollution of a nearby bay by the Chisso Company and a food chain involving shellfish taken from that bay, was continuing to fell its victims (there would be 46 deaths and 75 disabilities by 1970). In May the itai-itai ("ouch-ouch") disease, which is painful and causes bone crumbling, was the first illness to be designated as pollution-generated by the Japanese government. Observed on and off since 1920 at various sites, the latest outbreak was placed at the doors of a Mitsui Mining and Smelting facility which let cadmium escape into a river, whence the pollutant found its way into the food chain via rice paddies and rice. The thalidomide drug tragedy was still on everyone's mind. The severely malformed children of women who had taken that drug during their pregnancies were still not cared for by either the Dainippon Pharmaceutical Company or the government, and schools refused to accept them (Ui, 1972; Iijima, 1979).

Then, in the summer of 1968, a disease of unknown origin made its appearance in southern Japan. Victims suffered from disfiguring skin acne and discoloration, from fatigue, numbness, respiratory distress, vomiting, and loss of hair. Eventually 10,000 people were stricken and some died. What was the cause? An investigation of 121 cases was conducted, with 121 healthy individuals matched to the victims by age and sex being used as a control group. All 242 were questioned regarding their diets, personal habits, and places of work. When it was discovered that the only significant difference between the two groups was in the amount of fried foods eaten, the disease was traced to rice oil produced by the Kanemi Company (Grundy, Weisbrod, and Epstein, 1974, 111).

It took another 7 months to find the specific agent in the oil. Autopsies performed on victims revealed the presence of PCBs, polychlorinated byphenyls (Iijima, 1979, 268). Oil made from rice bran at the Kanemi plant was heated at low pressure to remove objectionable odors. The heating pipes were filled with hot Kanechlor, a PCB containing fluid, but the pipes were corroding and tiny pinholes in them allowed PCBs to leak directly into the oil. In fact, Kanemi had been in the habit of replenishing about 27 kg of lost PCBs a month for some time.

There are other, less direct paths by which the extremely toxic PCBs can reach humans. For example, the Kanemi rice oil had also been used as an additive to chicken feed. In early 1968, 1 million chickens were given this feed and every second chicken died. In the United States 140,000 chickens were slaughtered in the state of New York on one occasion when data collected by the Campbell's Soup Company revealed more than the permitted level of PCBs in chickens raised by certain growers. The source of PCBs was found to be plastic bakery wrappers mixed in with ground-up stale bread from bakeries used as feed. On another occasion PCBs leaked from a heating system into fishmeal in a North Carolina pasteurization plant. About 12,000 tons of fishmeal were contaminated and 88,000 chickens, already fed this fishmeal, had to be destroyed when the product was recalled (Grundy, Weisbrod, and Epstein, 1974, 112).

PCBs were not used only in heat exchangers. They are a good hydraulic fluid and an excellent dielectric. They also decompose very slowly. These properties made them a good choice for insulating oil in capacitors and transformers. But they are no longer considered suitable for such applications—or any other where they can find their way into the environment. Their hardiness accounts for the fact that they are found in the oceans in larger quantities than DDT, although they enter the oceans initially in much lesser amounts (Murdoch, 1979, 286–287). Even a total shutdown of all possible sources of PCB contamination—responsible for about 28,000 tons of PCBs escaping into air, land, and water every year—would not result in an immediate decrease in its presence. Meanwhile, even the present burden of PCBs on the environment is not fully understood (Grundy, Weisbrod, and Epstein, 1974, 112–113).

Asbestos in the Air; Asbestos in Drinking Water? Although the use of PCBs in food and feed-processing plants might not have been a problem in the above instances if the heat transfer systems employing the toxic oil had been secure and closed, one should never be as optimistic as to expect a leak-proof piping system; and when foodstuffs are involved, one should not even contemplate a substantial use of toxic materials for any reason. There are other processes, however, where even an approximation to a closed system could have great benefits. One such example is ore processing as practiced by the Reserve Mining Company, a subsidiary of Armco and Republic steel companies, when it first started its huge taconite (iron ore) plant at Silver Bay, Minnesota in 1955 (Bartlett, 1980; Lawless, 1977, 297–307).

Reserve was invited to open operations by the state of Minnesota; it employed over 3000 workers; and it was also making a contribution to the nation by utilizing abundant but low-grade ore to produce much needed iron and steel. Taconite contains only 25 to 30 percent iron oxide. It is crushed and washed in steps to form a powder from which the iron is extracted by electromagnets. Everything at Reserve was on a large scale—size, cost ($350 million), power consumption, and water usage. Rather than utilizing settling basins and recycling the used water as is done in smaller plants, fresh water was drawn from Lake Superior and wastewater was discharged into it at the rate of half a million gallons per minute. Despite assurances from the University of Minnesota's experiment station that the fine taconite powder would settle out, it did not. The lake became discolored by the discharge of about 67,000 tons of taconite tailings a day, and fishers complained about the disappearance of fish.

The company was not doing anything very different from what industry had always been doing. In Sweden, paper mills were polluting the Baltic Sea, in the Soviet Union unique Lake Baikal was being similarly disfigured. But the managers of Reserve had not counted on the fact that times—and public attitudes—had changed. In the 1960s concerns about the environment were gradually being taken more seriously. The federal government and eventually the state of Minnesota took the company to court since it would not voluntarily change its practices and retrofit its extraction process to a closed system with no discharge of wastes into the lake. Private citizens had become active and were influential in causing the government agencies to act.

The court proceedings, including appeals, were lengthy (14 years) and replete with contradictory testimony from experts representing the opposing sides. It was not until asbestoslike fibers were found to be entering the air and the lake from the Reserve operations that a more definite link to a real hazard could be established. However, the amounts entering the drinking-water supplies of cities ringing Lake Superior, and the levels considered to be harmful, could not be established with any degree of precision. Accordingly, Reserve's taconite beneficiation plant was not closed down but ordered to switch over to a land disposal site as soon as possible.

Land Subsidence

> Roads began cracking like china. Railroad tracks buckled. Sewer lines burst. Oil drillers saw their pipes twist and bend like spaghetti. Pacific tides washed over harbor bulkheads.... (Marx, 1977, 140)

An earthquake? Not quite. Just a gradual but steady land subsidence, inadvertently caused by human beings, played back at rapid speed. Wesley Marx continues as follows in his book *Acts of God, Acts of Man* with a description of the geologic subversion of Long Beach during 1941 to 1945:

> The Long Beach naval shipyard wondered if it could stay intact long enough to help end World War II. Some areas sank as much as twenty-nine feet, or enough to reach the groundwater table....Gravity no longer provided drainage free of charge....When would the sinking stop? Some industrialists didn't care to find out. They were prepared to climb out of their inverted landscape and never return (Marx, 1977, 140).

When it was discovered that extensive pumping of oil fields in the vicinity was draining the underground reservoirs and reducing underground pressure, water was reinjected into those reservoirs and the lowering of the land surface was greatly reduced. Water pumping can cause similar subsidence, as many cities and areas have experienced, from Tokyo, Osaka, and Nagoya in Japan to the 3000-square-mile area between Houston and Galveston,

Texas. Houses will then lean and lowered land will experience severe flooding during typhoons and hurricanes.

Water is injected underground not only to prevent subsidence, but also to force oil out of aging wells. In 1963, such high-pressure injection in oil fields near Baldwin Dam in Los Angeles caused the bottom of the reservoir to crack open along an old fault line. The crack reached all the way to the foundation of the dam. The 12-year-old dam had been designed to withstand earthquakes but not such surprises. The released waters killed five people and did $14 million worth of damage to property (Marx, 1977, 141; Sowers and Sowers, 1970, 535).

The examples of land subsidence illustrate how human activity can change our environment in many different ways. Some effects may be local and reparable, but loss of life and property can result when hidden changes lead to unexpected events.

Too Little Water for the Everglades The great marshes of southern Florida have attracted farmers and real estate developers since the beginning of the century. When drained, they present valuable ground. From 1909 to 1912 a fraudulent land development scheme was attempted in collusion with the then U.S. Secretary of Agriculture. Arthur Morgan blew the whistle on that situation, jeopardizing not only his own position as a supervising drainage engineer with the U.S. Department of Agriculture, but also that of the head of the Office of Drainage Investigation. An attempt to drain the Everglades was made again by a Florida governor from 1926 to 1929. Once more Arthur Morgan, this time in private practice, stepped in to reveal the inadequacy of the plans and thus discourage bond sales.

But schemes affecting the Everglades were not over yet. Beginning in 1949, the U.S. Army Corps of Engineers started diverting excess water from the giant Lake Okeechobee to the Gulf of Mexico to reduce the danger of flooding to nearby sugar plantations. As a result the Everglades, lacking water during the dry season, were drying up. A priceless wildlife refuge was falling prey to humanity's appetite. In addition, the diversion of waters to the Gulf and the ocean also affected human habitations in southern Florida. Cities which once thought they had unlimited supplies of fresh groundwater found they were pumping salt water instead as ocean waters seeped in (Morgan, 1971, 370–389).

Southern Florida represents a complex environmental unit with a delicate balance. Any intrusion by human engineering must be seen as an experiment which must be conducted with great care. Unfortunately too many public agencies view any change in plans as unacceptable once a course has been charted. As Arthur Morgan points out in his book *Dams and Other Disasters*, the Corps was particularly prone to such an attitude, which was fostered by the crisis-oriented training at West Point Military Academy (Morgan, 1971, 37).

Why should the Everglades be preserved? We leave this as a study question for later. But first let us set a wider context by considering some views of our place in nature.

Living Things and Views of Nature

A fundamental issue in environmental ethics is whether morality is exclusively *anthropocentric,* that is, human-centered. Western religions have generally regarded nature as an unrestricted gift to be ruled and exploited. Likewise, traditional secular worldviews have drawn a sharp line between the human and nonhuman world, making the latter mere means to satisfying human ends.

Increasingly, human-centered views of nature have shifted away from an exploitative to a conservationist attitude. Recognition of the limits of natural resources has inspired awareness of the need to conserve for the sake of both present and future generations of human beings. But the assumption remains that only human beings have intrinsic moral worth, possess rights, and are owed duties of respect. Conservation is justified in terms of human needs for a livable environment. Or it is justified by appeal to the human right to life, which implies a right to have a livable environment (Blackstone, 1974).

Nonanthropocentric or nature-centered views of nature, by contrast, begin with the premise that humans are only part of nature and that human worth must be understood from a holistic ecological perspective. One such approach views the natural environment as having intrinsic value—value in and of itself, quite apart from any value humans place on it. According to this view, there is a direct moral imperative to *preserve,* not just conserve, the environment. This approach was voiced by the naturalists Aldo Leopold (1887–1948) and John Muir (1838–1914, founder of the Sierra Club).

Aldo Leopold's "land ethic" was founded on the principle that acts are good insofar as they preserve the integrity, stability, and beauty of the community of natural things. John Muir's naturalism took a mystic direction in ascribing elements of the divine to nature.

> Watch the sunbeams over the forest awakening the flowers, feeding them every one, warming, reviving the myriads of the air, setting countless wings in motion—making diamonds of dewdrops, lakes, painting the spray of falls in rainbow colors. Enjoy the great night like a day, hinting the eternal and imperishable in nature amid the transient and material (Muir, 1980, 113).

Muir would have felt at home in Japan, where an inspiring, lively waterfall or a quiet rock may be accorded the respect due a deity.

A different approach to a nature-centered ethics focuses on sentient animals, rather than all of nature. Sentient animals are those which feel pain and pleasure and have desires. Utilitarians have often extended their theory (that right action maximizes goodness for all affected) to sentient animals as well as humans. Most notably, Peter Singer developed a utilitarian perspective in his influential book, *Animal Liberation.* (Somewhat ironically this book has been called the bible of the animal rights movement: Singer does not ascribe rights to animals, although other philosophers have [Regan, 1983].)

Singer insists that moral judgments must take into account the effects of our actions on sentient animals. Failure to do so is a form of discrimination akin to racism and sexism—what he labels "speciesism." Thus, in building a dam that will cause flooding to grasslands, engineers should take into account the impact on animals living there. Singer allows that sometimes animals' interests have to give way to human interests, but their interests should always be considered and weighed.

There is a gulf between the human-centered and nature-centered perspectives, but its extent should not be exaggerated. Most agree that human interests have some priority over those of animals and plants. Most agree that nature should not be denuded but instead left in a recoverable state as a safe exit and for the benefit of later generations to enjoy. Moreover, even if animal interests are not counted anywhere near on a par with human interests, it remains abhorrent to inflict needless suffering on animals who share our capacity for pain.

The point of all this is that not everything of importance fits neatly into cost-benefit analyses with limited time horizons; much must be accounted for by means of constraints or limits which cannot necessarily be assigned dollar signs.

The Commons and a Livable Environment

Aristotle once remarked that what is common to the most people gets the least amount of care. Common sense confirms the frequent tendency of people to be thoughtless about things they do not own and which seem to be in unlimited supply. William Foster Lloyd was an astute observer of this phenomenon. In 1833 he described what the ecologist Garrett Hardin would later call "the tragedy of the commons" (Hardin, 1968, 254).

Lloyd observed that cattle in the common pasture of a village were punier and more stunted than those kept on private land. The common fields were themselves more bare-worn than private pastures. His explanation began with the premise that each farmer is understandably motivated by self-interest to enlarge his or her herd by one or two cows, claiming that the overall effect is minuscule. Yet the combined effects of all the farmers behaving this way is overgrazing of the pasture, even though it is true that each act taken by itself does negligible damage.

In this century, increasing population and decreasing natural resources have prompted similar thinking about ourselves in relation to nature. The same kind of competitive, unmalicious, but unthinking exploitation arises with respect to all natural resources held in common: air, land, forests, lakes, and oceans. Indeed, increasingly we must regard the entire biosphere as our "commons."

Few today would endorse past arguments by land developers, road builders, forest harvesters, and manufacturers that they should be left completely unhindered because the damage they do to the environment is minuscule. Economic rights have limits when the public good has been put at risk by cumulative small effects. Some have even argued that fundamental democratic rights can be invoked to justify controls on environmental harms. The basic right to life entails a right to a livable environment at a time when pollution and resource depletion has reached alarming proportions (Blackstone, 1974).

In addition to voluntary conservation efforts, there is a need for a shared effort to exercise democratic and international controls responsibly. Such controls involve some coercion. Only the naive would think that the short-term, profit-seeking orientation of most corporations could otherwise be sufficiently influenced by environmental appeals. Indeed, within limits businesses are eager for controls which require all competitors to be guided by similar restrictions that maintain fairness in competition.

Democratic controls can take many forms. They include passing laws, internalizing costs, and relying on technology assessment in approving projects.

Guilty until Proven Innocent?

The examples just given cover barely a fraction of the many environmental issues which might arise in engineering practice. But they suffice to raise some key questions of ethical import: *Who* is affecting *whom*—and *where, when,* and *how?*

In the mercury, cadmium, and PCB pollution cases in Japan the question was: Who is releasing toxic substances? In all instances the polluting companies refused to acknowledge their mistakes and rejected claims by the victims. The government did not intercede on behalf of those affected. It took long and costly court battles to win partial victories. Without volunteer activists little progress would have been made. Is this the way it ought to be?

The victims of pollution in Japan were easily identified, the offenders less so. In the Reserve Mining Company episode, the reverse occurred. The taconite facility was an obvious polluter in some respects even though its more harmful emissions could not be detected with the naked eye and were not introduced as evidence until later in the trial. But there were no identifiable victims because the ill effects of asbestos take time to develop. And it would also be difficult to determine where around Lake Superior the contamination would be most serious in the long run.

These difficulties are multiplied many times in the case of acid rain. How can control of Ohio utilities be instituted when it is still impossible to link them to specific harms in New York and Canada? If it is believed that massive controls should be imposed now in order to prevent unmanageable disasters ahead, who should pay the costs, which by all estimates will be in the billions each year? Is it fair to pass laws that burden already heavily regulated utilities, or should the costs of installing more effective scrubbers be borne by us all?

Any search for answers as to how the responsibility for environmental degradation should be shared will invariably lead us to further questions about how polluters and pollutants should be controlled. The first recourse recommended in all these matters is usually the judicial system. Let the courts decide who is guilty and how they are to be stopped! We have already described some of the shortcomings of this approach in Chap. 3. To reiterate: The courts move slowly, few individuals (whether plaintiff or defendant) can afford the process, and overreliance on the law promotes minimal compliance. In environmental cases the difficulties are compounded. A judge cannot be expected to be a specialist on health, safety, and the environment, nor can he or she usurp the powers of the legislature and prescribe control mechanisms. The most a judge can do is to guarantee a fair legal process.

Assuring a fair legal process is in itself a major undertaking. Technical information supportive of either the plaintiff or the defendant must be provided by expert testimony. Invariably both sides in a dispute will succeed in marshaling expert witnesses whose testimony will be contradictory. This should not surprise us because in spite of a professed adherence to rationality and commitment to truth, the engineer or scientist testifying will rely on a personal value system in selecting and presenting his or her information, mistrusting other interpretations. Earlier we had occasion to stress the importance of stating one's biases when serving as a legal witness or consultant. In an adversary hearing, admissions of bias and—perhaps more significantly—admissions of uncertainty can be exploited by the opposing side to such an extent that many specialists have become hesitant about giving evidence, to the detriment of all concerned. Rachel Carson, the embattled author of *Silent Spring,* described the consequences of this problem in a letter to a friend:

> I'm convinced there is a psychological angle in all this that people, especially professional men, are uncomfortable about coming out against something, especially if they haven't absolute proof that "something" is wrong, but only a good suspicion. So they will go along with a program about which they privately have acute misgivings (quoted in Graham, 1970, 23).

Those experts who agree to testify open themselves to a barrage of hostile questions which can become personal in nature. Considerable embarrassment can occur if the truth is stretched or hidden. In the Reserve Mining Company trials some of the witnesses did not fare well. One expert who had served as a consultant to the plaintiffs later testified for the defendant, whom he could furnish with inside information. One expert claimed to have a Ph.D. degree from Purdue when in fact he did not. One consultant to Reserve was removed from a National Academy of Sciences committee studying the effects of asbestos in drinking water when it was disclosed he had agreed to provide Reserve with information on the confidential deliberations of the committee. There were conflicts of interest linking individual employees and offices of the State of Minnesota, the University of Minnesota, and Reserve, as well as the judges hearing the case. How believable can expert testimony be under such circumstances?

With so many uncertainties, one approach would be not to release a new product or process until it is shown to be risk-free. Industry and its engineers might claim that this is tantamount to declaring something to be guilty merely because innocence cannot be established. Is this not against our legal principles? Why does the U.S. Department of Agriculture declare poultry with 5 parts per million (ppm) or more of PCBs to be unfit for human consumption? The 5-ppm limit would be difficult to justify, just as a higher limit would be dangerous to defend, on strictly scientific grounds (Grundy, Weisbrod, and Epstein, 1974, 112). Where should the burden of proof lie? It is estimated that victims of Kanemi rice-oil disease had taken in at least 500 mg of PCBs. Translated into pounds of chicken contaminated with 5 ppm of PCBs, this would amount to 220 pounds of chicken or over 4 pounds of chicken consumed per week for a year. If this seems like a lot of chicken to eat, how much less would suffice to poison people with PCBs? In addressing the problem of where the burden of proof should lie, Garrett Hardin states:

> In criminal law, as practiced in Britain and America, a man is "innocent until proven guilty...." Scientists, however, see things otherwise. Science is an occupation in which most experiments fail.... Confronted with any new, untried, nostrum, a scientist, if called upon to place a bet, will bet that it won't work. Such is the conservative judgment (Hardin, 1968, 58–59).

Engineers have more confidence in their projects and will therefore chafe at such an interpretation. But they would have to be omniscient if they could foresee all the environmental effects of their work. And they would have to think themselves omnipotent if they hoped to control those consequences without redesigns or retrofits. Accordingly, they must adopt the viewpoint of engineering as experimentation. The guilty-until-proven-innocent approach will then appear less unreasonable. Indeed it is not threatening since it applies to things and not to persons. If the expected benefit of a project or product is so great that prolonged testing seems an unreasonable cost, a preliminary release could be authorized—provided that "safe exits" existed and funds were available for subsequent corrective action, should it become necessary.

Internalizing Costs of Environmental Degradation

When we are told how efficient and cheap many of our products and processes are—from agriculture to the manufacture of plastics—the figures usually include only the direct costs of labor, raw materials, and use of facilities.

If we are quoted a dollar figure, it is at best an approximation of the price. The true cost would have to include numerous indirect factors such as the effects of pollution, the depletion of energy and raw materials, and social costs. If these, or an approximation of them, were internalized—that is, added to the price—then those for whose benefit the environmental degradation had occurred could be charged directly for corrective actions. The problem with the "technofix" approach—using technology to repair the damages of technology—is not so much with physical realization as it is with the financial burden. As taxpayers are beginning to revolt against higher levies, the method of having the user of a service or product pay for all its costs is gaining more favor. The engineer must join with the economist, the natural and physical scientists, the lawyer, and the politician in an effort to find acceptable mechanisms for pricing and releasing products so that the environment is protected through truly self-correcting procedures rather than adequate-appearing yet often circumventable laws. But again, as we did in our discussion on designing for safety in Chap. 4, we wish to point out here that good design practices may in themselves provide the answers for environmental protection without added real cost. For example, consider the case of a lathe which was redesigned to be vibration-free and manufactured to close tolerances. It not only met occupational safety and health standards with regard to noise, which its predecessor had not, but it also was more reliable, more efficient, and had a longer useful life, thus offsetting the additional costs of manufacturing it (Melman, 1982, 176).

Technology Assessment

The Congress of the United States of America has an Office of Technology Assessment. It prepares studies on the social and environmental effects of technology in areas such as cashless trading (via bank card), nuclear war, or pollution. At the federal and state levels, many large projects must be examined in terms of their environmental impact before they are approved. The purpose of all this activity is praiseworthy. But how effective can it be?

Engineers, it is often said, tend to find the right answers to the wrong questions. The economist Robert Theobald made the following comment on education:

> The university is ideally designed to insure that you remain certain that you know the answers to questions that other people posed long ago. The problem today is that the questions we should be answering are not yet known. Unfortunately the process required for discovering the right questions is totally different from the process of discovering the right answers (quoted in Thrall and Starr, 1972, 17).

It should be quite apparent that it is not easy to know what questions to ask. And technology assessment and other forecasting methods suffer because of this.

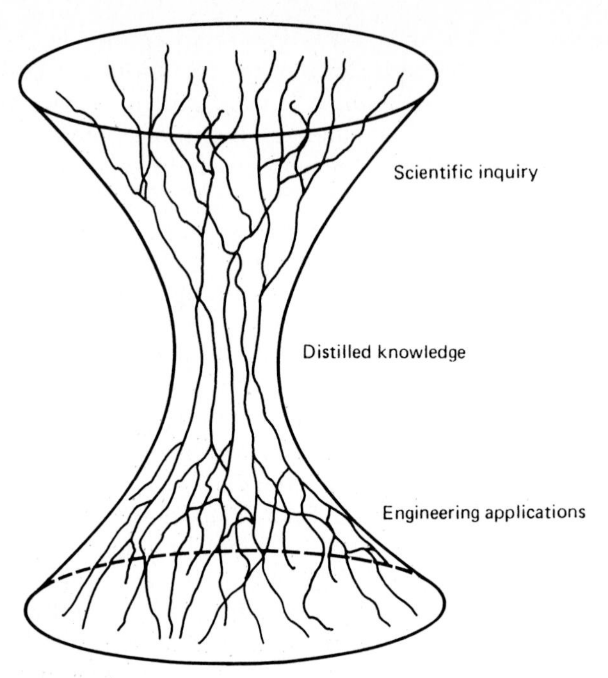

FIGURE 7-3
Distilling and applying knowledge.

When scientists conduct experiments, they endeavor to distill some key concepts out of their myriad observations. As shown in Fig. 7-3, a funnel can be used to portray this activity. At the narrow end of the funnel we have the current wisdom, the state of the art. Engineers make use of it to design and build their projects. These develop in many possible directions, as shown by the shape of the lower, inverted funnel. The difficult task of technology assessment and environmental impact analyses is to explore the extent of this spread and to separate the more significant among the possibly adverse effects.

The danger in any assessment of technology is that some serious risks can easily be overlooked while the studies and subsequent reports, properly authenticated by the aura of scientific methodology, assure the decision maker that nothing is amiss—or perhaps that perceived risks are more serious than they really are. We do not wish to belittle such efforts and we think that they are worthwhile, if only because of those questions they raise—and answers they uncover—that otherwise might not have surfaced. But there is a danger in believing that no further action is required once the reports have been approved and filed. Our contention remains that engineering must be understood as social experimentation and that the experiment continues, indeed enters a new phase, when the engineering project is implemented. Only by careful monitoring will it be possible to gather a more complete picture of the tangled web of effects encompassed in Fig. 7-3 within the inverted, lower funnel.

Study Questions

1 Describe how products or processes in an engineering field of interest to you can damage the environment, including human beings. Then examine pertinent laws designed to protect the environment at the local, state, and national levels and determine their effectiveness. Discuss alternatives to the regulatory approach.

2 Write an essay on the topic "Why Save Endangered Species?" or "Why Save the Everglades?" You may wish to discuss some of the following points: Is it reasonable to put much effort into saving endangered species from extinction? What, if any, moral theories apply to plant and animal life and our relation thereto? Are we duty bound to preserve all forms of natural life, or does such life enjoy an intrinsic right to exist? (Compare with Rescher, 1980, essay 7; and with Regan, 1984).

3 Are there any moral reasons why landowners may on occasion be constrained from doing to their properties what they please?

4 Phosphate detergents were alternately promoted and banned between 1947 and 1971. This caused great confusion in some communities (Lawless, 1977, 450). Critically examine the use of phosphate detergents as an experiment involving the environment.

5 Environmentalists are often accused of being elitists who wish to preserve the environment for their own enjoyment without regard for the needs of others. Granted that environmental controls may be necessary to preserve our habitat in the long run, do rich people (or nations) have the right to impose such controls now when they will harm poor people (or nations) more in the short run than they harm the rich?

6 Consider the following example of environmental side effects cited by Garrett Hardin:

> The Zambesi River...was dammed...to create the 1700-square-mile Lake Kariba. The effect *desired:* electricity. The "side-effects" *produced:* (1) destructive flooding of rich alluvial agricultural land above the dam; (2) uprooting of long-settled farmers from this land to be resettled on poorer hilly land that required farming practices with which they were not familiar; (3) impoverishment of these farmers...[and various other social disorders]; (6) creation of a new biotic zone along the lake shore that favored the multiplication of tsetse flies (Hardin, 1968, 68).

Similar problems have occurred when dams were built in the United States and when the Aswan Dam was erected on the Nile. One might ask if the original purpose may not itself begin to look like merely a side effect. If so, Hardin asks, can we *never* do anything? Describe under what conditions you think a dam such as the one on the Zambesi River should be built and operated. To whom is the engineer in charge of its construction ultimately responsible?

7 On January 15, 1919, a large wooden tank, 58 feet high and over 90 feet in diameter and holding over 2.5 million gallons of molasses, suddenly burst open. The immediate vicinity around this tank, which stood in the north end of Boston, was quickly

covered by a stream of molasses which reached up to the second floor of many buildings. Numerous people were trapped by the sticky fluid and drawn down into it. What simple measure could have prevented the fluid from spreading and creating such an unusual "molasses environment"? Even today there are tanks holding vast quantities of water, oil, or liquefied natural gas where no adequate measures have been taken to protect the public from the dangers they present.

COMPUTER ETHICS

Computers have become the technological backbone of our society. Their degree of sophistication, range of applications, and sheer numbers continue to increase. Through networks they span the globe. Yet electronic computers are still only a few decades old, and it is difficult to foresee all the moral problems which will eventually surround them.

The present state of computers is sometimes compared to that of the automobile in the early part of this century. At that time the impact of cars on work and leisure patterns, pollution, energy consumption, and sexual mores was largely unimagined. If anything, it is more difficult to envisage the eventual impact of computers since they are not limited to any one primary area of use comparable to a car's function in transportation.

It is already clear, however, that computers will cause or contribute to a variety of moral problems. To deal with these problems a new area of applied ethics called *computer ethics* is developing (Johnson, 1985; Johnson and Snapper, 1985). Computer ethics has special importance for the new groups of professionals emerging with computer technology—for example, designers of computers, programmers, systems analysts, and operators. To the extent that engineers design, manufacture, and apply computers, computer ethics is a branch of engineering ethics. But the many professionals who use and control computers share the responsibility for their applications.

Some of the issues in computer ethics concern shifts in power relationships resulting from the new capacities of computers. Others concern abuse of existing installations, as when computers are used in criminal activities and invasions of privacy. Still others relate to assessments of the desirable directions of new computer applications. As we shall see, the issues are sometimes similar to ones we have discussed earlier, but here the computer connection gives them a special flavor.

Power Relationships

Computers dramatically increase the ability of centralized bureaucracies to manage enormous quantities of data, involving multiple variables, and at astonishing speed. During the 1960s and 1970s social critics became alarmed at the prospect that computers would concentrate power in a few centralized bureaucracies of big government and big business, thereby eroding democratic systems by moving toward totalitarianism.

These fears were not unwarranted, but they have lessened due to recent developments in computer technology. In the early stages of computer development there were two good reasons for believing that computers would inevitably tend to centralize power (Simon, 1979). Early large computers were many times cheaper to use when dealing with large tasks than were the many smaller computers it would have taken to perform similar tasks. Thus it seemed that economics would favor a few large and centrally located computers, suggesting concentration of power in a few hands. Moreover, the large early computer systems could only be used by people geographically close to them, again implying that relatively few people would have access to them.

The invention, development, and proliferation of microcomputers changed all this. Small computers became increasingly powerful and economically competitive with larger models. Furthermore, remote access and time sharing allowed computer users in distant locations to share the resources of large computer systems. These changes opened new possibilities for decentralized computer power.

For example, it was once feared that computers would give the federal government far greater power to control nationally funded systems, such as the welfare and medical systems, lessening control by local and state governments. But in fact data systems have turned out to be two-way, allowing both small governments and individuals to have much greater access to information resources amassed at the federal level.

Computers are powerful tools which do not by themselves generate power shifts. They contribute to greater centralization or decentralization insofar as human decision makers so direct them.

This is not to say that computers are entirely value-neutral. For in addition to their hardware there is software—and programs can quite easily be biased, just as can any form of communication or way of doing things. For example, a computerized study of the feasibility of constructing a nuclear power plant can easily become biased in one direction or another if the computer program is developed by a group entirely pro- or anti-nuclear energy (Johnson, 1985, 81).

Computer Abuse

The term "computer abuse" does not refer merely to the damaging of computers, although it includes that kind of problem. As we will define it, computer abuse is any unethical or illegal conduct in which computers play a central role (whether as instruments or objects), as opposed to their playing an incidental or peripheral role. The distinction between central and incidental roles is somewhat rough, but a few examples will help to clarify it.

Computers are only incidentally involved when extortion is attempted via a phone which is part of a computerized telephone system (Kling, 1980, 408). By contrast, computers are centrally involved when an extortionist disguises his voice by means of a computer as he talks into a phone. And computers are even more centrally involved when an unauthorized person uses a telephone computer system to obtain private phone numbers, or when someone maliciously alters or scrambles the programming of a telephone computer, or when a private or governmental hit squad uses computers to identify fingerprints of its intended victims.

Finally there are "hackers" who compulsively challenge any computer security system. Some carry their art to the point of implanting "time bombs" or "Trojan horses" or "viruses" that will "choke networks with dead-end tasks, spew out false information, erase files, and even destroy equipment." (Marshall, 1988).

Computer Theft and Fraud

Some of the most commonly discussed cases of computer abuse are instances of outright theft and fraud. There are many forms of computer theft and fraud. For example: (1) stealing or cheating by employees at work; (2) stealing by nonemployees or former employees; (3) stealing from or cheating clients and consumers; (4) violating contracts for computer sales or service; (5) conspiring to use computer networks to engage in widespread fraud; (6) advertising "next-generation" computers in a misleading manner; (7) extortion practiced by the sole programmer or operator of a computer installation who has a unique knowledge of how it works or is programmed.

Public interest has often been drawn to the glamorous capers of computer criminals (Whiteside, 1978). Enormous sums of money have been involved. The average amount stolen in conventional embezzlement is $19,000. The amount for an average computer-related embezzlement is $430,000, and many millions are often involved. Yet the giant thefts uncovered are believed to be only a small fraction of computer theft—less than 1 percent (T. Logsdon, 1980, 163–164).

Crime by computer has proven to be unusually inviting to many. Computer crooks tend to be intelligent and to view their exploits as intellectual challenges. In addition, the computer terminal is both physically and psychologically far removed from face to face contact with the victims of the crimes perpetrated. Unlike violent criminals, computer criminals find it easy to deceive themselves into thinking they are not really hurting anyone, especially if they see their actions as nothing more than pranks. In addition there are often inadequate safeguards against computer crime. The technology for preventing crime and catching criminals has lagged behind implementation of new computer applications. Computers reduce paperwork, but this has the drawback of removing the normal trail of written evidence involved in conventional white-collar crime (forgeries, receipts, etc.) Finally, the penalties for computer crime, as for white-collar crime in general, are mild compared to those for more conventional crimes.

Computer crime raises obvious moral concerns related to basic issues of honesty, integrity, and trust. It also forces a rethinking of public attitudes about crime and its punishment. Is it fair that the penalty for breaking into a gas station and stealing $100 should be the same as for embezzling $100,000 from a bank? How should society weigh crimes of minor violence against nonviolent crimes involving huge sums of money?

The potential for computer crime should enter significantly into the thinking of engineers who design computers. In fact, protection against criminal abuse has become a major constraint for effective and successful design of many computer systems and programs. Engineers must envisage not only the intended context in which the computer will be used, but both likely and possible abuses.

For some time secret computer passwords have been used as a security feature. More recently introduced, and still of limited effectiveness, is data encryption. This technique is widely employed to prevent theft from funds transfer systems. In data encryption, messages are scrambled before transmission and unscrambled after reception according to secret codes. Such devices, of course, require special precautions in maintaining confidentiality and security, and engineers have a major role to play in making recommendations in these areas.

Privacy

Storage, retrieval, and transmission of information using computers as data processors has revolutionized communication. Yet this very benefit poses serious moral threats to the right to privacy. By making more data available to more people with more ease, computers make privacy more difficult to protect. Here we will discuss privacy and confidentiality in regard to individuals, but the issues are similar in relation to corporations.

In the Chap. 6 discussion of employee rights, we defined the right to privacy as the right to control access to and use of information about oneself. We indicated that the right was vitally important for a variety of reasons, and could be grounded within each of the major types of ethical theories. Here we will take note of some of the threats to that right from computerized data centers.

Imagine that you are arrested for a serious crime you did not commit—for example, murder or grand theft. Records of the arrest, any subsequent criminal charges, and information about you gathered for the trial proceedings might be placed on computer tapes easily accessible to any law enforcement officer in the country. Prospective employers doing security checks could gain access to the information. The record clearly indicates that you were found innocent legally. Nevertheless that computerized record could constitute a standing bias against you for the rest of your life, at least in the eyes of many people with access to it.

The same bias could exist if you had actually committed some much less serious crime—say a misdemeanor. If you were arrested when you were 15 for drinking alcohol or swearing at an officer, for example, the record could stay with you. Or imagine that medical data about your visits to a psychiatrist during a period of depression could be accessed through a data bank. Or that erroneous data about a loan default were placed in a national credit data bank to which you had limited access. Or merely suppose that your tastes in magazine subscriptions were known easily to any employer or ad agency in the country.

The potential abuses of information about us are unlimited and become more likely with the proliferation of access to that information. For this reason a series of new laws has been enacted (Rule, McAdam, Stearns, and Uglow, 1981). For example, the 1970 Fair Credit Reporting Act restricted access to credit files. Information can be obtained only by consumer consent or a court order, or for a limited range of valid credit checks needed in business, employment, and insurance transactions or investigations. The act also gave consumers the right to examine and challenge information about themselves contained in computerized files.

The Privacy Act of 1974 extended this right of inspection and error correction to federal government files. It also prohibited the information contained in government files from being used for purposes beyond those for which it was originally gathered unless such use was explicitly agreed to by the person whose file it is. Numerous other laws have been passed and are being considered to extend protection of individual privacy within private business and industry.

Such laws are expensive to implement, sometimes costing tens and hundreds of millions of dollars to enforce. They also lessen economic efficiency. In special circumstances they can have harmful effects on the public. There is little question, for example, that it would save lives if medical researchers had much freer access to confidential medical records. And it would be much more convenient to have one centralized National Data Center. This idea was proposed in the mid-1960s and is still alive in the minds of many. But privacy within a computerized world can apparently be protected only by making it inconvenient and expensive for others to gather data bank information on us.

Earlier we mentioned that inadvertent errors in data banks can also constitute a serious problem. Consider the following episode reported in the French press in 1979 and recounted by Jacques Vallee.

The three young men who bought gas at a filling station outside of Paris, France, did not look trustworthy to Mr. Nicholas, the station owner. Something about their clothes, the mended license plate, and the hurriedly scrawled signature on the check spurred him to call the police after the trio had left. Sure enough, police files queried by a computer revealed that the car had been stolen. A special police team was dispatched, and the stolen car was intercepted at a red light. Two plainclothes police officers approached the suspects, the only one in uniform remaining in the patrol car. One of the young men, Marcel Seltier, reported, "We did not understand anything. We saw the one with the gun aim at Claude. A moment later a shot rang out.

The bullet went through the windshield and hit Claude's face just under the nose. We thought they were gangsters."

This is the case of the trigger-happy police who in 1979 dispensed justice on the spot under the erroneous impression that the suspects were dangerous. Undoubtedly the lawmen discharged their weapons in response to some inadvertently suspicious movements on the part of the three young men still in the car. The car had indeed been stolen several years earlier. In the meantime, it had been recovered and resold by the insurance company (Vallee, 1982; Schinzinger, 1986).

The error of not recording the final sale cannot be blamed on the computerized data file as such. Nevertheless, a system which combines rapid remote access to information with instant life and death decisions is ill conceived if its users are not trained to expect occasional human or machine errors and act accordingly. A sobering thought concerns error corrections, even when they are undertaken in good faith. To quote Vallee:

> When data is "expunged" or deleted from a computer (following a judge's order to purge a certain record, for example), it is generally *not* true that the data can also be deleted from the older "back-up" versions of that computer's memory, which are kept in a secure vault somewhere in case the information ever needs to be reconstructed after a system failure. Thus, the "expunged" data could still be examined if someone really had the inclination to search for it (Vallee, 1982, 30).

Instant reaction to wrong information—computer-generated and imbued with the aura of authenticity—should also make us worry about military reliance on fully automated missile warning and delivery systems, regardless of where in the world we may live.

Further Social Issues

Crime, privacy, and power shifts have been the focus of most discussions about computer ethics. More attention, however, is now being given to a variety of other social and professional issues. Let us survey some of these, beginning with social issues.

1 Selective disclosure of their views by politicians has always occurred. In a speech to a conservative group a candidate for senator will tend to say very different things from what he or she tells a liberal group. Different topics may be discussed, different emphases given, and inconsistent remarks made. Computers make it possible to turn this political maneuver into a science. Dozens of types of letters can be worded to appeal to different kinds of groups. Each type reveals only a fraction of the candidate's views—just that portion which will be most attractive to a selected sociopolitical group.

The information about these groups of people is obtained by computer from public records. The characterizations of the groups' attitudes and norms are computer-generated. The letters sent to them are personalized by computer. And the mailing process is also computerized. With electronic accuracy and efficiency, politicians are enabled to have many different faces when viewed by different groups.

Several moral issues are raised by this possible application of technology: (1) Does such selective disclosure constitute deception? (2) Does filtering the truth about a politician's views undermine the autonomy of voters in making decisions? (3) Since use of computers is expensive, is it fair that the rich have more extensive access to this technology, or should equal-time laws for television be extended to computers? (Maner, 1980, 2–8).

2 Many new occupations have been created by computers: data processing, programming, information science, key punch operating, computer systems analysis, computer sales positions, and so forth. Should the importance of responsible behavior by these groups lead the public to seek their full professionalization and require licensing, registration, or continuing education? What should the public demand by way of accountability from professional organizations representing these groups?

3 Most existing professional groups are finding their work altered by computers. This will have subtle effects on areas of personal accountability. For example, it was once clear that doctors were fully responsible for the diagnoses they made: Errors in judgment were *their* errors. Increasingly, however, diagnoses and other medical decisions are and will come to be made with varying degrees of computer assistance. Health data on patients may be collected with computerized instruments and then entered into a computer which is programmed to diagnose illness on the basis of the medical symptoms entered. Under such conditions, when would a doctor be responsible for a mistaken diagnosis that resulted from a computer error? When would a doctor be blameworthy for overreliance on a mistaken computer diagnosis?

4 Computers have led and will continue to lead to elimination of jobs. What employer attitudes are desirable in dealing with this situation? No employer, of course, can afford to pay people for doing no work. Yet especially within large corporations it is often possible to readjust work assignments and work loads and to wait for people to retire or to change jobs voluntarily before laying off employees. Such benign employment practices have often been embraced from prudential motives to prevent a public and employee backlash against introduction of computer technologies which eliminate jobs (Kling, 1979, 10), but moral considerations relating to human costs should be weighed even more heavily.

5 There are also questions concerning public accountability of businesses using computer-based services. It can be made very difficult or relatively simple for a consumer to notice and correct computer errors or computer-printed errors. For example, a grocery-store receipt can itemize items by obscure symbols or simple words understandable to a customer. Here again moral reasons reinforce long-term good business sense in favoring policies beneficial to consumer needs and interests.

6 Program trading is the automatic, hands-off trading by computer of stocks, futures, and options on the stock market. Did this practice contribute to the "meltdown on Black Monday" (October 19, 1987), when the U.S. stock market took a precipitous plunge, and should it be controlled?

7 The U.S. Department of Defense is supporting the creation of autonomous weapons which can be aimed and fired by on-board computers which make all necessary decisions, including enemy identification. Computer scientists and engineers are divided over the advisability of such a major step toward automation of the battlefield.

8 There is a dangerous instability in computerized defense systems even if they are working perfectly. Let us assume then that all the nuclear warning software works without error, and that the hardware is fail-safe. Nevertheless, the combination of two such correctly functioning but opposing systems together is unstable. This is because secrecy prevents either system from knowing exactly what the other is doing, which means that any input which could be interpreted as a danger signal must be responded to by an increase in readiness on the receiving side. That readiness change, in turn, is monitored by the opposing side which then steps up its readiness, and so on. This feedback loop triggers an escalating spiral. Does the possibility of an entirely unprovoked attack triggered by the interaction of two perfectly operating computer-based systems as described by Raushenbakh (1988) enhance security?

Professional Issues

Many of the issues in engineering ethics which we have dealt with earlier arise again within the context of computer work. New variations or new difficulties may be involved, often owing to the high degree of job complexity and required technical proficiency introduced by computers. Such was the case, for example, in the whistle-blowing case of Virginia Edgerton given as a study question in Chap. 6. A number of interesting case studies have been collected by Donn Parker in his excellent book, *Ethical Conflicts in Computer Science and Technology*. The following examples are based on some of Parker's cases.

Safety Dependence on computers has intensified the division of labor within engineering. For example, civil engineers designing a flood control system have to rely on information and programs obtained from systems analysts and implemented by computer programmers. Suppose the systems analysts refuse to assume any moral or legal responsibility for the safety of the people affected by the flood control plans, arguing that they are merely providing tools whose use is entirely up to the engineers. Should the civil engineers be held accountable for any harm caused by poor computer programs? Presumably their accountability does extend to errors resulting from their own inadequate specifications which they supply to the computer experts. Yet should not the engineers also be expected to contract with computer specialists who agree to be partially accountable for the end-use effects of their program? (Parker, 1979, 34–38)

Ownership Rights Consider an engineer who develops a program used as a tool in developing other programs assigned to her. Subsequently she changes jobs and takes the only copy of the first program with her for use on her new job. Suppose first that the program was developed on company time under the first employer's explicit directives. Taking it to a new job without the original employer's consent would be a violation of that employer's right to the product (and possibly a breach of confidentiality). As a variant situation, however, suppose the program was not written under direct assignment from the first employer, but was undertaken by the engineer at her own discretion to help her on her regular work assignments. Suppose also that to a large extent the program was developed on her own time on weekends, although she did use the employer's facilities and computer services. Did the employer own or partially own the program? Was she required to obtain the employer's permission before using it on the new job? (Parker, 1979, 72–74)

Whistle-Blowing An engineer working as a computer programmer played a minor role in developing a computer system for a state department of health. The system stored medical information on individuals identified by name. Through no fault of the engineer, few controls had been placed on the system to limit easy access to it by unauthorized people. Upon learning of this the engineer first informed his supervisor and then higher management, all of whom refused to do anything about the situation because of the anticipated expense required to correct it. In violation of the rules for using the system, the programmer very easily obtained a copy of his own medical records. He then sent them to a state legislator as evidence for his claims that the right of citizens to confidentiality regarding such information was threatened by the system. Was his behavior improper? Was his subsequent firing justified? (Parker, 1979, 90–93)

Employer Authority and Professional Rights A project leader working for a large retail business was assigned the task of developing a customer billing and credit system. The budget assigned for the project appeared at first to be adequate. Yet by the time the system was half completed it was clear the funds were not nearly enough. The project leader asked for more money, but the request was denied. He fully informed management of the serious problems which were likely to occur if he had to stay within the original budget. He would be forced to omit several important program functions relating both to convenience and to safety: for example, efficient detection and correction mechanisms for errors, automatic handling and reporting of special customer exceptions, and audit controls. Management insisted that these functions could be added after the more minimal system was produced and installed in stores. Working under direct orders, the project leader completed the minimal system, only to find his worst fears realized after it was installed. Numerous customers were given incorrect billings or ones they could not understand. It was easy for retail salespersons to take advantage of the system to steal from the company, and several did so. Within a year the company's profits and business were beginning to drop. This led to middle-level management changes, and the project leader found himself blamed for designing an inadequate system.

Did the project leader have an obligation either to clients or to the company to act differently than he did? Did he have a moral right to take further steps in support of his original request or later to protect himself from managerial sanctions? (Parker, 1979, 109–111)

Parker's example comes close to illustrating a common failing among computer experts: Selling the customer (or the boss) on a more complex system than is warranted or promising more capacity and faster delivery than is achievable. The latter was the case with a revolutionary, computer based system for trust accounts that the Bank of America had to abandon after buying it for $20 million, spending another $60 million trying to make it work, and wasting several years in the process. (Frantz, 1988)

Informed Consent A team of engineers and biomedical computer scientists develop a system for identifying people from a distance of up to 200 meters. A short tube attached to a sophisticated receiver and computer, and aimed at a person's head, reads the individual's unique pattern of brain waves when standard words are spoken. The team patents the invention and forms a company to manufacture and sell it. The device is an immediate success within the banking industry. It is used to secretly verify the identification of customers at tellers' windows. The scientists and engineers, however, disavow any responsibility for such uses of the device without customer notification or consent. They contend that the companies which buy the product are responsible for its use. They also refuse to be involved in notifying public representatives about the product's availability and the way it is being used.

Does employing the device without customer awareness violate the right to privacy or to informed consent? Do the engineers and scientists perhaps have a moral obligation to market the product with suggested guidelines for its ethical use? Should they be involved in public discussions about permissible ways of using it? (Parker, 1979, 126–128)

Study Questions

1 Present and defend your answers to the questions raised about the cases based on Parker's examples.

2 Look up the following terms in a legal dictionary: fraud, theft, extortion, sabotage, vandalism, burglary. Then imagine and briefly describe an example of a computer crime falling under the definition of each.

3 Write a short research paper exploring the threats to privacy posed by data banks. In your essay comment on some specific advantages and disadvantages of having one centralized national data bank which pools all available government information on citizens.

4 The U.S. Office of Technology Assessment has reported that new rules may soon be needed for computer services concerning the prediction of criminal and financially negligent behavior. People are likely in the future to be denied credit, employment, and insurance because they fall into certain categories based on statistical correlations developed by computers.

Comment on who you think should establish these new rules, and give any examples of rules you would favor. For example, do you think it would be fair to allow race or sex to be used as a variable in this context?

5 Tracy Kidder's Pulitzer Prize–winning book, *The Soul of a New Machine,* recounts the human drama behind the development of a minicomputer by Data General Corporation (Kidder, 1981). Read the book, and as you do so, identify any moral issues which may arise in the course of creating a new computer system and bringing it to market.

6 The following warning to parents whose children use home computers was carried by the Associated Press (*Los Angeles Times,* 25 Dec. 1987, p. I-47): "In recent years more sexually oriented materials have been showing up for home computers—some on floppy disks with X-rated artwork and games, and other accessed by phone lines from electronic bulletin boards... with names like Cucumber,... Orgy, Nudepics, Porno, Xpics, and Slave."

Discuss the ethical issues raised by pornography in this new medium. Should there be controls? How can access be denied to children? Are there any parallels with constraints on the use of the postal service?

7 In 1985 computer scientist David L. Parnas resigned from an advisory panel on computing in support of battle management, a nine-member group established by the Strategic Defense Initiative office in the Department of Defense. He explained his position in a position paper (Parnas, 1985) and a year later, in response to criticism, in a letter in which he writes:

> I resigned... when it became clear to me that the panel would not give serious consideration to the question of feasibility. Even at that time I did not go public. I sent copies of my position to Government officials, not to the press. After copies had been leaked to the press by others, it became clear to me that the SDI management's response was damage control, not serious consideration of the issues that I had raised. Only then did I agree to explain my position in public (Parnas, 1986).

Seek further information on this case and write a report on Parnas's actions. Did he act as whistle-blower? If so, as a responsible one?

WEAPONS DEVELOPMENT

Much of the world's technological activity has a military focus. Based just on size of expenditures, direct or indirect involvement of engineers, and startling new developments, military technology would deserve serious discussion in these pages. The moral problems it raises, however, are of a magnitude which makes the other cases we have treated so far pale in comparison. The continuing automation of the battlescene has made military activity seem less troublesome and more appealing to technologically advanced nations, while the real horrors of total war—whether fought with classical weapons, hydrogen bombs, or biochemical agents—are too far beyond the comprehension of many diplomats, engineers, and younger generals to guide their decisions.

Let us first take a leap back into history and see what has not changed, which lessons we have not learned. When Xerxes reached the Hellespont on his westward march, he had a bridge built over the waterway. It is said that the bridge succumbed to a storm, the engineers were beheaded, the water was given a ceremonial flogging, and a second bridge was ordered. This time no chances were taken as 674 galleys in two rows were tied together with flax ropes weighing 50 pounds to the foot. The roadbed of planks, brush, and earth withstood the crossing of 150,000 soldiers and several times that number of noncombatants (De Camp, 1963, 89).

There are several reasons for an engineer to do his or her best on a military job. High among them are patriotism and prudential interest. The latter can be occasioned by threats from an easily displeased ruler or by the lure of commercial success. Xerxes' bridge builders certainly had compelling reasons to build a good bridge. With their experience they might even have benefited from staying on at the site, offering to build bridges for other armies who came along. (Indeed, the next that came along was Xerxes' army on retreat.)

The Weapons Seesaw

The trade in arms and military know-how has a long tradition. Today military expenditures throughout the world total about $900 billion dollars annually. Of this amount, one-quarter is earmarked for purchases of weapons and related equipment, 17 percent of which are traded internationally.

Among the world's most successful arms merchants and manufacturers was the family Krupp. Krupp was an expert at exploiting the defensive-offensive weapons seesaw. This is how it worked, according to William Manchester in his book *The Arms of Krupp:*

> Having perfected his nickel steel armor, Fritz advertised it in every chancellery. Armies and navies invested in it. Then he unveiled chrome steel shells that would pierce the nickel steel. Armies and navies invested again. Next—this was at the Chicago fair, and was enough in itself to justify the pavilion—he appeared with a high-carbon armor plate that would resist the new shells. Orders poured in. But just when every general and admiral thought he had equipped his forces with invincible shields Fritz popped up again. Good news for the valiant advocates of attack: it turned out that the improved plate could be pierced by "capped shot" with

> explosive noses, which cost like the devil. The governments of the world dug deep into their exchequers, and they went right on digging. Altogether thirty of them had been caught in the lash and counterlash. Fritz showed the figures to the emperor, who chuckled. He would have gagged if he had known the truth: he himself was being trapped in a variation of the seesaw (Manchester, 1970, 248).

In the Far East the firm of Krupp (Germany) was joined by Vickers (Great Britain) and Schneider (France) in supplying arms to the Chinese, the Japanese, and the Russians. Sir William White, chief designer of the British Admiralty and later director of warship construction at Armstrong, Whitworth and Co., was said not to be unwilling

> to play the part of the *honnete coutier* by pointing out the growth of the Japanese navy to his Chinese clients, or of the Chinese to their indomitable rivals.... By such means he was able to increase the profits of the great company which employed him, and to extend what is, perhaps, the most important of our national industries, and to kindle in the hearts of two Asiatic peoples the flames of an enlightened and sacred patriotism (Seldes, 1934, 36).

Everything worked out quite well for the arms merchants when after both the Japanese and Russian fleets had been outfitted, the latter was badly battered during the Russo-Japanese War in 1905, requiring new orders to rehabilitate it.

Japan won the war, but its troops had suffered terrible bloodbaths on land caused by Russian use of Maxim's machine guns. The American inventor Hiram S. Maxim had by then joined forces with the Swedish weapons manufacturer Nordenfeldt and its master salesperson, Boris Zaharoff, who was known as the Mystery Man of Europe and the Merchant of Death. It was Zaharoff who set off an arms race in the Balkans. Among other feats he sold one of Nordenfeldt's clumsy, 1881 model steam-powered submarines to the Greek government as a novelty, "then two to the Turks as a counter to the Greek threat, then two more to the Greeks, and so on" (Slade, 1986).

But back to the Krupps in Essen, Germany. While Krupp men were busy with their steel, Krupp wives did good deeds among the workers' families. This was good stuff for satire, superbly used in George Bernard Shaw's portrayal of the Krupps in *Major Barbara.* Berta Krupp once expressed horror at the fact that the Essen Works would manufacture "bomb cannons." She is said to have been assured that few would dare get into the way of such a cannon—thus it would promote peace. By World War I, heavy ordnance took the form of 420-millimeter howitzers lofting 1-yard-long shells weighing 1 ton each over 9 miles, and of 210-millimeter cannons which fired 264-pound shells at Paris from 76 miles away. Both types of guns were occasionally dubbed Dicke (Big) Berta, some say in honor of Berta Krupp, others say in memory of the monk Berthold Schwarz, the reinventor of gun powder in the western world.

Big Berta was superseded in World War II by Big Dora, two stories high with a barrel 90 feet long and a bore of 840 millimeters. This monster could fire 8-ton shells up to 30 miles. But by then the big guns were no longer as useful. Mobile delivery of destructive force by tanks and aircraft had been perfected, to be followed by rockets and nuclear weapons.

This change in technology was accompanied by a change in strategy. Where armies once faced each other in battle, bombs could now be delivered far beyond the front lines on the enemy's population centers. The scene of battle was thus moved, and the change is reflected in the work of Erich Maria Remarque, whose *All Quiet on the Western Front* dealt with the sufferings of the soldier in the front lines of World War I, but whose World War II novel *A Time to Live, a Time to Die* centered on the home front and dealt with the effects of war on the civilian population.

In the early months of World War II the chiefs of staff of the warring countries would still agonize over the question of whether or not to bomb targets in civilian population centers at night. Toward the end of the war, night raids had become common practice and civilians themselves had become the targets. The agonizing decisions were now up to them: Which child or children do you leave behind if the asphalt roads are so soft from the heat that you cannot possibly escape the flames if you carry all?

To us the atom bombs dropped on Hiroshima and Nagasaki are horrible not only because of the many deaths they caused (single air raids during World War II had killed larger numbers of people), nor because of the ghastly medical consequences for survivors, nor because they were unnecessary (Cousins, 1987, 40-50), but mostly because they ushered in the age of rapid, irretrievable delivery of destructive power in immense concentrations. To deliver the equivalent explosive effect of the Hiroshima bomb would have required 20,000 tons of explosives (TNT). Loaded on railroad hopper cars this would have filled 267 cars making a train 2 miles long, just for one bomb. It would have taken 740 of today's B-52 bombers to carry this load in conventional explosives.

In the early 1960s the Soviet Union exploded hydrogen bombs in the 50- and 60-megaton range for test purposes. Each of those bombs had two to three thousand times the destructive power of the Hiroshima bomb, and it would take a train 4000 to 6000 miles long to carry the TNT required to produce an explosion equivalent to that produced by any one of them. You would have to wait at a railroad crossing for close to 100 hours to let such a train pass. Or if carried by bombers, you would see the sky darkened with one and a half million planes. The total megatonnage of nuclear weapons available to the two major superpowers for instant departure toward targets is about 3300 megatons for the United States (in 10,800 warheads on 2000 launchers, including bombers) and 5800 megatons for the U.S.S.R. (in 9500 warheads on 2700 launchers, including bombers). (Mayers, 1986, 63)

Today's bombs are smaller (though often clustered), mostly carried by missiles, and targetable with great exactness. Countermeasures are being developed and the offensive-defensive seesaw is exercised in the same manner as a century ago. The Office of Technology Assessment of the U.S. government estimates that a large-scale nuclear exchange between the United States and the U.S.S.R. could kill more than 250 million people outright in those two countries alone, let alone those who would suffer lingering illnesses and later deaths from starvation, exposure, or disease. The effects of a nuclear winter created by the reduction in sunlight reaching the earth could spread disaster to all parts of the world, thus affecting nonwarring nations as well.

Those consequences are not disputed. What occupies the best minds are questions regarding national policy: Is the threat of mutual annihilation an effective deterrent for the future? Would unilateral disarmament, or disarmament without adequate safeguards, be an effective indicator of trust which could bring about a deescalation of the arms race and a reduction of the potential for nuclear war? In the meantime the nuclear arsenal becomes bigger and more technically perfect. A comparison of the total tonnage of explosives unleashed by all sides during World War II with what is ready to be used now in an all-out war is graphically portrayed in Fig. 7-4. Over 6000 times more destructive power than was used during the 6 years of World War II could be delivered within hours during a full exchange of hostilities between today's members of the nuclear club.

The Engineer's Involvement in Weapons Work

Instruments of torture leave little to the imagination. Descriptions of the use of the rack, the thumbscrew, and the electric prod convey instant sensations of pain and suffering. Our modern weapons of war, however, at first seem more remote. Unless one has experienced or studied their consequences, they appear quite guiltless in themselves. Even those who deliver their deathly charges against the enemy mostly do so at a distance, making their use seem more acceptable—or at least less unsettling.

How do the men and women who design weapons, manufacture them, and use them feel about their work? Most have reservations about it, but those who stay in it also have reasons to support their continued involvement. The following cases involve real weapons. The people are composites who represent the various positions taken by typical citizens who are also engineers.

1 Bob's employer manufactures antipersonnel bombs. By clustering 665 guava-sized bomblets and letting them explode above ground, an area covering the equivalent of ten football fields is subjected to a shower of sharp fragments. Alternatively the bombs can be timed to explode hours apart after delivery. Originally the fragments were made of steel; now less easily detected plastics are sometimes used, making the treatment of wounds, including the location and removal of the fragments, more time-consuming for the surgeon. Recently another innovation was introduced: By coating the bomblets with phosphorus the fragments could inflict internal burns as well. Thus the antipersonnel bomb does its job quite well without necessarily kill-

Firepower to Destroy a World...Plus

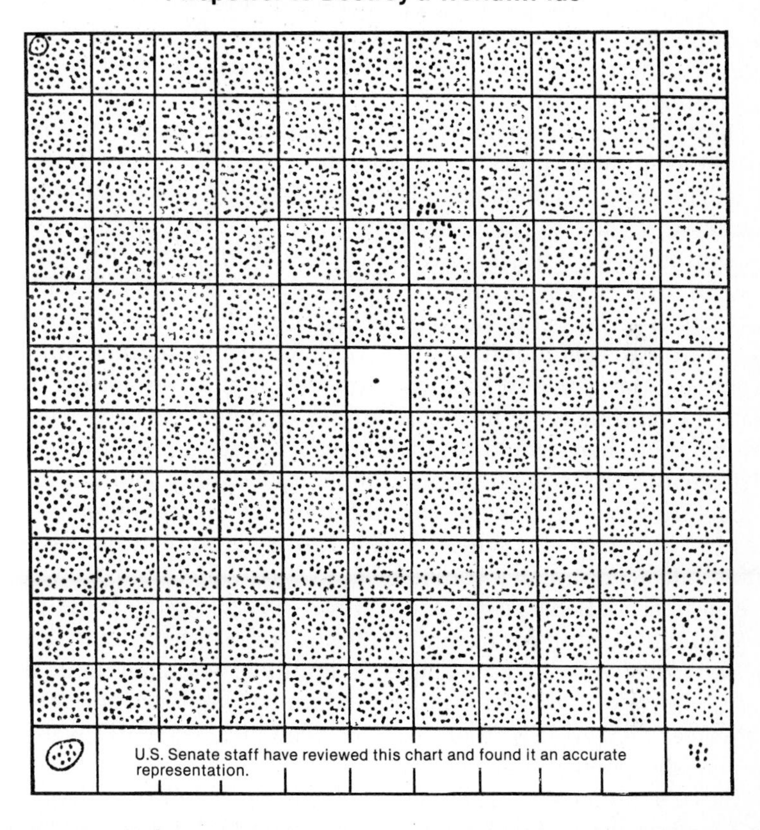

The dot in the center square represents all the firepower of World War II—3 megatons. The other dots represent the firepower in existing nuclear weapons—18,000 megatons (equal to 6,000 WW IIs). About half belong to the Soviet Union, the other half to the U.S.

The top left circle represents the weapons on just one Poseidon submarine—9 megatons (equal to the firepower of 3 WW IIs)—enough to destroy over 200 of the largest Soviet cities. The U.S. has 31 such subs and 10 similar Polaris subs. The lower left circle represents one new Trident sub—24 megatons (equal to the firepower of 8 WW IIs)—enough to destroy every major city in the northern hemisphere. The Soviets have similar levels of destructive power.

Another interpretation: The single dot is the bomb on Hiroshima. If a defensive shield repels only 99 percent of incoming missiles or planes, all the other dots are Hiroshima-rated warheads which get through in an all-out exchange among all nuclear powers.

Researched and drawn by James Geler and Sharyl Green for Parents and Teachers for Social Responsibility. May be copied freely.

FIGURE 7-4

ing in that it ties up much of the enemy's resources in treating the wounded who survive its explosion.

Bob himself does not handle the bombs in any way, but as an industrial engineer he enables the factory to run efficiently. He does not like to be involved in making weapons, but then he tells himself that someone has to produce them. If he does not do his job, someone else will, so nothing would change. Furthermore, with the cost of living being what it is, he owes his family a steady income.

2 Mary is a chemical engineer. A promotion has gotten her into napalm manufacturing. She knows it is nasty stuff. She remembers Professor Wald, a Nobel laureate in biology from Harvard, berating the chemical industry for producing this "most brutal and destructive weapon that has ever been created." But this was when she was in college, during the Vietnam war. Civilians were forever not leaving the fighting zone and then there were complaints about them being hurt or killed. She abhors war like most human beings, but she feels that the government knows more than she does about international dangers and that it is better to fight a war abroad than on our shores. If everyone were to decide on her or his own what to do and what not to do, then there would be utter chaos. Perhaps society can tolerate a few oddballs with their own ideas, but companies certainly should be prepared to manufacture the weapons our armed forces need. Incidentally, if Mary continues to perform well on her job she will be promoted out of her present position into working on a commercial product with much growth potential.

3 Ron is a specialist in missile control and guidance. He is proud to be able to help his country through his efforts in the defense industry. The missiles he works on will carry single or multiple warheads with the kind of dreadful firepower which, in his estimation, has kept any potential enemy in check since 1945. At least there has not been another world war—the result of mutual deterrence, he believes.

4 Marco's foremost love is physical electronics. He works in one of the finest laser laboratories. Some of his colleagues do exciting research in particle beams. That the laboratory is interested in developing something akin to the "death ray" described by science fiction writers of his youth is of secondary importance. More bothersome is the secrecy which prevents him from freely exchanging ideas with experts across the world. But why change jobs if he will never find facilities like those he has now?

5 Joanne is an electronics engineer whose work assignment includes avionics for fighter planes which are mostly sold abroad. She has no qualms about such planes going to what she considers friendly countries, but she draws the line at their sale to potentially hostile nations. Joanne realizes that she has no leverage within the company, so she occasionally alerts journalist friends with news she feels all citizens should have. "Let the voters direct the country at election time"—that is her motto.

6 Ted's background and advanced degrees in engineering physics gave him a ready entry into nuclear bomb development. As a well-informed citizen he is seriously concerned with the dangers of the ever growing nuclear arsenal. He is also aware of the possibilities of an accidental nuclear exchange. In the meantime he is working hard to reduce the risk of accidents such as the thirty-two "broken arrows" (near catastrophic accidents) reported by the Pentagon—or the many others that he knows have occurred worldwide. Ted continues in his work because he believes that only specialists, with firsthand experience of what modern weapons can do, can eventually turn around the suicidal trend represented by their development. Who else can engage in meaningful arms control negotiations?

Our names are fictitious, as are the specific jobs described. But all over the world talented people engage in weapons work. Surely everyone who accepts a job in a war-related industry should also seriously consider his or her motives in doing so. Prudential self-interest is not sufficient to guarantee responsible participation in what must be regarded as humankind's most crucial engineering experiment. Only those who have arrived at morally autonomous, well-reasoned positions for either engaging in or abstaining from weapons work can be counted on to carefully monitor the experiment and try to keep it from running a wild course.

Defense Industry Problems

Across the globe we find nations which confer special privileges on their defense industries without giving sufficient thought to the problems which can accompany large military buildups. Unethical business practices, for instance, occur as in all massive projects, but the urgency of completing a weapons system before it becomes obsolete and the secrecy which surrounds it makes proper oversight particularly difficult.

This is one of the examples we describe briefly below. The other cases address problems which are more serious because they are not as easily recognized. Thus the second example has to do with "technology creep"—the development of new weapons, such as the cruise missile, which can alter diplomatic arrangements even as they are being negotiated. We then examine the issue of secrecy in different contexts. As a last example we take up the overall effect of defense spending on a nation's economy and the distortions it can produce. Our discussion may reveal a United States perspective, but the underlying problems are of a global nature.

The following cases illustrate the various problems just enumerated and should give pause for ethical reflection by both critics and ardent supporters of weapons manufacturing and development.

1 The problem of waste and cost overruns is a continuing one in the defense industry (A. E. Fitzgerald, 1972; Gansler; 1980; Melman, 1970). In Chap. 6 we mentioned one example—the $2 billion cost overruns on development of the C5-A transport plane, overruns reported to the public by Ernest Fitzgerald. Fitzgerald, who is a deputy for management systems in the Pentagon, has been a significant critic of how the defense industry has operated at efficiencies far below commercial standards. He has described how contractors' work forces were swelled with underutilized engineers and high-salary sales personnel, resulting in lavish overhead fees. Or how small contractors were willing to comply with cost-cutting plans, but large suppliers felt secure in not complying. (At present, twenty-five firms hold 50 percent of all defense contracts in the United States, and eight firms conduct 45 percent of all defense research.)

High cost and poor quality were encouraged in various ways: Planned funding levels were leaked to prospective contractors. Cost estimates were based on historical data, thus incorporating past inefficiencies. Costs were cut when necessary by lowering quality, especially when component specifications were not finalized until the contract was completed. Sole-supplier policies gave a contractor the incentive to "buy in" with an artificially low bid, only to plead for additional funds later on. And those funds were usually forthcoming, since the Department of Defense has historically accepted what it knows to be optimistically low development-cost estimates because they stand a better chance of being approved by Congress (Gansler, 1980, 296).

2 In Goethe's poem *Der Zauberlehrling,* the sorcerer's apprentice employs his master's magic incantation to make the broom fetch water. When he cannot remember the proper command to stop the helpful broom, however, he comes near to drowning before the master returns. Technology, as the slave in service of the arms race, resembles the sorcerer's broom. Not only has the world's arsenal grown inordinately expensive (even without graft), and not only does it contribute to a steadily worsening inflation, it has also gained a momentum all its own. George Kennan, a former ambassador to the Soviet Union, has stated:

> I see this competitive buildup of armaments conceived initially as a means to an end but soon becoming the end itself. I see it taking possession of men's imagination and behavior, becoming a force in its own right detaching itself from the political differences that initially inspired it, and then leading both parties, invariably and inexorably, to the war they no longer know how to avoid (Kennan, 1981).

The arsenal is not only growing in size, it is also getting "better." Diplomats may be striving to avert a major conflict, but all the while an exuberance for new developments creates a technology creep which can at times postpone or even upset all negotiations. Nations are suddenly seen to shift to new positions as new devices to more accurately target missiles, or perhaps an entirely new weapon, are reported to be in the offing. This can destabilize (and occasionally stabilize) the political process. Meanwhile, as Jerome Wiesner and Herbert York have written:

> . . . both sides in the arms race are . . . confronted by the dilemma of steadily decreasing national security. *It is our considered professional judgment that this dilemma has no technical solution.* If the great powers continue to look for solutions

> in the area of science and technology only, the result will be to worsen the situation (Wiesner and York, 1964, 27).

The technological imperative that innovations must be implemented should give advocates of preparedness for conventional, limited war some cause for concern as well. Giving in to the excitement of equipping and trying out weapons employing the latest in technology may provide added capability to sophisticated, fully automatic systems such as intercontinental ballistic missiles. But if tactical, humanly operated weapons fall prey to the gadget craze, a less than optimal system may result. The F-15 fighter illustrates this problem of preoccupation with prestige-boosting modernism. The plane was the fastest and most maneuverable of its kind, yet 40 percent of the F-15s were not available for service at any one time because of defects, difficulty of repair, and lack of spare parts. We observe that faddishness is not restricted to the world of fashion—it occurs in technology as well. The engineer must constantly be on guard not to fall prey to it.

3 This example concerns peacetime secrecy in work of military import. Norbert Wiener, the founder of cybernetics, was moved to write in answer to a request for information on some of his wartime work in missile control, "that to provide information is not necessarily an innocent act." He refused to give the information because too often the scientist places "unlimited powers in the hands of people he is least inclined to trust with their use" (Wiener, 1947, 46).

Secrecy poses problems for engineers in various ways. Should discoveries of military significance always be made available to the government? Can they be shared with other researchers, with other countries? Or should they be withheld from the larger scientific and public community altogether? If governmental secrecy in weapons development is allowed to become all-pervasive, on the other hand, will it also serve to mask corruption or embarrassing mistakes within the defense establishment? Can secrecy contribute to the promotion of particular weapon systems, such as the x-ray laser, without fear of criticism? (Adam, 1988) There are no easy answers to these questions and they deserve to be discussed more widely.

4 Of particular importance is that we ask ourselves how long a nation can divert tremendous resources (funds, materials, talent) into an economically noncontributing sector without overburdening the economy (Dumas, 1986). Every dollar, ruble, or cruzeiro spent on defense produces fewer jobs than an equal allocation for typically neglected sectors such as education or roads. At a time when most nations' true security lies in a stronger economy, redirection or "economic conversion" becomes mandatory. This entails retraining of defense industry engineers and managers so their designs, manufacturing processes, and sales techniques can bring reasonably priced, competitive civilian goods on an open market with a chance of success. A changeover requires careful planning so no communities suffer from major dislocations. The key to success is to start now while such a "safe exit" still exists.

Study Questions

1 Is it right to ask whether a weapons system is cost-effective? What does it matter how expensive it is when the nation's security is at stake and the weapons work as intended?

2 In a farewell address to the nation in 1961, outgoing President Dwight Eisenhower sounded a warning about the emerging military-industrial complex. He told his listeners that we cannot immediately tool up for massive weapons production should another major war occur, that our arms must thus be ready and mighty, and that therefore we must have a large peacetime defense establishment. But he went on to say:

> This conjunction of an immense military establishment and a large arms industry is new in the American experience. The total influence, political, even spiritual, is felt in every city, every State house, every office of the Federal Government. We recognize the imperative need for this development. Yet we must not fail to comprehend its grave implications. Our toil, resources and livelihood are all involved; so is the very structure of our society.

What implications did Eisenhower have in mind? Are there ethical implications as well? Of what kind?

3 Earl Louis Mountbatten spoke the following words shortly before his assassination:

> Next month I enter my eightieth year. I am one of the few survivors of the First World War who rose to high command in the second and I know how impossible it is to pursue military operations in accordance with fixed plans and agreements. In warfare the unexpected is the rule and no one can anticipate what an opponent's reaction will be to the unexpected.... There are powerful voices around the world who still give credence to the old Roman precept—if you desire peace, prepare for war. This is absolute nuclear nonsense and I repeat—it is a disastrous misconception to believe that by increasing the total uncertainty one increases one's own certainty (address on the occasion of being awarded the Louise Weiss Foundation prize; delivered at Strasbourg in 1979 to the Stockholm International Peace Research Institute).

Interpret Mountbatten's words from the viewpoint of those who see modern war as a technological experiment.

4 Consider the following blunt pronouncements by statesmen and generals (Schwartz, 1971, 203). Is there any message in them for engineers?

a Douglas MacArthur (commencement speech at Michigan State University, 1961): "Global war has become a Frankenstein to destroy both sides. If you lose you are annihilated. If you win, you stand only to lose. No longer does it possess even the chance of the winner of a duel. It contains only the germs of double suicide."

b Dwight D. Eisenhower, from a speech before the American Society of Newspaper Editors, April 16, 1953: "Every gun that is made, every warship launched, every rocket fired signifies, in the final sense, a theft from those who hunger and are not fed, those who are cold and are not clothed. This world in arms is not spending money alone. It is spending the sweat of its laborers, the genius of its scientists, the hopes of its children."

c Dwight D. Eisenhower, after the collapse of the 1960 summit meeting: "All of us know that, whether started deliberately or accidentally, global war would leave civilization in a shambles. In a nuclear war there can be no victors—only losers."

d Nikita Khrushchev, when Premier of the U.S.S.R.: "Only madmen and lunatics can now call for another world war. As for the men of sound mind—and they account for the majority even among the most deadly enemies of communism—they cannot but be aware of the fatal consequences of another war."

e John F. Kennedy, addressing the United Nations: "The weapons of war must be abolished before they abolish us."

6 The Just-War theory considers a war to be acceptable when it satisfies several stringent criteria: The war must be fought for a just *cause*, the *motives* must be good, it must follow a call from higher *authority* to legitimize it, and the use of *force* must be based on necessity (Hehir, 1980, 368–369). Central to notions of a just war are the principles of noncombatant immunity and proportionality. Noncombatants are those who will not be actively participating in combat and therefore do not need to be killed or restrained. Proportionality addresses the extent of damage or consequences allowable in terms of need and cost. Describe a scenario for the conduct of a just war and describe the kinds of weapons engineers might have to develop to wage one.

7 On what ethical grounds could the nuclear stockpile and/or the Strategic Defense Initiative (SDI, or "Star Wars") be justified or found not to be justified? You may include what you consider stabilizing or destabilizing effects. Possible sources: U.S. Catholic Bishops (1982), Kavka (1984); both are reprinted in Sterba (1985).

8 The following problem is taken from an article by Tekla Perry in the *IEEE Spectrum* [although it involves the National Aeronautics and Space Administration rather than the Defense Department, many of the actors (companies and government) involved in space research are also involved in weapons development]:

> Arthur is chief engineer in a components house. As such, he sits in meetings concerning bidding on contracts. At one such meeting between top company executives and the National Aeronautics and Space Administration, which is interested in getting a major contract, NASA presents specifications for components that are to be several orders of magnitude more reliable than the current state of the art. The components are not part of a life-support system, yet are critical for the success of several planned experiments. Arthur does not believe such reliability can be achieved by his company or any other, and he knows the executives feel the same. Nevertheless, the executives indicate an interest to bid on the contract without questioning the specifications. Arthur discusses the matter privately with the executives and recommends that they review the seemingly technical impossibility with NASA and try to amend the contract. The executives say that they intend, if they win the contract, to argue midstream for a change. They remind Arthur that if they don't win the contract, several engineers in Arthur's division will have to be laid off. Arthur is well-liked by his employees and fears the lay-offs would affect some close friendships. What should Arthur do? (Perry, 58)

After you have prepared your answer, you might wish to consult Perry's article for various other reactions to the dilemma it presents.

9 Prepare a case study on any one of the following weapons, all of which faced serious development problems: the M-16 rifle (Squires, 1986), Sergeant York Gun

(Adam 1987), Bradley infantry vehicle (Cousins, 1987). The references are listed only to get you started; seek more material on your own.

10 Discuss the topic "Technology and war—which promotes which?" One reference might be the anthology *Military Enterprise and Technological Change* recently assembled by M. R. Smith (1985) and discussed by S. W. Leslie in *Science* (17 Jan. 1986, pp. 277–278). Another is an essay by David Noble (1985).

11 To what extent is the economy helped or hurt by military spending? Read and report on *Our Depleted Society* (Melman, 1965) or *The Overburdened Economy* (Dumas, 1986) and present opposing views as expressed in publications of the U.S. Department of Defense.

SUMMARY

Engineering is increasingly a social experiment on an international scale, requiring engineers to achieve wider perspective on their endeavors as employees of multinational corporations, and in dealing with the environment, computers, and weapons development.

The maxim "When in Rome do as the Romans do" is inadequate as a guide to conduct in foreign countries. It implies Ethical Conventionalism, which is the false view that morality is merely a matter of local customs—as if it were all right to be a racist when working in South Africa, or to allow the unsafe practices which led to the disaster at Bhopal. Yet foreign customs often are morally relevant factors which should be taken into account in making moral judgments, especially about how to appropriately transfer technology from one setting to another. As Moral Relationalism says, moral judgments need to be made in relation to many factors which can vary from situation to situation, and moral rules which are both simple and exceptionless are rare.

The world is an ecosphere, an international "commons" of air, water, and other basic resources. The natural environment is not tightly segmented by national boundaries. Acid rain and other pollutants cause problems that require international cooperation. Even when the environmental impacts are more localized, their subtlety and complexity demand that the social experimenter exercise an imaginative and cautious vision. Human-centered and nature-centered visions of the world offer alternative ways of approaching environmental ethics.

Computer ethics is the branch of engineering ethics dealing with moral issues in computer technology. It is becoming increasingly important as computers become the technological backbone of contemporary society. Many issues arise over the possibility of computer abuse—unethical or illegal conduct in which computers are made to play a central role, as in theft, fraud, and violation of privacy rights. Some issues are merely special examples of those we have dealt with earlier: for example, whistle-blowing, safety, informed consent, ownership rights, and conscientious refusal to engage in unethical activity. But still others have to do with the effects of the widespread use of computers in our society: for example, the elimination and transformation of jobs, or the use of computers in surveillance and political campaigns.

There are special moral problems intrinsic to the defense industry, such as those related to planned cost overruns, uncritical proliferation of new weapons, and secrecy in military work. Yet those issues appear insignificant compared with the problem of war and peace. And overshadowing all other issues dealt with in this book is the possibility of a nuclear holocaust. Certainly the decision to enter or avoid weapons development as a career is among the most important confronting engineers. It is also one of the most deeply personal decisions one can make and should involve a searching examination of both one's individual conscience and the social and political issues of weapons technology.

CHAPTER **8**

CAREER CHOICE AND PROFESSIONAL OUTLOOK

On February 2, 1976, three engineers made news headlines when they resigned from General Electric (G.E.) to protest the nuclear energy industry in which they had worked for years. Each of the engineers was married and had three children. Each left a comfortable job to work full time and without pay for a nuclear protest group. Greg Minor had worked 16 years for General Electric and was manager of advanced control and instrumentation for the nuclear energy division of G.E.'s San Jose plant. Dale G. Bridenbaugh had worked 23 years with General Electric and was manager of performance evaluation and improvement. And Richard Hubbard, at age 38, was manager of quality assurance in the control and instrumentation department.

The three gave as the reason for their resignations the extreme danger they believed to result from the way nuclear energy was being developed. In his resignation letter Greg Minor expressed their shared sentiments:

> My reason for leaving is a deep conviction that nuclear reactors and nuclear weapons now present a serious danger to the future of all life on this planet. I am convinced that the reactors, the nuclear fuel cycle and waste storage systems are not safe (quoted in Barnett, 1976, 34).

Beyond the element of public protest involved, the episode illustrates a dimension of ethics in engineering with which we have not yet dealt. Moral considerations often do and should play a central role in decisions about career choice and career changes. In some ways career decisions are among the most morally important decisions a person makes. At the same time they are very personal decisions. Just as some engineers have decided on moral grounds that they ought not to participate in nuclear energy development, many others have seen supporting it as a moral imperative, considering it as the best solution to the energy crisis and the best way to remove dependence on foreign oil, both of which have wide repercussions for the well-being of our society.

In this chapter we begin by exploring some general connections between morality and career decisions, stressing how one's work affects one's sense of self-worth as a professional and as a human being. We then move on to an examination of how engineering work is fragmented through separation by functions, how those who are affected (favorably or unfavorably) by engineered products also fall into different groups, and why it is important that engineers and their managers maintain a global, integrative outlook.

This book stresses the ethical issues faced by the employed engineer, but in this chapter we devote a section to the consulting engineer. The problems under discussion here are related closely to business practices; accordingly we cover some of what engineering ethics was all about several decades ago.

In a section on the responsibility of and to the profession we depart from our earlier emphasis on the individual engineer and his or her options as an autonomous ethical agent. Here we examine the role of the profession as a whole and its collective responsibilities.

The chapter ends with a short summary of the main themes we have presented in this book.

ETHICS AND VOCATION

At first sight it may seem odd to think of career decisions as having any special connection with morality. In choosing careers we consider prospects of job offers, advancement, security, salary, prestige, life style, personal challenge, and personal satisfaction. These factors apparently have more to do with prudence in seeking personal happiness than with morality.

Perhaps, as Kant thought, there is a kind of indirect moral duty to achieve our own happiness (Kant, 18). If we are depressed and unhappy, we may be unwilling to pursue our moral duties. Or as Erich Fromm, Abraham Maslow, and other recent humanistic psychologists have said, people are unable to be genuinely concerned about others if they are unable to care about themselves. But this line of thought would establish only a very tenuous connection between careers and morality.

Before identifying more direct connections, however, let us elaborate on the aspects of engineering mentioned above which relate to seeking personal happiness.

Existential Pleasures of Engineering

Happiness can be viewed, following Aristotle, as self-realization, rather than mere contentment (Aristotle, 1104). Self-realization, in turn, comes through the exercise of one's highest talents, interests, skills, and virtues. Within limits, the greater the complexity and challenge to one's talents, the greater one's happiness tends to be. Now engineering is a complex and challenging discipline. Moreover, the undergraduate curriculum for engineering is acknowledged to be more rigorous and difficult than the majority of academic disciplines. Combining these generalizations, we might guess that students are attracted to engineering at least in part because of the challenge it offers to intelligent people.

Do empirical studies back up this somewhat flattering portrayal? To a significant extent, yes. Typical students are motivated to enter engineering primarily by a desire for interesting and challenging work. They have an "activist orientation" in the sense of wanting to create concrete objects and systems—to make them and to make them work. They are more intelligent than average college students, although they tend to have a low tolerance for ambiguities and uncertainties which cannot be measured and translated into figures (Perrucci and Gerstl, *Profession Without Community: Engineers in American Society*, 27–52).

What is it that is so appealing and challenging in making technological products? Perhaps no one has so elegantly conveyed the excitement of engineering as Samuel Florman in his book *The Existential Pleasures of Engineering*. By "existential pleasures" Florman means deep-rooted and elemental satisfactions. He portrays engineering as essentially an attempt to obtain and apply an understanding of the universe so as to fulfill human needs and desires. This attempt calls forth some of the finest aspirations and deepest impulses of human beings.

The first existential pleasure Florman distinguishes resides in the act of personally changing the world. By nature, humans are compelled to improve the world. There is no end to the possibilities for achieving improvements, and the allure of "endless vistas bewitches the engineer of every era" (Florman, 1976, 120–121).

Changing the world brings with it the second type of existential pleasure: the joy of creative effort. This includes planning, designing, testing, producing, selling, constructing and maintaining. In contrast with the scientist's, whose main interest is in discovering new knowledge, the engineer's greatest enjoyment derives from creatively solving practical problems (Florman, 1976, 143).

Yet the engineer shares, as a third type of pleasure, the scientist's joy in understanding the laws and riddles of the universe. Both may experience "quasi-mystical moments of peace and wonder" (Florman, 1976, 141).

A fourth pleasure relates specifically to size in the world. The magnitude of natural phenomena—oceans, rivers, mountains, and prairies—both intimidates and inspires. In response, engineers conceive immense ships, bridges, tunnels, and other "mammoth undertakings [which] appeal to a human passion that appears to be inextinguishable" (Florman, 1976, 122).

A fifth pleasure relates to regularly being in the presence of machines. A mechanical environment can generate a comforting and absorbing sense of a manageable, controlled, and ordered world. "For a period of time, personal concerns, particularly petty concerns, are forgotten, as the mind becomes enchanted with the patterns of an orderly and circumscribed scene" (Florman, 1976, 137).

Concern for Humanity

Florman concludes his inventory of the existential pleasures of engineering with what he says is the primary and most important one of all: "a strong sense of *helping,* of directing efforts toward easing the lot of one's fellows" (Florman, 1976, 145). "The main existential pleasure of the engineer," he writes, "will always be to contribute to the well-being of his fellow man" (Florman, 1976, 147).

This paramount source of existential pleasure and motivation suggests one straightforward connection between morality and career decisions in engineering. Virtually every traditional characterization of the nature and goals of engineering has emphasized the contribution it makes to improving human life. This is not mere propaganda unrelated to what really concerns engineers on the job. The same theme is sounded in the personal testimony of many engineers as they reflect on what their careers have meant to them (Florman, 1976, 94–95).

This needs to be borne in mind when interpreting empirical studies of students' main motives for entering engineering. The technical challenge it offers is always seen by them against the background of participating in a socially useful and important enterprise. Engineering would not attract students if it were viewed by them as generally directed toward immoral ends.

Our interest here is in the possibility of this implicit background of moral concern becoming more explicit in thinking about career decisions. One might well ask what moral aims one's career has or might have. From such a perspective, ethics does not need to be viewed negatively as a burden or a constraint on one's career. Rather it offers positive ideals for expressing natural moral interests in concrete ways in one's professional life. We are led to consider what has been called the ethics, or moral philosophy, of vocation.

Philosophy of Vocation

William Frankena, a leading contemporary ethicist, has attempted to state the fundamental tenets of a moral philosophy of vocation. Four of his suggestive ideas may be summarized as follows (Frankena, 1976, 393–408).

First, according to Frankena, each person has the important duty of selecting a vocation. Most vocations are jobs which enable people to earn a living, whether or not a given individual needs the income from his or her job. But they are also, and more essentially, types of work which enable people to find senses of identity, personal worth, and meaning, and which promote feelings of contributing to the good of others. Thus the work of raising children qualifies as a vocation, even though in fact no income is usually obtained from it. Selecting a vocation is a duty, according to Frankena, both because of the moral importance of seeking self-fulfillment and because of the good that can be achieved for others through pursuing a vocation.

Second, vocations carry with them a set of specific duties. Fulfilling those is the main way in which a person fulfills the general duty of doing good for others. Utilitarians might say that the amount and kind of good one has a duty to produce is largely determined by the nature of one's vocation. This could be called "vocation-utilitarianism," in which good consequences are weighed largely in respect to the role of a vocation rather than in respect to individual acts (act-utilitarianism) or general rules (rule-utilitarianism).

Frankena is not a utilitarian, but as a duty ethicist he admits that one fundamental duty is the principle of beneficence: to create good and prevent harm. The amount of good people are obligated to create is partly determined by the nature of their vocations. This is because concrete moral obligations are specified in part by a direct appeal to professional duties, as well as by an appeal to foundational ethical theories. Presumably professional duties are themselves grounded in the more general ethical principles, as we have suggested throughout this book.

Third, while there is a duty to have a vocation, there is normally no specific vocation one is obligated to pursue. Rather, people should be free to select their own careers as an aspect of their autonomous self-expression. This is because both personal self-realization (what Frankena calls "the good life") and the morally good life are based on autonomous decision making.

Fourth, and last, there are objective considerations which ought to be weighed in making career choices. Most important are the prospects of happiness and self-fulfillment. One must ask which available vocation, or type of vocation, it would be rational to choose in order to best express one's main talents and basic interests. Also very important, however, are the moral values to which one is committed. For it is largely in and through one's vocation that those values will be expressed and realized.

These four tenets can be developed with respect to the choice of engineering as a vocation. No one is obligated to become an engineer, for that should be a matter of free personal decision. Yet the decision to enter engineering should be made on the basis of a searching assessment of one's talents and interests, and also on the basis of relating engineering to one's basic moral values.

Engineering involves a particular set of professional duties specified in terms of practical ways to prevent harm and promote good. A reasonable decision to enter engineering entails a prior weighing of those duties against one's general values and ideals. Does an engineering career promise an avenue for self-realization and is it compatible with one's moral concern?

Frankena's line of thought can also be extended to decisions about entering branches of engineering and areas of engineering work. To use an example already mentioned, people contemplating a career having to do with the development of nuclear energy ought to investigate the types of moral issues and dilemmas they might have to confront. A serious investigation may involve coursework related to the social ramifications of nuclear engineering.

This last point relates to the implications of a philosophy of vocation for education. Frankena says his views require that education be both liberal and vocational. It should be vocational in the sense of preparing people for the vocations he thinks they are obligated to pursue. But it should be a liberal arts education as well, in order to enable them to choose their vocations autonomously and with wisdom. Education should foster both self-knowledge and serious reflection upon the nature and function of values.

Work Ethics

Frankena's views provide the theoretical basis for what is commonly called a *work ethic*. This idea calls for comment. There is, in fact, no one work ethic. There are many different ones which different people embrace, and no doubt there are several morally legitimate and some unjustified perspectives on work (Cherrington, 1980, 19–30).

One version was identified by the sociologist Max Weber in *The Protestant Ethic and The Spirit of Capitalism*. The Protestant work ethic, according to Weber, provides the psychological explanation for the rise of modern capitalism. In brief, it was the idea that financial success is a sign that predestination has ordained one as favored by God. This was thought to imply that making maximal profits is a duty mandated by God. Profit becomes an end in itself rather than a means to other ends. It is to be sought rationally, diligently, and without compromise with other values.

Carried to an extreme, this idea can lead to the life and ideology of the workaholic. Whether out of religious or secular concerns, workaholics are compulsive overworkers. Their entire energies are devoted to work, to the point of unhealthy addiction. As with other addicts, workaholics are said to be pushed by uncontrolled and unconscious urges: anxiety, guilt, insecurity, or a sense of inferiority. As a result their behavior is irrational. They neglect their families, have few personal relationships not tied to their work, and seem unable to fill leisure time with meaningful activity. Death tends to come early, often from heart attack (Cherrington, 1980, 253–274).

Focusing on this irrational extreme should not lead to a rejection of the possibility of a reasonable work ethic. There is a difference between a compulsive workaholic and a healthy hard worker who also has a wide range of additional interests. Major professions, like engineering, make substantial demands upon those engaged in them which require the virtues of discipline, initiative, and keeping up to date in regard to knowledge and skills. To this extent professional ethics merges with a work ethic which emphasizes attitudes and virtues needed for fulfilling professional obligations.

Beyond this point, there is a range of general attitudes which may enter into an individual's personal work ethic. Several of those attitudes might be summarized as follows:

1 Work is intrinsically valuable to the extent that it is enjoyable or meaningful in allowing personal expression and self-fulfillment. Meaningful work is worth doing for its own sake and for the sense of personal identity and self-esteem it brings.

2 Work is the major instrumental good in life. It is the central means for providing the income needed to avoid economic dependence on others, for obtaining desired goods and services, and for achieving status and recognition from others.

3 Work is a necessary evil. It is the sort of thing one must do in order to avoid worse evils, such as dependency and poverty. But it is mind-numbing, degrading, and a major source of anxiety and unhappiness.

As we noted earlier, most people enter engineering for its inherent challenge. This presupposes that they seek a vocation warranting attitudes in the first category, although the second type of attitude is also important. They hope to find in engineering a vocation which evokes the pride and self-esteem of the craftsperson. Meaningful tasks are to be accomplished because of their intrinsic importance, not because a supervisor is watching.

Alienation, Integrity, and Self-Respect

Engineers, like all professionals and other workers, are vulnerable to periodic alienation from their work. To be alienated from one's work is to be separated from it in ways harmful to one's identity, integrity, and self-respect. Usually this involves painful feelings, such as depression, anxiety, anger, and hatred, although it can also occur without these emotions. Essentially it is the state of not identifying with what one does, not seeing the work as a personal expression of oneself.

The concept of *worker alienation* was introduced by Karl Marx (1818–1883) as part of his critique of capitalism. Alienation occurs when work becomes a commodity sold to an employer, rather than an expression of one's personal nature. Work becomes a humiliating "necessary" evil, as the third attitude above suggests.

> What constitutes the alienation of labor? First, that the work is *external* to the worker, that it is not part of his nature; and that, consequently, he does not fulfill himself in his work but denies himself, has a feeling of misery rather than well being, does not develop freely his mental and physical energies but is physically exhausted and mentally debased. The worker therefore feels himself at home only during his leisure time, whereas at work he feels homeless. . . . It is not the satisfaction of a need, but only a *means* for satisfying other needs (K. Marx, 98).

For Marx, work should be central to our identity—to our sense of who we are. The many hours invested in it should be experienced overall as evoking and developing our talents and highest potentials. Alienation from work activities instead leads to self-estrangement—to not being able to identify with ourselves insofar as we are our work. Furthermore, it is usually accompanied by two other forms of alienation: alienation from the product of our labor and from other workers. Product alienation is the inability to identify with and affirm the worth of the products or services we provide. Alienation from other workers means not identifying with other human beings engaged in related activities. Whereas fulfilling work involves seeing oneself embodied in the products of one's labor and also feeling a part of a group effort in which one is unified with other workers, alienation severs these connections with the physical world and with the community of workers.

Marx argued that alienation from products, productive activity, and other workers was an inevitable result of capitalism with its emphasis on consumerism and mass accumulation of private property and power in the hands of the middle class. In our view, these claims had some plausibility from the perspective of nineteenth-century industrial society, in which worker exploitation was widespread. Today, however, the sources of worker alienation are more complex, and are apparent in communist societies as well, although Marx's clarification of the concept of alienation is still useful.

Certainly one major cause of worker alienation involves morality: a gap between one's fundamental moral values and work. This is a loss of moral integrity—a loss of the unity of moral concern in one's life. It occurs when work becomes flagrantly cut off from one's primary moral ideals. Thus, persons committed to a humane world where people show concern for other people will experience a loss of moral unity in their lives when they spend their working hours manufacturing needlessly polluting and hazardous plastics for use in making dangerous toys (Nielsen, 1987). This will also bring lowered self-respect, that is, a lessened sense of one's own worth as a human being.

Most alienation arises on a smaller scale and may be temporary as one rethinks particular roles and job assignments that threaten one's integrity. This rethinking is a periodic need for all professionals concerned with maintaining a sense of self-respect throughout a career. It is made complicated by the fact that few (if any) jobs allow a perfect mesh between one's personal ideals and one's work activities. Most professionals testify that some *compromise* is an inescapable, and even desirable, aspect of their careers.

"Compromise" can mean two things. In one sense it means to undermine integrity by violating one's fundamental moral principles. In another sense, the one intended here, it means to settle differences by mutual concessions or to reconcile conflicts through adjustments in attitude and conduct. In this sense, compromises are sometimes good and sometimes bad. They may be reasonable ways to sustain relationships in the face of deep differences, or valuable ways to carry on with a life in the face of hardship and difficulty. Or they may lead to so severe a conflict between the working and private life that the engineer had better seek work elsewhere.

Models of Professional Roles

Models and metaphors often serve to organize thinking and crystallize attitudes. It was with this in mind that we earlier suggested the model of engineering as social experimentation and the engineer as an experimenter. Many other models have been advocated in connection with the social and moral roles of engineering and engineers. In concluding this section, we list some of those models which at times have seemed attractive to some engineers. We leave for a study question the assessment of the models in connection with thinking about career goals. Those who wish to examine the influence of several of them in the first half of this century are referred to Edwin Layton's masterful study *The Revolt of the Engineers.*

1 *Savior.* Plato believed a philosopher-king was needed in order to create the ideal society. Others have believed that engineers hold the key to creating a utopian society. In part this is to be achieved through technological developments which lead to material prosperity. In part it is to arise through creation of a technocracy in which engineering ways of thinking are applied to large-scale social planning. The representative engineer is a savior who will redeem society from poverty, inefficiency, waste, and the drudgery of manual labor.

2 *Guardian.* Perhaps engineers cannot usher in utopia. Yet it is they who know best the directions in which, and pace at which, technology should develop. Accordingly, they should be given positions of high authority based on their expertise in determining what is in the best interests of society (Veblen, 1965, 52–82).

3 *Bureaucratic servant.* Within the corporate setting in which engineers work, management should make the decisions about the directions of technological development. The proper role of the engineer is to be the servant or handmaiden who receives and translates the directives of management into concrete achievements. The engineer is the loyal organization man or woman whose special skills reside in solving problems assigned by management, within the constraints set by management.

4 *Social servant.* The role of engineers lies exclusively in obedient service to others, but their true master is society. Society expresses its interests either directly through purchasing patterns, or indirectly through government representatives and consumer groups. Engineers in cooperation with management have the task of receiving society's directives and satisfying society's desires.

5 *Social enabler and catalyst.* This is a variation on model 4. Service to society includes, but is not exhausted by, carrying out social directives. Ultimate power and authority lie with management, but nevertheless the engineer plays a vital and active role beyond mere order-following. Sometimes engineers are needed to help management and society understand their own needs and to make informed decisions about desirable ends and means of technological development (Fruchtbaum, 1980, 258).

6 *Game player.* Engineers are neither servants nor masters of anyone. Instead they play by the economic game rules that happen to be in effect at a given time. Their aim, like that of managers, is to play successfully within organizations, enjoying both the pleasures of technological work and the satisfaction of winning and moving ahead in a competitive world (Maccoby, 1978).

Study Questions

1 The following widely discussed case study was written by a leading British philosopher, Bernard Williams. While the case is about a chemist, the issues it raises are equally relevant to engineering.

> George, who has just taken his Ph.D. in chemistry, finds it extremely difficult to get a job. He is not very robust in health, which cuts down the number of jobs he might be able to do satisfactorily. His wife has to go out to work to keep [i.e., to support] them, which itself causes a great deal of strain, since they have small children and there are severe problems about looking after them. The results of all this, especially on the children, are damaging. An older chemist, who knows about this situation, says that he can get George a decently paid job in a certain laboratory, which pursues research into chemical and biological warfare. George says that he cannot accept this, since he is opposed to chemical and biological warfare. The older man replies that he is not too keen on it himself, come to that, but after all George's refusal is not going to make the job or the laboratory go away; what is more, he happens to know that if George refuses the job, it will certainly go to a contemporary of George's who is not inhibited by any such scruples and is likely if appointed to push along the research with greater zeal than George would. Indeed, it is not merely concern for George and his family, but (to speak frankly and in confidence) some alarm about this other man's excess of zeal, which has led the older man to offer to use his influence to get George the job.... George's wife, to whom he is deeply attached, has views (the details of which need not concern us) from which it follows that at least there is nothing particularly wrong with research into CBW. What should he do? (Williams, 97–98)

In defending your answer, make reference to several ethical theories, including act-utilitarianism and virtue ethics.

2 Do you agree with Frankena that there is a duty to have a vocation, in his sense of the term "vocation"? Respond to the criticism that individuals have a moral right to do whatever they want with their lives, as long as they do not hurt others, and that this makes it all right for people who inherit wealth not to have vocations.
3 Formulate a work ethic which expresses your own attitudes about work.
4 Assess the positive and negative implications of each of the following models for engineers: savior, guardian, bureaucratic servant, social servant, social enabler and catalyst, and game player. Which best captures the type of engineer you would like to be or would like engineers to be? Develop your answer, tracing its social implications, by applying the model(s) to one of the main topics discussed in the last chapter: multinational corporations, environmental ethics, computer ethics, or weapons development.

5 Read and report on "Individual Choices: What You Can Do," Chapters 8-11 in *Career Development for Engineers and Scientists* by Morrison and Vosburgh (1987).

OVERCOMING FRAGMENTATION

Engineers can be classified by the products they create as chemical, civil, electrical, or mechanical engineers, and so on. This is also the way they are educated, certified, and grouped in major professional societies. More important for our purposes is a division of engineers by function, because it is within their functions and where those functions overlap that engineers face most of their ethical dilemmas. This division, which is practiced by most engineering organizations, can also result in a fragmentation of awareness to the point where few engineers retain a global outlook on their products and on the settings in which these products will be put to use.

A classification according to function will typically divide engineering activities into design, manufacture, operation, and phase-out. These are listed as column headings in Table 8-1. In our tabulation the design function incorporates research, development, de novo design, and modification of earlier designs. Manufacturing also includes repair, while operation of equipment includes maintenance and periodic inspection. We have left out sales engineering because the ethical issues associated with it are generally not unique to engineering. Service or field engineers are not listed separately because their main activities of maintenance and repair are subsumed under operation and manufacture. On the other hand, Table 8-1 introduces an activity rarely seen as a separate classification: phase-out. We have added it because of the growing need for planning orderly shutdown of major projects. This can include such tasks as decommissioning a nuclear power plant, stocking spare parts for a popular product which is being replaced by a newer model, and disposing of toxic substances which are the by-product of a manufacturing process or which leave the plant as an inherent component of the product.

Orthogonally to the ordering by function or activity one could also classify engineers according to the mix of clients and employers they serve, yielding groups such as industry engineers, public works engineers, consulting engineers, teaching engineers, and so forth. We have chosen instead to concentrate on a broader classification by groups of people which are affected by engineering projects. These are the parties interested in the safe progress and outcome of a project. The Table 8-1 row headings list these groups. First there are the workers in the manufacturing plant or on the construction site. They can be hurt by unsafe equipment, collapsing structures, and toxic substances. But they are not users of the product as yet. The user/client is listed next, also any worker employed by the user/client who operates or is exposed to the engineered product. The user can be an active consumer (homemaker, hobbyist, or car owner or driver) or a passive consumer (an air traveler who has little choice as to the particular plane being used). Finally we have the "innocent" bystanders: people and the environment. The environ-

TABLE 8-1
A TAXONOMY OF MALFUNCTIONS, THEIR CAUSES AND EFFECTS

		Origin of deficiency			
Affected party		**1 Design**	**2 Manufacture**	**3 Operation**	**4 Phase-out**
A	Worker (Production/ construction)	(a) Mechanization (314) (b) Chernobyl (151)	(a) Milford Haven (66) (b) Quebec Bridge (114)	(a) Grain silo (118) (b) Train wrecks	Radioactive isotopes (78)
B	Active consumer (Client's worker)	(a) Sugar mill (315) (b) Ford Pinto (143)	(a) Heart valve (315) (b) Oil rig (118)	(a) *Challenger* (185) (b) TMI (146)	Spare parts shortage
C	Passive client (No choice)	(a) DC-10 (43) (b) *Titanic* (63)	Building collapse (313)	Transport (air, land, sea)	Declining water quality (318)
D	Bystander (Near or distant)	Loss of privacy through data bank	(a) Buffalo Creek (102) (b) Molasses (277)	(a) Bhopal (258) (b) Police file (282)	(a) Air pollution (b) Love Canal (315)
E	Nature (Fauna, flora, view)	(a) Everglades (269) (b) Eiffel Tower (315)	Loss of topsoil to development	(a) Acid rain (264) (b) Ozone layer (265)	(a) River pollution (b) North Sea (315)

Examples described in the text are followed by parentheses in which the respective page numbers are indicated. The other examples are mentioned in this chapter or are self-explanatory.

ment stands for nature and the rights of future generations for a livable natural world.

A Taxonomy of Malfunctions

Most engineering projects are successful and serve their purposes well without malfunctions or adverse effects on the social and natural ecologies. Among those that fail, the majority fail safely without hurting people physically (though possibly financially) because defects are being caught early on through periodic checks—which are part of the monitoring responsibility of engineers as experimenters. Listed as entries in Table 8-1 are engineering projects (experiments) which failed because of insufficient knowledge (gaps in the state of the art), lack of concern for safety, or too much concern for institutional posture and profit.

Each malfunction is related to the activity primarily responsible for its occurrence by the column in which it appears, and to the group of people affected (interested party) by its row. The examples used are from cases which appeared earlier in this book or which will be explained briefly below. They include both dramatic and everyday failures.

The construction industry has gathered interesting data on the causes of structural failures. A study by the Building Research Association (Great Britain) revealed that 58 percent of failures could be attributed to faulty design and 35 percent to faulty execution. Other reasons (note that there is some overlap between these and the former two) were that materials or components failed to meet accepted standards (12 percent of cases) and that users expected more than the designers had anticipated (11 percent of cases) (Ransom, 1981).

Another study, performed at the Swiss Federal Institute of Technology in Zurich (ETH), indicated that the causes of 800 structural failures—responsible for 504 deaths and 592 people injured—were primarily the results of "human mistakes, errors, carelessness, and so on. . . ." Where engineers were at fault the causes were mostly insufficient knowledge (36 percent), underestimation of environmental influence (16 percent), ignorance, carelessness, and/or negligence (14 percent), forgetfulness and/or error (13 percent), and relying upon others without sufficient control (9 percent) (Matousak, 1977, cited in Florman, 1987, 102–103).

The uncertainties which led us to propose our model of engineering projects as experiments are reflected in the studies cited above by the categories of insufficient design knowledge (but not plain ignorance), unexpected influences of application and environment, and variations in materials. The human failings, such as ignorance, negligence, carelessness, forgetfulness, and errors reflect a certain lack of "virtue" which managers sensitive to the issues could correct by good supervision of technical tasks and training so as to turn safety concerns into a habit.

Before we further examine the entries in Table 8-1, we should mention that the classifications we have adopted have a certain arbitrariness in that clear boundaries between functions cannot be drawn except on organization charts. But it is precisely because such artificial, institutional boundaries exist in industry, where they are supposed to provide efficient division of labor and make managing easier, that we have drawn them here as well. To some extent the same can be said about the classifications outlined in the rows in Table 8-1. There the arbitrariness stems from our desire to avoid clutter in the table.

A Look at the Examples

In the following we shall refer to entries in Table 8-1 by row-column-position. We begin with the examples labeled "Mechanization" (A-1-a) and "Sugar Mill" (B-1-a) in the left upper part of the tabulation and then move directly to "Love Canal" (D-4-b) and "North Sea" (E-5-b) in the lower right corner. This will take us from one of the earliest cases in the industrial revolution to one of the currently pressing problems.

The introduction of machines into the textile industries of England created angry reaction from those who saw their jobs, livelihoods, and ways of life (hard but free) threatened. Here are the numbers reported in the British Parliamentary Papers of 1840 of the men replaced by various textile machines. Spinning jenny: 9 of 10 warp spinners and 13 of 14 weft spinners; scribbling engine: 15 of 16 scribblers; gig mill: 11 of 12 shearmen; shearing frame: 3 of 4 shearmen. The effect on the communities can be fathomed only when one reads that scribblers constituted about 10 percent of the preindustrial adult work force and shearmen about 15 percent (Randall, 1986). Early and continuing opposition known as the Wiltshire Outrage was followed by the more forceful tactics of the Luddites in Yorkshire. The workers and the textile mill owners both insisted on the moral foundations of their respective positions, but the execution of some of the followers of the legendary Ned Ludd showed who had the power. In addressing the problem of Ludd's destruction of frames, Lord Byron asked Parliament in 1812: "Can you commit a whole country to their own prisons? Will you erect a gibbet in every field, and hang up men like scarecrows?" (Quoted by Winner, 1977, 127)

Today the methods of maintaining industrial peace have improved, but the debate continues as computers and robots take over many routine jobs. In the long run the lot of the worker should improve, but little is done to remedy the short-term dislocations which hurt many communities because too few employers assume any responsibility for such matters. Another area in which early industrial technology exacted a great price was in work-related injuries. At first they were ignored because there was an unlimited supply of labor. The second step was to do everything possible to prevent injuries from slowing production. It is said that in eighteenth-century Caribbean sugar mills, axes were kept ready "to amputate a slave's arm should he be caught in the inrunning nip point of the rollers used to crush the sugar cane"

(Roberts, 1984). In England, where children's working conditions were regulated in 1802, safety guards on belt transmissions were not required until 1844.

Slowly those threats which could be perceived by our senses were contained, until today we are finally addressing the problem of toxic wastes, which escape easy detection or jurisdiction. Examples are the Love Canal case and North Sea pollution. Around 1946 the Hooker Chemical Co. stopped depositing pesticide wastes in a landfill which was subsequently sold. In the early 1970s it was found that 82 different chemicals were leaking from their buried 55-gallon drums into the basements of houses which had been built on the landfill. This situation gave rise to a major controversy because of uncertainty as to the extent of damage, questions regarding responsibility for resettlement, and withholding of information. With respect to the latter it was revealed that Hooker officials had learned of leakages as early as 1958—by which time the land had been sold to the Niagara Falls school board—but claimed they did not want to subject the new owner to any litigation. The North Sea, meanwhile, is being polluted by the effluents carried into it by many rivers, some of which pass through several countries where they pick up toxic wastes. Jurisdictional difficulties led the adjoining countries to adopt in 1987 an informal compact to do the best they can to reduce further deterioration.

The Eiffel Tower (E-1-b) is listed as a scenic despoiler only to remind us that tastes will differ. When it was built, many Parisians consoled themselves with the thought that it would not last longer than 20 years. De Maupassant disliked the tower but often ate in its restaurant (although the food was not good) because "it's the only place in Paris where I don't have to see it" (Meisler, 1987). Today no one would think of demolishing the Eiffel tower except for safety reasons.

"Heart Valve" (B-2-a) refers to a widely used, and initially very successful, mechanical heart valve developed and manufactured by Shiley Inc. (now part of Pfizer). The valve has a tilting disk which is held in place by two metal struts. In 147 patients a strut has broken, preventing normal operation of the valve; death has resulted in 65 percent of the cases where this has happened. The strut breaks at a weld that would be unnecessary if both struts were fabricated as a single piece. Statistically the risk of strut fracture is less than the risk of death from surgery to replace a valve, but this is not much of a consolation for valve implant recipients. It is also unsettling for them to know that problems were recognized a long time before this type of valve was withdrawn (Steinbrook, 1985).

One interesting case for which there is no suitable slot in Table 8-1 is that of the stun gun. The stun gun is used by police in place of a club or firearm to subdue suspects who resist arrest. It produces an electric shock wherever it touches the victim's body. Unlike the laser gun, which administers a single shock via a projectile connected to a conducting wire, the stun gun is capable of repeated discharges. While this nonlethal weapon was at first welcomed by most legal and law enforcement professionals, there are now second thoughts after reports of its misuse as a device of torture. We mention this example to underline two facts: (1) no arrangement of things or activities can encompass all of them satisfactorily and (2) there are many more examples than we can possibly cover. They are left for the reader to find in the current and archival literature.

Global View: A Management Priority

An examination of Table 8-1 reveals that malfunctions of engineered products or systems can originate in one or more engineering functions and can affect one or more groups of people. Where the errors will occur and who will be hurt cannot be predicted easily. It is therefore necessary that engineers and their managers adopt a broad outlook which permits them to scan the totality of activities within the manufacturing establishment, as well as outside the establishment among the users and the general public.

We have already referred to the arbitrariness in the lines of demarcation between the columns and rows of Table 8-1. It is one of the important roles of engineering management to remove these boundaries as much as possible. All too often, engineers in design are isolated from engineers in manufacturing and sales; all too often, the architect-engineer is not represented at the construction site. Only firms with project-oriented teams show some promise of integration. And even then their view of the product environment is frequently too narrowly focused on the client's point of view, allowing scant attention to the interests of other parties who may be affected by the product.

Earlier we mentioned management's responsibility for creating a work environment in which the "virtues" of engineering can blossom: responsibility, acquisition of knowledge, truthfulness, and so forth. Here we urge management to take on a more difficult, but also more structured duty—the duty to provide overall integration of the processes leading from design to manufacture to operation of a product, an integration which results in fewer serial and more parallel operations. Engineers must be given the opportunity to reach beyond their cubicles so safety concerns can be looked at jointly and passed on more naturally. Otherwise one has a situation similar to what happens when different regulatory agencies share responsibility over a product but act independently (see Study Question 12, p. 145).

Fortunately many engineers see the need for wider concerns. If they find safety issues that need addressing, they must be provided the avenues for bringing them to the attention of management at whatever level necessary to get action. In many organizations there are ombudspeople to provide such access. Procedures of this type serve prudential ends as well, since they "not only protect management from lawsuits, but also help it retain capable, ethical employees" (Matley et al., 1987).

We do not consider it sufficient, however, merely to enact organizational changes to remedy failure-proneness. Structural changes could be necessary but may not be sufficient. When we asked what might be amiss in academe when so many professors are aware of or suspect scientific fraud in some of their colleagues' work (Study Question p. 224), we did not intend to single out that profession. As Matley and coauthors found through a survey of chemical engineers (Matley et al., 1987), 16 percent of respondents admitted to having done something unethical in their work, but 49 percent indicated knowing of others who did. Engineers should feel free to actively participate in the reduction of ethical problems where they exist, and management must give them the opportunity to do so.

Optimization

It is often said that engineers can do for ten dollars what everyone else can do for a thousand. In other words, engineers not only provide needed products and services, they do so at the best possible price. "Price" should be seen as a conglomerate of attributes: monetary value, safety, reliability, aesthetics, and other performance criteria. Not all desiderata can be met simultaneously. When engineers practice "optimization," it is therefore not the single-minded pursuit of a narrow goal but a search for the best possible solution under constraints imposed by society and nature.

One of the greatest dangers in engineering is suboptimization. This is the practice of finding the best solution for only part of the problem. Such a solution may force the rest of the problem into a less than satisfactory solution mode. It is the habit of suboptimization which is partly responsible for the saying that engineers find all the right answers to the wrong problems. It is our view that the kind of management which integrates engineering functions and assesses the effects of an engineered product on the social and natural ecologies is best suited to combat the suboptimization habit. Allowing its professional employees to adopt a global outlook which detects more than the obvious constraints should be one of management's top priorities.

Let us dwell a moment on the subject of constraints. A number of constraints were discussed implicitly in earlier chapters. Safety is one of them. It should be recalled that an important feature of a safe product or system is its feature for safe exit of workers, users, or innocent bystanders. Another constraint had to do with the disposal of harmful wastes which accrue during manufacture or use. We have also pointed to economic loss as a possible effect of product malfunction. It must not be overlooked since loss of livelihood can be as hurtful as (and lead to) loss of health. The level of acceptability which sets the limit in a constraint inequality (say, speed to be no greater than 65 mph) must be established with the consent of those who become the human subjects of the experiment.

The two constraints which we have not emphasized so far have to do with the natural environment. The world's resources are limited. Even if there seems to be enough around for us now, the rising expectations of developing nations will lead to stiff competition later on. We cannot always count on a technofix which will get us out of a bind through new discoveries or inventions. If nothing else, wasteful use of raw materials and depletable energy sources now will drive their prices unaffordably high for future generations.

Drawing on nature's bounty should be viewed as an economic experiment, or as the economic portion of an engineering experiment. Part of the experimental procedure would be to regard natural resources as capital rather than as commodities (Schumacher, 1973). When it comes to water, we often regard it only as a readily available service. The case of water introduces another type of constraint which must not be overlooked: what is abundantly available and of good quality now may not be so in the future. Water, for instance, may remain available in a given location, but it may become unusable in the future because of poor quality. This is happening to some groundwater basins.

Another aspect of the economic model is life-cycle cost. *Life-cycle cost* includes the cost of operation and maintenance over the life of the product and the cost of its disposal afterwards. All this may be difficult to calculate when exact interest figures are not available, but there is no excuse for leaving out of consideration those costs or constraints which cannot be evaluated easily. It is appropriate to mention in this connection the need to internalize the costs of pollution, waste production, and other side effects which burden the social and natural ecologies. All too often they are not ascribed to the process which produced them—i.e., they are not counted as part of the total cost for that process.

The human-centered view of nature as a treasure trove into which we may dip at pleasure to satisfy all our wants is of Faustian dimensions. Faust's deal with the devil provided for satisfaction now, payment later. Goethe's version of Faust has been described as an alchemistic drama from beginning to end (C. G. Jung, 1946). The economist Binswanger has successfully interpreted Goethe's drama as a critique of our economic system which strives for constant expansion. In the first part of the play, the alchemist Mephistopheles (the devil) provides Faust with the golden drink (the love potion) which rejuvenates him, but love turns out to have only momentary and not lasting value. In the second part of the drama, Faust finally achieves satisfaction through a land reclamation project of major proportions (complete with risks of flooding) through the machinations of Mephistopheles who opens the way to mortgaging the future for present gain. The present gain is capital—in the form of paper money. "To accomplish this one needs the vision of the inventor, the engineer, the entrepreneur, who through new ideas, projects,and investments changes the world to make it an object of trade and monetary value" (Binswanger, 1985). Here Binswanger speaks of the modern alchemist. Too late Faust realizes that he has created a future in which there is no room for concern and caring.

A final set of constraints—one which should remain in clear view and never be hidden—is composed of standards, rules, and laws. These constraints should be interpreted broadly and adhered to in their intended spirit.

The minimalist's view that no more should be done than absolutely required cannot ensure true safety and will eventually breed more, stricter, and less palatable regulations. Once again we encounter a cost which should be taken into account now but is forgotten until it haunts us later.

Optimization under constraints is not unlike reaching the best possible compromise with society and nature acting as your "adversary." A compromise in which hidden conditions are overlooked is not a good compromise. So it is with optimization when it turns into suboptimization because competing objectives or constraints have been neglected.

Study Questions

1 Are there any entries (examples) in Table 8-1 which are misplaced or for which you can offer better sample cases?
2 How do you think engineering education does, or should, prepare engineers to think globally? Are the prescriptions laid out in this section unrealistic? Too stiff or too ambiguous? In a student's program, what is the proper allocation of effort to analysis, design, experimentation, and nontechnical subjects? Discuss this with professors, other students, and practicing engineers.
3 Discuss what were (seem to have been and should have been) the criterion of optimality and the set of constraints in one of the following projects: space shuttle *Challenger,* Ford Pinto, St. Lawrence Seaway (not discussed in this book), or a case selected by you.
4 Select and discuss an engineering product or service that appears to have been "suboptimized."
5 Alasdair MacIntyre says that law should be used only as a last resort. Continuous resort to law is a sign that some deeper moral relationship has broken down (A. MacIntyre, 1980). At what level of law (standards, regulations, statutory law, litigation) would you say this applies to engineering?

CONSULTING ENGINEERING

Consulting engineers operate in private practice. They are compensated by fees for the services they render, and not by salaries received from employers. Because of this, they tend to have greater freedom to make decisions about the projects they undertake. Yet their freedom is not absolute: They share with salaried engineers the need to earn a living.

Here we will raise questions in four areas—advertising, competitive bidding, contingency fees, and provisions for resolution of disputes—which illustrate some of the special responsibilities of consulting engineers. We will also note how in safety matters consulting engineers may have greater responsibility than salaried engineers, corresponding to their greater freedom.

Advertising

Some corporate engineers are involved in advertising because they work in product sales divisions. But within corporations, advertising of services, job openings, and the corporate image are left primarily to advertising executives and the personnel department. By contrast, consulting engineers are directly responsible for advertising their services, even when they hire consultants to help them.

Prior to a 1976 Supreme Court decision, competitive advertising in engineering was considered a moral issue and was banned by professional codes of ethics. As in law and medicine, anything beyond a tasteful notification of the availability of one's services was thought to be "unprofessional." It was deemed unfair to colleagues to win work through one's skill as an advertiser rather than through one's earned reputation as an engineer. It was also felt that competitive advertising caused friction among those in the field and lessened their mutual respect and that vigorous advertising damaged the profession's public image by placing engineering on a par with purely money-centered businesses.

However, the Supreme Court disagreed with that view. According to its ruling, as well as other rulings by the Federal Trade Commission, general bans on professional advertising are improper restraints of competition. They serve to keep prices for services higher than they might otherwise be, and they also reduce public awareness of the range of professional services available.

These rulings have shifted attention away from whether professional advertising is moral or not toward whether it is honest or not. Deceptive advertising normally occurs when products or services are made to look better than they actually are. This can be done in many ways: (1) by outright lies, (2) by half-truths, (3) through exaggeration, (4) by making false innuendos, suggestions, or implications, (5) through obfuscation created by ambiguity, vagueness, or incoherence, (6) through subliminal manipulation of the unconscious, etc. (Leiser, 1979).

There are notorious difficulties in determining whether specific ads are deceptive or not. Clearly it is deceptive for a consulting firm to claim in a brochure that it played a major role in a well-known project when it actually played a very minor role. But suppose the firm makes no such claim and merely shows a picture of a major construction project in which it played only a minor role? Or, more interestingly, suppose it shows the picture along with a footnote which states in fine print the true details about its minor role in the project? What if the statement is printed in larger type and not buried in a footnote?

As another example, think of a photograph of an engineering product (say, an electronics component) used in an ad to convey the impression that the item is routinely manufactured and available for purchase, perhaps even "off the shelf," when in actuality the picture shows only a preliminary prototype or mockup and the item is just being developed. To what extent, then, should the buyer—as the subject or participant of an "experiment" conducted by the manufacturer—be protected from misleading information about a product?

Advertisers of consumer products are generally allowed to suppress negative aspects of the items they are promoting and even to engage in some degree of exaggeration or "puffery" of the positive aspects. Notable exceptions are ads for cigarettes and saccharin products, which by law must carry health warnings. By contrast, norms concerning the advertising of professional services are much stricter. For example, the code of the National Society of Professional Engineers (NSPE) forbids all of the following:

> ...the use of statements containing a material misrepresentation of fact or omitting a material fact necessary to keep the statement from being misleading; statements intended or likely to create an unjustified expectation; statements containing prediction of future success; statements containing an opinion as to the quality of the Engineer's services; or statements intended or likely to attract clients by the use of showmanship, puffery, or self-laudation, including the use of slogans, jingles, or sensational language format (*NSPE Code of Ethics*, Sec. 3b).

Is there sufficient warrant for these tougher restrictions? Do they perhaps suppress vigorous competitive advertising which could be beneficial to the public? How are reasonable restrictions on the manner of advertising to be justified in terms of ethical theories? These are questions which deserve to be addressed within engineering ethics.

Active solicitation of clients through advertising or personal contacts has been considered especially unprofessional when it takes work away from other engineers. Yet here again one may ask whether some degree of solicitation serves the public interest by encouraging healthy competition. The dangers of allowing it, of course, must also be weighed carefully. Does it open the door to those who are dishonest and who might in very subtle ways unfairly criticize the work of other engineers whom they seek to supplant, or to those who might exaggerate the merits of their work? Certainly strong restrictions on misleading advertising in this area are especially important.

Competitive Bidding

For many years codes prohibited consulting engineers from engaging in competitive bidding, that is, from competing for jobs on the basis of submitting proposed fees. The following statement, for example, formerly appeared in the code of the American Society of Civil Engineers:

> It shall be considered unprofessional and inconsistent with honorable and dignified conduct and contrary to the public interest for any member of the American Society of Civil Engineers to invite or submit priced proposals under conditions that constitute price competition for professional services (quoted in Alger, Christensen, and Olmsted, 1965, 35).

It was considered permissible for industrial and construction firms to use competitive bidding because they could formulate cost estimates with some accuracy based on fixed design specifications. By contrast, the job of the consulting engineer is generally to develop creative designs for solving novel problems. Often there is no way to make precise bids. Allowing competitive bidding in such cases, it was felt, would open the door to irresponsible engineering in that inaccurate bids would encourage either cutting safety and quality (in the case of low bids) or padding and overdesigning (in the case of high bids).

However, in 1978 the Supreme Court ruled that professional societies were unfairly restraining free trade by banning competitive bidding. The ruling still left several loopholes, though. In particular, it allowed state registration boards to retain their bans on competitive bidding by registered engineers. It also allowed individual consulting firms to refuse to engage in competitive bidding. Thus fee competition where creative design is involved has remained a lively ethical issue. Is it in the best interests of clients and the public to encourage the practice?

If the use of competitive bidding is widely rejected by engineering firms, clients will have to rely almost exclusively on reputation and proven qualifications in choosing between them. This raises the problem of how the qualifications are to be determined in an equitable way. Is the younger, but still competent, consulting engineer placed at an unfair disadvantage? Or is it reasonable to view this disadvantage as justifiable, given the general importance of experience in consulting work?

Contingency Fees

Consulting engineers play the primary role in making arrangements about payment for their work. Naturally this calls for exercising a sense of honesty and fairness. But what is involved specifically?

As one illustration of the kinds of problems which may arise, consider the following entry in the code of the National Society of Professional Engineers:

> An Engineer shall not request, propose, or accept a professional commission on a contingent basis under circumstances in which his professional judgment may be compromised, or when a contingency provision is used as a device for promoting or securing a professional commission (*NSPE Code of Ethics,* Sec. 11b).

A *contingency fee* or *commission* is one which is dependent upon some special condition beyond the normal performance of satisfactory work. Typically, under a contingency-fee arrangement the consultant is paid only if she or he succeeds in saving the client money. Thus a client may hire a consultant to uncover cost-saving methods which will save 10 percent on an already contracted project. If the consultant does not succeed in doing so, no fee is paid. The fee may be either an agreed-upon amount or a fixed percentage of the savings to be realized.

In many contingency-fee situations it is easy for the consultant's judgment to become biased. For example, the prospects of winning the fee may tempt the consultant to specify inferior materials or design concepts in order to cut construction costs. Hence the point of the NSPE code entry. But even allowing for this problem, is the thoroughgoing ban on such fees in the NSPE code warranted? There is, after all, a point to their use. They are intended to help stimulate imaginative and hopefully responsible ways of saving costs to clients or the public, and presumably this consideration deserves some weight.

Resolving the issue calls for balancing out the potential gains against the potential losses that result from allowing or banning the practice. In this respect it is like many other issues in engineering ethics which call for reasonable judgments based on both past experience and foresight. And philosophical ethical theories can be useful in making those judgments by providing a general framework for assessing the morally relevant features of the problems under consideration.

Safety and Client Needs

The greater amount of job freedom enjoyed by consulting engineers as opposed to salaried engineers leads to wider areas of responsible decision making concerning safety. It also generates special difficulties.

Very often, for example, consulting engineers have the option of accepting or not accepting "design-only projects." A design-only project is one where the consultant contracts only to design something, but not to have any role—even a supervisory one—in its construction. Design-only projects are sometimes problematic because of difficulties encountered in implementing the designing engineer's specifications and because that engineer is often the only individual really well qualified to identify the areas of difficulty (Alderman and Schultz, 8–12). For example, clients or contractors may lack adequately trained inspectors of their own. In fact, when novel projects are being undertaken, clients may not even know that their own inspectors are unsatisfactory. Again, a contractor may be unable or unwilling to spot areas where the original design needs to be modified so as to best serve the client. The designer is often the person best able to ensure that the client's needs are met, as well as the safety needs of the project, yet he or she may not be around to do so.

The importance of having the designer involved in on-site inspection is illustrated by the following example:

> An engineering firm designed a flood control project for the temporary retention of storm water in a nearby city. Included in the project were some high reinforced concrete retaining walls to support the earth at the sides of the retention basin. Although the consulting engineer had no responsibility for site visits or inspection, one of the designers decided to visit the site, using part of his lunch hour to see how it was progressing. He found that the retaining wall footings had been poured and the wall forms were placed. He was shocked, however, to find that the reinforcing steel extended from the footings into the walls only a small fraction of the specified distance. He immediately returned to his office and the client was notified of the situation. The inspectors responsible were disciplined and correc-

tive measures taken with respect to the steel reinforcement. There is no question that in the first heavy rain the walls would have collapsed had the designer not discovered this fault. The result would have been heavy property loss, waste of resources, environmental damage and possible injury, even loss of life (Alderman and Schultz, 11).

It is thus a significant area of inquiry to determine when consulting engineers should or should not accept design-only projects. And when they do accept them, are they not obligated to make at least occasional on-site inspections later, in order "to monitor the experiment" they have set in motion? That is, are there at least some minimal moral responsibilities in this context which reach beyond the legal responsibilities specified in the contract?

In the course of making on-site inspections consulting engineers may notice unsafe practices which endanger workers. For example, they may notice the absence of a sufficient number of secondary support struts for a building or bridge. They may know from their past experience that this could cause a partial collapse of the structure while construction workers are on the job. Of course job safety is the primary responsibility of the contractor who has direct control over the construction. Yet for the consultant to do nothing would be negligent, if not callous. But how far do the consultant's responsibilities extend? Is a letter to the construction supervisor sufficient? Or is the consultant morally required to follow through by checking to see that the problem is corrected? It should be noted that an engineer who does point out construction deficiencies on one occasion—even if not contractually required to—but refrains from doing so on other occasions can be held liable for complicity in any damages resulting from unreported deficiencies. (Review Study Question 3, pp. 50-51.)

Provision for Resolution of Disputes

Large and complex engineering projects involve many participants at different levels of responsibility within the organizations representing the owner, the consulting engineer, and the construction firm. Overlapping responsibilities, fragmented control, indecision, delays, and an inability to resolve disputes quickly and amicably characterize many projects. To forestall potential liabilities in such situations, the various parties involved usually devote much time to protecting themselves when it could more profitably be used to improve the quality of the project. Resolution of disagreements is made more difficult when construction lasts several years and personnel changes occur during that period, because mutual trust and understanding are not easily nurtured under such conditions.

It has been observed by engineers engaged in construction projects that

> ...owners often initiate polarization from the outset by placing liability on engineers, vendors, and constructors without categorization or regard to which, or under what conditions faults may arise.... The engineer extends the chain by prepar-

> ing tighter specifications, employing every exculpatory phrase at his command, and inviting the constructor to nominate his best "sea-lawyer" as project manager. All this leads to the compounding of heavy contingency factors—where possible—or further assumption of risk by the implementing entities (P. Smith, 1978, 33).

The position of owners is not difficult to understand, as they have the most to lose. Thus the tendency to shift risks onto others. Engineers are tied to the contract provisions. They will be reluctant to innovate, preferring instead to stick to the tried and true ways of doing things. Contractors are used to taking large risks, but they are quick to recognize conditions "beyond their control."

Litigation has increased considerably in recent decades, and the character of litigation has changed as well. The construction industry is no exception, and its experience, as described by a panel of experts, serves as a good illustration of the kinds of legal problems consulting engineers now sometimes face:

> Traditionally law suits were fairly clear cut and involved matters directly related to the construction process; suits by owners were relatively uncommon; the design professional had to contend with virtually no litigation; and disputes were almost entirely confined to participants in the construction process.
>
> Today not only has the number of lawsuits dramatically increased but the nature of the lawsuits and the participants also have changed. Third forces, historically external to the process, today are the motivating factors behind a great many suits (Buehler, 1978, 94).

The third forces mentioned above typically are citizen intervenors or regulatory commissions.

Since litigation is time-consuming and costly, the consulting engineer should arrange contractually for methods of resolving conflicts. Quite apart from defining how risks are to be apportioned and payment of fees to be made, there should be contractual provisions for dispute-solving vehicles (designed to avoid costly court battles) such as mediation-arbitration in which a mediator attempts to resolve a dispute first, and if that is not fruitful, to act as the final, binding arbitrator. It should also be specified contractually that the National Joint Board for Settlement of Jurisdictional Disputes will be called upon to provide a hearing board and appeals board.

Engineering practice today does not regularly provide for clear-cut arbitration or conflict resolution. We believe that the consulting engineer is the proper party to assume the obligation of assuring that such clauses are included in contracts and are adhered to by everyone involved. This responsibility arises from the engineer's close contacts with both the owner and the constructor of a project and from the "social experimentation" nature of engineering. While contractual arrangements cannot solve problems caused by third-party intervenors, they can forestall a great many conflicts the origins of which can easily be foreseen and provided for.

Study Questions

1 Locate three advertisements: one for a technological product; one for the services of a consulting engineer; and one for positions in an engineering firm. Critique each ad in terms of whether the information or pictures included are misleading or deceptive in any way (be specific).

2 State why you agree, or why you disagree, with the following positions regarding advertising in local telephone directories. The first three examples comprise Case No. 72-1 of the *NSPE: Opinions of the Board of Ethical Review* (vol. IV, 1976). The fourth is our adaptation of a case involving a dentist and the dentists' association in San Francisco.

a "Are bold face listings in the classified section of local telephone directories consistent with the Code of Ethics?" The NSPE says "no."

b "Are bold face listings in the regular section of local telephone directories consistent with the Code of Ethics?" The NSPE says "yes."

c "Are professional card-type listings (set off by lines or blank space) in the classified section of local telephone directories consistent with the Code of Ethics?" The NSPE says "no."

d Mr. Zebra is a consulting engineer whose firm, Zebra Associates, appears last in the telephone directory's classified listing of engineers. In order to gain a more advantageous position in the yellow pages and in other directories, he changes the name of his firm to Antelope and Zebra. Antelope is a purely fictitious partner. Is this ethical?

In a similar case a dentists' board of review said "no."In considering the first three cases you may wish to consult the reasoning of the NSPE Board of Ethical Review. Also, examine some of the other numerous rulings by that board.

3 Is there anything unethical about the conduct of the engineer employee described in the following?

> A firm in private practice handles many small projects for an industrial client, averaging 20 to 30 projects a year. The firm has a signed agreement with the industrial client which does not obligate the client to give the firm any work, but does establish the respective responsibilities, terms of payment and other contractual details when the client does use the firm's services. The actual assignments are made by means of purchase orders referring to the agreement. An engineer employee of the firm resigns his employment and establishes his own firm and then actively solicits the industrial client of his former employer without any prior indication of interest by the client (*NSPE: Opinions of the Board of Ethical Review*, Case No. 73-7, vol. IV, 1976).

4 Is the decision of the consulting firm in the following example the morally obligatory one?

> A large civil engineering consulting firm completed a comprehensive arterial highway plan for a large, midwestern metropolitan area. The area appropriated funds for the next phase of the program: the preparation of a design report covering the recommended highways. The officials concerned, not knowing what a reasonable fee would be for the design report work, felt that it was necessary to invite proposals from various consultants. The consulting firm explained that it could not participate if the selection were to be made on the basis of engineering fees. The officials replied that, although price would not be the only consider-

ation, it would be a very important one. They could not explain to their constituents why the work was awarded to Engineer X if Engineer Y offered to do it for 2 percent less. The civil engineering company declined to participate (Alger, Christensen, and Olmsted, 1965, 49).

5 In the following case are Doe's presentation and offer entirely ethical?

John Doe, P.E., a principal in a consulting engineering firm, attended a public meeting of a township board of supervisors which had under consideration a water pollution control project with an estimated construction cost of $7 million. Doe presented a so-called "cost-saving plan" to the supervisors under which his firm would work with the engineering firm retained for the project to find "cost-saving" methods to enable the township to proceed with the project and thereby not lose the federal funding share because of the township's difficulty in financing its share of the project.

Doe further advised the supervisors that his company contemplated providing his "cost-saving" services on the basis of being paid ten percent of the savings; his firm would not be paid any amount if it did not achieve a reduction in the construction cost. Doe added that his firm's value engineering approach would be based on an analysis of the plans and specifications prepared by the design firm and that his operation would not require that the design firm be displaced (*NSPE: Opinions of the Board of Ethical Review*, Case No. 77-10, *Professional Engineer*, vol. 48, no. 6, June 1978, p. 52).

6 Should Engineer A in the following example take any further action, and if so what type?

During an investigation of a bridge collapse, Engineer A investigates another similar bridge, and finds it to be only marginally safe. He contacts the governmental agency responsible for the bridge and informs them of his concern for the safety of the structure. He is told that the agency is aware of this situation, and has planned to provide in next year's budget for its repair. Until then, the bridge must remain open to traffic. Without this bridge, emergency vehicles such as police and fire apparatus would have to use an alternate route which would increase their response time about twenty minutes.

Engineer A is thanked for his concern and asked to say nothing about the condition of the bridge. The agency is confident that the bridge will be safe (unpublished case study written up by and used with permission of L. R. Smith and Sheri Smith).

7 Examine and assess some recent disputes over whether fee competition among consulting engineers is ethical. A good place to begin is with the articles by Crawford Greene and Gerald Swenson listed in the Bibliography.

RESPONSIBILITIES OF AND TO THE PROFESSION

Throughout this book we have focused on the responsibilities of engineers *in* their work—that is, the responsibilities they have for the projects on which they work and for related aspects of their jobs. Yet it is important to avoid any suggestion that we have made an exhaustive survey of the issues in engineering ethics. Hence we conclude with a sketch of a few of the topics we have either omitted or mentioned only in passing, and which can be gathered under the unifying theme of responsibilities of and to the profession. Many of the points we raise here will be stated in the form of questions in order to challenge readers to continue further inquiry.

There are important issues pertaining to the responsibilities of the engineering profession as a whole. Many of those issues have to do with the collective responsibilities of professional and technical societies. There are also related issues having to do with the responsibilities of engineers to their profession, to peers, and to the world in which they live as citizens. In addition, there are questions about moral ideals that should be promoted in order to inspire desirable conduct that goes beyond the call of duty.

Collective Responsibilities of the Profession

It makes sense to praise or blame a profession as a whole for either establishing or failing to establish appropriate standards, and for either fostering or discouraging appropriate ideals. It also makes sense to praise or criticize its general role in society. This presupposes that there are general obligations and ideals which a profession as a collective body should pursue.

For one thing, making general appraisals of professions entails examining their "macroeconomics": that is, examining how they do and should function as a group within contemporary society. For example, to what extent is it desirable for the engineering profession as a whole to set standards in such areas as disposal of toxic wastes? (Ladd, 1980, 158)

Or, to take another kind of topic, is the trend toward increasing rule-making on behalf of professionalism within engineering in the public interest? Here many issues are involved, at least given the model of professionalism derived from developments in medicine and law:

1 Should the engineering profession be allowed to have the authority to decide which students and how many students will be admitted to schools of engineering? Should laypersons representing the public have a say?

2 Should registration and licensing of all engineers be mandatory, as they are for doctors and lawyers? There would be potential benefits: for example, greater assurance that minimal standards of training and skill would be met by all engineers. But there would also be drawbacks, if only that bureaucratic red tape would increase.

3 Should continuing education be mandatory for all engineers?

These and related issues are given regular, detailed examination at professional conferences. A sample of how they are typically approached can be found in *Ethics, Professionalism, and Maintaining Competence,* edited by Russel C. Jones et al., which is a collection of papers presented at a civil engineering conference. And many professional journals regularly carry articles treating these topics: for example, *Engineering Education, Professional Engineer, Engineering Issues (ASCE),*

Chemical Engineering, and *IEEE Spectrum*. A stimulating study of historical and sociological backgrounds for dealing with these topics is *The Credential Society* (Collins, 1964). However, there have as yet been few attempts to approach the issues within a framework of philosophical ethical theory.

Responsibilities of Professional Societies

Many of the large-scale social and economic issues involved in engineering ethics relate directly to professional societies. Professional societies help unify a profession. They also act on behalf of the profession, or at least on behalf of a large segment of the profession. A number of responsibilities ascribed to professions as collectives transfer directly to existing professional societies.

Many of the current tensions in professional societies exist because of uncertainties about their involvement in moral issues. This was illustrated in the BART case, as we saw in Chap. 6. One chapter of the California Society of Professional Engineers felt it should play a role in supporting the efforts of the three engineers who sought to act outside normal organizational channels in serving the public. Another chapter felt it was inappropriate for the society to do so.

It is unlikely that existing professional societies will, and perhaps undesirable that they should, take any univocal proemployee or promanagement stand. Their memberships, after all, are typically a mixture of engineers in management, supervision, and nonmanagement. Yet professional societies can, should, and are increasingly playing a role in conflicts involving moral issues, although rank and file engineers remain skeptical because they still consider the societies management-dominated (Flores, 1982, 80). Through membership participation on committees, they provide a sympathetic and informed forum for hearing opposing viewpoints and making recommendations. Through their guidelines for employment practice and conflict resolution they can help forestall debilitating disputes within corporations. Details concerning the desirable extent and form of such activities deserve ongoing discussion within engineering ethics.

Certainly we might expect professional societies increasingly to foster the study of engineering ethics. They are ideal groups for sponsoring ethics workshops, conducting surveys on matters of ethical concern, informing their members of developments related to ethics, and encouraging schools of engineering to support regular and continuing education courses in engineering ethics. Unfortunately a study of the activities of professional societies concluded that "little attention and only minimal resources have been directed toward professional ethics matters" (Chalk, Frankel, and Chafer, 1980, 101). As at least two other observers have pointed out, progress needs to be made before a genuine community based on a sense of shared values exists in the engineering profession (Perrucci and Gerstl, *Profession without Community: Engineers in American Society*, 176).

Another set of issues concerns the role professional societies should play in political decisions related to engineering and technological development. Should there be lobbying groups at the state and national levels seeking legislation on issues like engineering registration, continuing education requirements, and engineers' rights? Opinions are still sharply divided.

There is also the question of the direction and course of existing and possible new professional organizations. Perhaps it is desirable to have greater unity among engineering societies and thus to encourage newer and higher-level umbrella organizations to arise (such as the recently established American Association of Engineering Societies). Yet there are risks in seeking more unified power and action. Would a single powerful engineering society comparable to the American Medical Association or the American Bar Association be in the public interest?

Ultimately these "macro" issues lead back to the "micro" issues related to individual responsibility. For it is individuals involved in their professional societies who are the ultimate loci of action. An interesting area of engineering ethics would be to examine the typical situations in which individuals might be called upon to perform special services within their professional societies, and how far they are obligated to do so.

This leads us to the related topic of the obligations engineers have to their profession.

Obligations to the Profession

The code of ethics of the Accreditation Board for Engineering and Technology suggests that engineers should obey the code in order to "uphold and advance the integrity, honor and dignity of the engineering profession." Similarly, the preamble to the code of the National Society of Professional Engineers suggests that the code should be followed in part "to uphold and advance the honor and dignity of the engineering profession." Should such statements be dismissed as representing remnants of the natural esprit de corps of an emerging major profession? Or are there special professional obligations to the engineering profession which engineers should recognize?

It can be argued that engineers have an obligation to participate in technical societies in order to keep themselves up to date in their field and to help assure the continuation of such societies. That obligation derives from another obligation to clients, employers, and the public to advance their own and their colleagues' skills and knowledge. And at the very least it is highly desirable for engineers to be active in the ethics-related activities of their professional societies.

Are there additional obligations to defend the honor of the engineering profession against attacks from its critics? Would not the public and the profession itself be better served through a constant effort to seek truth and fairness in assessments of the engineering profession, and through a willingness to raise trenchant criticisms as conscience dictates? The legitimate honor of any profession as a whole can only be earned through its members acting with integrity in meeting their professional responsibilities.

Surely something can also be said in defense of the idea of a duty to respect and defend the honor of the profession. Effective professional activity, whether in engineering or any other profession, requires a substantial degree of trust from clients and the public. Total absence of such trust would undermine the possibility of making contracts, engaging in cooperative work, exercising professional autonomy free of excessive regulation, and working under humane conditions. Building and sustaining that trust is an important responsibility shared by all engineers.

This does not mean glossing over the genuine problems and faults one may encounter as the member of a profession. Nor does it mean always endorsing the more optimistic view of one's profession. But it does mean taking seriously the fact that appearances and reputation are important and should be given balanced and fair attention. Just what this amounts to in practice for engineers is a legitimate area of study for engineering ethics.

We might add that there is always the danger that the idea of an obligation to one's profession can become perverted into a narrow, self-interested concern. Such would be the case, for example, if a profession deliberately limited the number of its practitioners so as to create a greater demand for them and hence manipulate their salaries or fees upward. For this reason it may be preferable to understand talk about "obligations to the profession" as shorthand for certain obligations to the public. Engineers as individuals and as a group owe it to the public to sustain a professional climate conducive to meeting their other obligations to the public.

Collegiality and Obligations to Peers

A significant and neglected topic in professional ethics is that of relationships with peers, both coworkers and other members of one's profession. While engineering codes of ethics mention this topic, they often mention only the negative aspects. The NSPE code, for example, states that

> Engineers shall not attempt to injure, maliciously or falsely, directly or indirectly, the professional reputation, prospects, practice or employment of other engineers, nor indiscriminately criticize other engineers' work. Engineers who believe others are guilty of unethical or illegal practice shall present such information to the proper authority for action (Sec. III-8).

Insistence on not defaming colleagues unjustly and on not condoning unethical practice is important, since professionals are often overly reluctant to criticize peers responsibly. But this insistence needs to be balanced with an emphasis on the positive role of *collegiality* among professionals in promoting the moral aims of professions.

What is collegiality? In a recent essay which approaches collegiality as a fundamental professional virtue, Craig Ihara offers the following definition:

"Collegiality is a kind of connectedness grounded in respect for professional expertise and in a commitment to the goals and values of the profession, and..., as such, collegiality includes a disposition to support and cooperate with one's colleagues" (Ihara, 1988, 60). In other words, the central elements of collegiality are respect, commitment, connectedness, and cooperation.

Respect is the positive attitude involved in valuing one's peers for their professional expertise and their devotion to the social goods promoted by the profession. In the case of engineering this means affirming the worth of other engineers engaged in producing socially useful and safe products. Like friendship, collegial respect ought to be reciprocal (that is, mutual, rather than one-sided), but unlike friendship it need not involve personal affection.

Commitment means sharing a devotion to the moral ideals inherent in the practice of engineering. Even where there is fierce competition among professionals working for profit-making corporations, there should be a sense that other engineers share a concern for the overall good made possible through this competition. This is analogous to how members of competing teams in sports (hopefully) maintain a sense of underlying values beyond winning.

Connectedness is "an awareness of being part of a cooperative undertaking created by shared commitments and expertise" (Ihara, 58). It is more than acting in ways that show respect for peers. One must do so with an appropriate attitude of affirming their worth and with a sense of being united with them in an enterprise defined by common goals. This sense of unity with other engineers evokes *cooperation* and mutual support.

Why is collegiality a virtue—that is, a valuable trait of character which ought to be encouraged among engineers and other professionals? Ihara offers two answers. Viewed from the perspective of society, collegiality is an instrumental value; it is good as a means to promoting professional aims. By enlivening one's sense of shared commitment with others, collegiality supports personal efforts to act responsibly in concert with colleagues. Hence it strengthens one's motivation to live up to professional standards.

Viewed from the perspective of professionals, collegiality is intrinsically valuable. It is part of what defines the *professional community*—in this case, the engineering community—as comprised of many individuals jointly pursuing the public good. Such a community cannot continue without some shared awareness of mutual commitment to professional ideals.

Understood in this way, collegiality is a professional virtue deserving further attention. There is a need to explore how it can be distorted and misused, such as when peers appeal to it in urging silence about corporate corruption. Collegiality is not an excuse nor a justification for shielding irresponsible conduct. In fact, as Ihara suggests, peers who engage in gross misconduct cease to be "colleagues" in the moral sense analyzed above, for they are no longer worthy of respect and support. Again, collegiality can degenerate into mere group self-interest, rather than shared devotion to the public good. Preoccupation with cutthroat competitiveness also threatens collegiality, as does a narrow focus on the corporate goal of maximizing profit in disregard of the public good.

Responsibilities of Engineer-Citizens

Do engineers as a group and as individuals have special responsibilities as citizens which go beyond those of nonengineer-citizens? For example, should they participate more actively than others in social debates concerning industrial pollution, automobile safety, and disposal of nuclear waste?

Answering this question would require a clarification of the obligations citizens in general have concerning public policy issues. But even here there is considerable disagreement. One view holds that no one is strictly obligated to participate in public decision making. Instead, such participation is a moral ideal which it is desirable for citizens to embrace and pursue as their time allows. A contrasting view holds that all citizens have an obligation to devote some of their time and energies to public policy matters. Minimal requirements for everyone are to stay informed about issues that can be voted on, while stronger obligations arise for those who by professional background are well grounded in specific issues as well as for those who have the time to train themselves as public advocates.

Engineers are not as well represented on many legislative and advisory bodies as they might be. Perhaps they are too modest about offering their services or maybe they see a number of complications arising from service of this kind.

For engineers in private practice, the latter consideration can be particularly troubling. For example, there is the matter of advertising: While an engineer who is employed by a company will bring recognition and honor to the company through volunteer activities, any such efforts on the part of a self-employed engineer could be interpreted as self-serving attempts to gain publicity and perhaps even to secure valuable inside information.

Ideals of Voluntarism

Should the engineering profession encourage the pro bono, voluntary giving of engineering services without fee or at reduced fees to especially needy groups? Is this an ideal which is desirable for engineering professional societies to embrace and foster among individuals and corporations?

Voluntarism of this sort has long been encouraged in medicine, law, and education. By sharp contrast, engineering codes of ethics have either been silent on this question or taken stands that discourage voluntarism. For example, the ECPD (now ABET) code was revised during the 1960s to state: "Engineers shall not undertake nor agree to perform any engineering service on a free basis." Most other codes also insisted that engineers are obligated to require "adequate" compensation for their work—meaning compensation at the present fee scale. Such statements are now being revised in light of Supreme Court rulings suggesting they restrain free trade. Nevertheless, there continues to be a sentiment against encouraging engineers to donate their services without full compensation.

Robert Baum has challenged this sentiment (Baum, 1985). He acknowledges that engineers have fewer opportunities to donate their services as individuals than do doctors and lawyers. This is because engineering services tend to require shared efforts and to demand the resources of the corporations for which most engineers work. But this merely shows that engineers might best help the needy through group efforts. (It is also true that increasing numbers of doctors and lawyers work for corporations).

Is the providing of engineering services to the needy an important matter, as is the providing of legal and medical services? Yes, for two reasons. On the one hand, meeting some legal and medical needs requires supplementation by engineering services. For example, Baum argues that Native Americans (American Indians) often lack the resources for the engineering studies needed in dealing with the Bureau of Land Management, which has authority to grant leases on Native American land. There is money for lawyers, but no money for costly environmental impact studies required for, say, challenging a proposed government project that is harmful in the view of a Native American group. There are similar problems associated with health issues raised by polluted water and soil on reservations.

On the other hand, there are financially disadvantaged groups, especially the elderly and some minorities living in both urban and rural areas, whose minimal needs are at present not met—needs for running water, sewage systems, electrical power, and inexpensive transportation. This could be remedied if access to engineering services were made available at lower than normal costs.

There are many options that the profession of engineering might explore. These include encouraging engineers to serve in government programs like VISTA, urging government to expand the services of the Army Corps of Engineers, encouraging engineering students to focus their senior projects on service for disadvantaged groups, and encouraging corporations to offer 5 percent or 10 percent of their services free or at reduced rates for charitable purposes.

Is there an obligation for professional societies to foster voluntarism among engineers in providing engineering services as reduced fees? Baumleaves the question open. His main concern was to argue that needy groups ought to have access to engineering services, but not to resolve the question of who should provide them (groups of engineers, corporations, local government, federal government, etc.). He suggests, however, that engineers do have one important duty concerning services for the needy: "to participate in dialogues concerning the needs of specific individuals and groups and the possible ways in which these needs might be met" (Baum, 1985, 133).

Baum feels that through initiating discussion between representatives of the engineering profession and of disadvantaged groups solutions may be found.

We would add that a morally concerned engineering profession should recognize the rights of corporations and individual engineers to voluntarily engage in philanthropic engineering service. Furthermore, it would be desirable for professional societies to endorse the voluntary exercise of this right as being a desirable ideal—an ideal of generosity that goes beyond the call of duty. While good deeds beyond the scope of one's primary work cannot compensate for unethical conduct inside it, a profession fully dedicated to the public good should recommend participation by engineers in all aspects of community life. Many individual engineers and some engineering societies are already engaged in such volunteer services. They range from tutoring disadvantaged students in mathematics and physics, to "urban technology" interest groups and senior engineering students who advise local governments on their engineering problems.

Study Questions

1 Most states do not require engineers employed by industrial corporations to be registered or licensed by the state (based on meeting certain minimal requirements of education, knowledge, and experience), although some require registration of engineers in charge of design and manufacturing processes that affect the public health and safety. This "industrial exemption" has come under increasing criticism. Do you agree or disagree with the following reasons for abolishing the industrial exemption and for requiring industrial engineers to be licensed? Are they all good moral reasons for requiring registration?

a Registered engineers assure the company of the services of those who have met the prescribed statutory requirements of the law enacted to protect the public health, safety, and welfare.

b An engineering staff composed of registered professional engineers enhances the prestige and public relations potential of the firm.

c Engineering registration improves morale of the engineer by attesting to his qualifications, competence, and professional attitude. It also encourages the engineer to take full responsibility for his work.

d Engineering registration improves company-client relations by attesting to engineering staff competence and satisfies the legal requirements of many states and municipalities requiring project control under a registered engineer.

e Engineering registration promotes high standards of professional conduct, ethical practice, integrity, and top-quality job performance (Kettler, 1983, 534).

2 Currently only state, not national, registration is possible for engineers, and even that is frequently optional. Assess the following views concerning the desirability of requiring national registration of all engineers. Are there other reasons for and against national registration?

a "The method of obtaining national registration as well as the approach would act as a vehicle for elevating the engineering profession to standards that doctors and lawyers now enjoy" ("Comments from Professional Engineers: National Registration," 1964, 26).

b "I feel that a national registration law administered by a Federal agency would be very detrimental to the engineering profession because it would transfer our local responsibilities and authority to a Federal power.... [In] practice I think it would become a dictorial agency, administering the law for its own purposes, with very little regard for the engineering profession" ("Comments from Professional Engineers: National Registration," 1964, 26).

3 An engineer learns that a friend and coworker has for several years gone to professional society meetings primarily for vacation purposes. The trips are billed to the employer, and the friend attends at most one or two sessions at the meetings (to put in an appearance) before going out on the town for a relaxing time. This clearly violates company policy. What should the engineer do?

4 Identify and discuss any moral duties, rights, and ideals pertinent to the following example.

> An engineer who had also had experience as a carpenter-contractor was asked by his church to assist in the construction of a new building. He finally served as the general contractor, the engineer-inspector, and the construction foreperson. He used labor donated by members of the church, including many carpenters, painters, cement finishers, and so on, even though they were not members of the skilled trades. No salaries were paid to any church members, but credit for labor was allowed against a pledge made by each member.... Before the engineer accepted the multiple responsibilities, bids from general contractors had been taken, and all were out of reach of the church group (Alger, Christensen, and Olmsted, 1965, 236).

5 Defend your view as to whether engineers have special obligations beyond those of nonengineers to enter into public debates over technological development. If you think they do not have special obligations, is it nevertheless especially desirable (as a moral ideal) for them to contribute to these debates? Should professional societies encourage such participation?

6 Examine several recent issues of professional engineering journals to locate articles on ethics. Report on the types of issues being addressed in those articles.

SUMMARY OF CHAPTER 8

Engineering offers a variety of "existential pleasures": for example, the delights of changing the world, being creative, understanding the physical universe, and promoting the good of others. It also offers economic rewards. Yet in addition there is a direct moral dimension related to the decisions to enter engineering as a profession, to specialize in a particular field of engineering, and to pursue a specific type of work within engineering. How best to meet the demands of one's personal conscience, as well as how best to secure one's personal happiness, should be considered in deciding.

Thus the notion of an ethics of vocation, or a moral philosophy of work, is relevant to career decisions. William Frankena has argued that there is a duty to pursue a vocation. Whether or not that is so, there clearly are special moral duties attached to specific professional roles within engineering. Career decisions involve relating one's basic moral convictions to such professional duties and also to the specific types of activities one will pursue.

Related to an ethics of vocation is the notion of a personal work ethic: that is, the values and ideals according to which one pursues one's career. Another aspect, as applied to engineering, is the assessment of various proposed models for the social role of engineers: for example, savior, guardian, bureaucratic servant, social servant, social enabler and catalyst, and game player.

Engineering is a process with stages (design, manufacture, sales, operation) which have no clear demarcations. As in ethics, where there are no sharp boundaries between conceptual and normative inquiries or between the foundational ethical principles, we must work our way through the forest of possible actions and outcomes. Clarity is achieved through this process more than from the final resting place in the forest. But to be successful one must not lose sight of the forest for the trees. It is management's task to structure engineering functions in such a way that the global outlook is never lost.

Consulting engineers have greater freedom than most salaried engineers. They also must deal with a wider variety of moral concerns since they are responsible for their own business decisions. Honesty in advertising may be difficult to maintain in the face of pressures to promote their business. Controversy still exists concerning the legitimacy of competitive bidding and accepting contingency fees. And special problems arise in determining the extent of responsibility in such areas as making safety inspections and arriving at ways to settle disputes.

Engineers are responsible to society not only through their individual activities but also through their profession. Their professional societies support the furtherance of technical knowledge, represent engineers collectively, and encourage volunteer assistance. But they can only be as effective as engineers will help them be through their support.

OVERVIEW AND CONCLUDING REMARKS

Before he was elected President of the United States, Herbert Hoover was the engineer-president of the American Institute of Mining Engineers and also of the former Federated American Engineering Societies. Serving in those capacities he had a significant impact on the development of the profession of engineering (Layton, 1971, 179–218). In his memoirs he describes the honors and liabilities of engineering:

> It is a great profession. There is the fascination of watching a figment of the imagination emerge through the aid of science to a plan on paper. Then it moves to realization in stone or metal or energy. Then it brings jobs and homes to men. Then it elevates the standards of living and adds to the comforts of life. That is the engineer's high privilege.
>
> The great liability of the engineer compared to men of other professions is that

his works are out in the open where all can see them. His acts, step by step, are in hard substance. He cannot bury his mistakes in the grave like the doctors. He cannot argue them into thin air or blame the judge like the lawyers. He cannot, like the architects, cover his failures with trees and vines. He cannot, like the politicians, screen his shortcomings by blaming his opponents and hope that the people will forget. The engineer simply cannot deny that he did it. If his works do not work, he is damned (Hoover, 1961, 132–133).

Hoover is reflecting on a time when engineering was still dominated, at least in outlook, by the independent craftsperson and consultant. When a bridge fell or a ship sank, the particular engineers responsible could be more easily identified. This made it easier to endorse Hoover's vision of individualism in regard both to creativity and personal accountability within engineering.

Today the products of engineering are as much "out in the open" as in Hoover's time. In fact, mass communication assures that mistakes are given even closer public scrutiny. And there are more engineers than ever. Yet despite their greater numbers, the engineers of today are less visible to the public than those of that earlier era. Technological progress is taken for granted as being the norm, and technological failure is blamed on corporations. And in the public's eye, the representative of any corporation is its top manager, who is often far removed from the daily creative endeavors of the company's engineers.

This "invisibility" of engineers makes it difficult for them to retain a sense of mutual understanding with and accountability to the public. The dominant image of engineers, even for many engineers themselves, has become that of a servant to organizations rather than a public guardian.

We have explored a number of the additional threats to a sense of personal responsibility coming from within organizations. Some of those arise inevitably from the context of contemporary competitive enterprise: fragmentation of work, pressured time schedules, tight budgets. But others are not inevitable: overly rigid lines of authority, disregard of the individual's conscience, suppression of free expression of professional judgment. The latter can be and are being changed as both employee rights and the public responsibilities of corporations receive greater attention.

There is no need to review here the many issues we have explored concerning the responsibilities of engineers. A rereading of the chapter summaries will suffice for that. Yet it is worthwhile to summarize briefly a few of the main themes which have surfaced throughout this book.

We have emphasized the personal moral autonomy of individuals. We did so with the conviction that responsible conduct is important in all aspects of living, not the least of which is work, and that it cannot be transferred away to employers, code writers, or lawmakers. Yet moral autonomy, as we have treated it, is vastly different from unrestrained individual whim. Autonomous moral agents are sensitive to the legitimate claims of employer authority and the need for reasonable laws and some codified ethical rules. At the same time they are sufficiently concerned to critically assess laws, codes, and directives in the light of reasoned value principles.

We have also emphasized the obligations of engineers to the public based on the public's right to make informed decisions about the risks affecting it. Meeting this obligation requires that engineers working within corporations and other large organizations be given the freedom to express and act on their professional judgment. Thus we were led to stress the importance of employee rights for engineers.

Those rights are limited by the legitimate authority of employers. While we explored several morally relevant aspects of that authority, we offered no simple algorithm for precisely determining its limits. Here, as elsewhere, there is room for disagreement among reasonable people. Engineers would be dogmatic if they always insisted their views were the correct ones. Managers would be equally dogmatic and abusive of their authority if they gave no recognition to the conscientious concern engineers often display and action they sometimes take in disputed areas. There is a need for appeal groups, both within the employing organizations (ombudspersons, ethics committees, etc.) and outside them (confidential review committees of professional societies or the government). Above all there is the need for mutual understanding among engineers and management about the need to cooperatively resolve value conflicts.

These themes can be integrated within a conception of engineering as an experiment on a societal scale involving human subjects. We can restate the points previously made within the framework of this paradigm: As subjects of the "experiment," clients and the public have a right to make informed decisions concerning the risks they will be subject to. As central participants in the "experiment," engineers have a special responsibility to respect those rights. To do so they must be allowed the freedom to conduct the "experiment" in a professional manner, which includes freedom from actions on the part of managers and supervisors that would seriously violate their consciences. On the other hand, managers play an equally central role in engineering "experiments." They have the legitimate authority to guide many areas of the work lives of engineers, although they should avoid undue restrictions on work and (as just noted) refrain from violating the consciences of those they supervise. But it must be kept in mind that moral dilemmas involving competing obligations to employers and the public are virtually inevitable for engineers. Resolving those dilemmas requires both autonomous moral reasoning about and tolerance and sensitivity to different views concerning areas of disagreement that can arise among responsible moral agents.

SAMPLE CODES AND GUIDELINES

ABET : CODE OF ETHICS
GUIDELINES
FAITH OF THE ENGINEER

AAES : MODEL GUIDE FOR
PROFESSIONAL CONDUCT

NSPE : CODE OF ETHICS

IEEE : CODE OF ETHICS

Accreditation Board for Engineering and Technology*

CODE OF ETHICS OF ENGINEERS

THE FUNDAMENTAL PRINCIPLES

Engineers uphold and advance the integrity, honor and dignity of the engineering profession by:

I. using their knowledge and skill for the enhancement of human welfare;

II. being honest and impartial, and serving with fidelity the public, their employers and clients;

III. striving to increase the competence and prestige of the engineering profession; and

IV. supporting the professional and technical societies of their disciplines.

THE FUNDAMENTAL CANONS

1. Engineers shall hold paramount the safety, health and welfare of the public in the performance of their professional duties.
2. Engineers shall perform services only in the areas of their competence.
3. Engineers shall issue public statements only in an objective and truthful manner.
4. Engineers shall act in professional matters for each employer or client as faithful agents or trustees, and shall avoid conflicts of interest.
5. Engineers shall build their professional reputation on the merit of their services and shall not compete unfairly with others.
6. Engineers shall act in such a manner as to uphold and enhance the honor, integrity and dignity of the profession.
7. Engineers shall continue their professional development throughout their careers and shall provide opportunities for the professional development of those engineers under their supervision.

345 East 47th Street New York, NY 10017

*Formerly Engineers' Council for Professional Development. (Approved by the ECPD Board of Directors, October 5, 1977)

AB-54 2/85

Accreditation Board for Engineering and Technology*

SUGGESTED GUIDELINES FOR USE WITH THE FUNDAMENTAL CANONS OF ETHICS

1 Engineers shall hold paramount the safety, health and welfare of the public in the performance of their professional duties.

a Engineers shall recognize that the lives, safety, health and welfare of the general public are dependent upon engineering judgments, decisions and practices incorporated into structures, machines, products, processes and devices.

b Engineers shall not approve nor seal plans and/or specifications that are not of a design safe to the public health and welfare and in conformity with accepted engineering standards.

c Should the Engineers' professional judgment be overruled under circumstances where the safety, health, and welfare of the public are endangered, the Engineers shall inform their clients or employers of the possible consequences and notify other proper authority of the situation, as may be appropriate.

(1) Engineers shall do whatever possible to provide published standards, test codes and quality control procedures that will enable the public to understand the degree of safety or life expectancy associated with the use of the design, products and systems for which they are responsible.

(2) Engineers will conduct reviews of the safety and reliability of the design, products or systems for which they are responsible before giving their approval to the plans for the design.

(3) Should Engineers observe conditions which they believe will endanger public safety or health, they shall inform the proper authority of the situation.

d Should Engineers have knowledge or reason to believe that another person or firm may be in violation of any of the provisions of these Guidelines, they shall present such information to the proper authority in writing and shall cooperate with the proper authority in furnishing such further information or assistance as may be required.

1 They shall advise proper authority if an adequate review of the safety and reliability of the products or systems has not been made or when the design imposes hazards to the public through its use.

2 They shall withhold approval of products or systems when changes or modifications are made which would affect adversely its performance insofar as safety and reliability are concerned.

e Engineers should seek opportunities to be of constructive service in civic

*Formerly Engineers' Council for Professional Development.

affairs and work for the advancement of the safety, health and well-being of their communities.

f Engineers should be committed to improving the environment to enhance the quality of life.

2 Engineers shall perform services only in areas of their competence.

a Engineers shall undertake to perform engineering assignments only when qualified by education or experience in the specific technical field of engineering involved.

b Engineers may accept an assignment requiring education or experience outside of their own fields of competence, but only to the extent that their services are restricted to those phases of the project in which they are qualified. All other phases of such project shall be performed by qualified associates, consultants, or employees.

c Engineers shall not affix their signatures and/or seals to any engineering plan or document dealing with subject matter in which they lack competence by virtue of education or experience, nor to any such plan or document not prepared under their direct supervisory control.

3 Engineers shall issue public statements only in an objective and truthful manner.

a Engineers shall endeavor to extend public knowledge, and to prevent misunderstandings of the achievements of engineering.

b Engineers shall be completely objective and truthful in all professional reports, statements, or testimony. They shall include all relevant and pertinent information in such reports, statements, or testimony.

c Engineers, when serving as expert or technical witnesses before any court, commission, or other tribunal, shall express an engineering opinion only when it is founded upon adequate knowledge of the facts in issue, upon a background of technical competence in the subject matter, and upon honest conviction of the accuracy and propriety of the testimony.

d Engineers shall issue no statements, criticisms, nor arguments on engineering matters which are inspired or paid for by an interested party, or parties, unless they have prefaced their comments by explicitly identifying themselves, by disclosing the identities of the party or parties on whose behalf they are speaking, and by revealing the existence of any pecuniary interest they may have in the instant matters.

e Engineers shall be dignified and modest in explaining their work and merit, and will avoid any act tending to promote their own interests at the expense of the integrity, honor and dignity of the profession.

4 Engineers shall act in professional matters for each employer or client as faithful agents or trustees, and shall avoid conflicts of interest.

a Engineers shall avoid all known conflicts of interest with their employers or clients and shall promptly inform their employers or clients of any business association, interest, or circumstances which could influence their judgment or the quality of their services.

b Engineers shall not knowingly undertake any assignments which would knowingly create a potential conflict of interest between themselves and their clients or their employers.

c Engineers shall not accept compensation, financial or otherwise, from more than one party for services on the same project, nor for services pertaining to the same project, unless the circumstances are fully disclosed to, and agreed to, by all interested parties.

d Engineers shall not solicit nor accept financial or other valuable considerations, including free engineering designs, from material or equipment suppliers for specifying their products.

e Engineers shall not solicit nor accept gratuities, directly or indirectly, from contractors, their agents, or other parties dealing with their clients or employers in connection with work for which they are responsible.

f When in public service as members, advisers, or employees of a governmental body or department, Engineers shall not participate in considerations or actions with respect to services provided by them or their organization in private or product engineering practice.

g Engineers shall not solicit nor accept an engineering contract from a governmental body on which a principal, officer or employee of their organization serves as a member.

h When, as a result of their studies, Engineers believe a project will not be successful, they shall so advise their employer or client.

i Engineers shall treat information coming to them in the course of their assignments as confidential, and shall not use such information as a means of making personal profit if such action is adverse to the interests of their clients, their employers or the public.

(1) They will not disclose confidential information concerning the business affairs or technical processes of any present or former employer or client or bidder under evaluation, without his consent.

(2) They shall not reveal confidential information nor findings of any commission or board of which they are members.

(3) When they use designs supplied to them by clients, these designs shall not be duplicated by the Engineers for others without express permission.

(4) While in the employ of others, Engineers will not enter promotional efforts or negotiations for work or make arrangements for other employment as principals or to practice in connection with specific projects for which they have gained particular and specialized knowledge without the consent of all interested parties.

j The Engineer shall act with fairness and justice to all parties when administering a construction (or other) contract.

k Before undertaking work for others in which Engineers may make improvements, plans, designs, inventions, or other records which may justify copyrights or patents, they shall enter into a positive agreement regarding ownership.

l Engineers shall admit and accept their own errors when proven wrong and refrain from distorting or altering the facts to justify their decisions.

m Engineers shall not accept professional employment outside of their regular work or interest without the knowledge of their employers.

n Engineers shall not attempt to attract an employee from another employer by false or misleading representations.

o Engineers shall not review the work of other Engineers except with the knowledge of such Engineers, or unless the assignments/or contractual agreements for the work have been terminated.

(1) Engineers in governmental, industrial or educational employment are entitled to review and evaluate the work of other engineers when so required by their duties.

(2) Engineers in sales or industrial employment are entitled to make engineering comparisons of their products with products of other suppliers.

(3) Engineers in sales employment shall not offer nor give engineering consultation or designs or advice other than specifically applying to equipment, materials or systems being sold or offered for sale by them.

5 Engineers shall build their professional reputation on the merit of their services and shall not compete unfairly with others.

a Engineers shall not pay nor offer to pay, either directly or indirectly, any commission, political contribution, or a gift, or other consideration in order to secure work, exclusive of securing salaried positions through employment agencies.

b Engineers should negotiate contracts for professional services fairly and only on the basis of demonstrated competence and qualifications for the type of professional service required.

c Engineers should negotiate a method and rate of compensation commensurate with the agreed upon scope of services. A meeting of the minds of the parties to the contract is essential to mutual confidence. The public interest requires that the cost of engineering services be fair and reasonable, but not the controlling consideration in selection of individuals or firms to provide these services.

(1) These principles shall be applied by Engineers in obtaining the services of other professionals.

d Engineers shall not attempt to supplant other Engineers in a particular employment after becoming aware that definite steps have been taken toward the others' employment or after they have been employed.

(1) They shall not solicit employment from clients who already have Engineers under contract for the same work.

(2) They shall not accept employment from clients who already have Engineers for the same work not yet completed or not yet paid for unless the performance or payment requirements in the contract are being litigated or the contracted Engineers' services have been terminated in writing by either party.

(3) In case of termination of litigation, the prospective Engineers before accepting the assignment shall advise the Engineers being terminated or involved in litigation.

e Engineers shall not request, propose nor accept professional commissions on a contingent basis under circumstances under which their professional judgments may be compromised, or when a contingency provision is used as a device for promoting or securing a professional commission.

f Engineers shall not falsify nor permit misrepresentation of their, or their associates', academic or professional qualifications. They shall not misrepresent nor exaggerate their degree of responsibility in or for the subject matter of prior assignments. Brochures or other presentations incident to the solicitation of employment shall not misrepresent pertinent facts concerning employers, employees, associates, joint ventures, or their past accomplishments with the intent and purpose of enhancing their qualifications and work.

g Engineers may advertise professional services only as a means of identification and limited to the following:

(1) Professional cards and listings in recognized and dignified publications, provided they are consistent in size and are in a section of the publication regularly devoted to such professional cards and listings. The information displayed must be restricted to firm name, address, telephone number, appropriate symbol, names of principal participants and the fields of practice in which the firm is qualified.

(2) Signs on equipment, offices and at the site of the projects for which they render services, limited to firm name, address, telephone number and type of services, as appropriate.

(3) Brochures, business cards, letterheads and other factual representations of experience, facilities, personnel and capacity to render service, providing the same are not misleading relative to the extent of participation in the projects cited and are not indiscriminately distributed.

(4) Listings in the classified section of telephone directories, limited to name, address, telephone number and specialties in which the firm is qualified without resorting to special or bold type.

h Engineers may use display advertising in recognized dignified business and professional publications, providing it is factual and relates only to engineering, is free from ostentation, contains no laudatory expressions or implication, is not misleading with respect to the Engineers' extent of participation in the services or projects described.

i Engineers may prepare articles for the lay or technical press which are factual, dignified and free from ostentations or laudatory implications. Such articles shall not imply other than their direct participation in the work described unless credit is given to others for their share of the work.

j Engineers may extend permission for their names to be used in commercial advertisements, such as may be published by manufacturers, contractors, material suppliers, etc., only by means of a modest dignified notation acknowledging their participation and the scope thereof in the project or product described. Such permission shall not include public endorsement of proprietary products.

k Engineers may advertise for recruitment of personnel in appropriate publications or by special distribution. The information presented must be displayed in a dignified manner, restricted to firm name, address, telephone number, appropriate symbol, names of principal participants, the fields of practice in which the firm is qualified and factual descriptions of positions available, qualifications required and benefits available.

l Engineers shall not enter competitions for designs for the purpose of obtaining commissions for specific projects, unless provision is made for reasonable compensation for all designs submitted.

m Engineers shall not maliciously or falsely, directly or indirectly, injure the professional reputation, prospects, practice or employment of another engineer, nor shall they indiscriminately criticize another's work.

n Engineers shall not undertake nor agree to perform any engineering service on a free basis, except professional services which are advisory in nature for civic, charitable, religious or non-profit organizations. When serving as members of such organizations, engineers are entitled to utilize their personal engineering knowledge in the service of these organizations.

o Engineers shall not use equipment, supplies, laboratory nor office facilities of their employers to carry on outside private practice without consent.

p In case of tax-free or tax-aided facilities, engineers should not use student services at less than rates of other employees of comparable competence, including fringe benefits.

6 Engineers shall act in such a manner as to uphold and enhance the honor, integrity and dignity of the profession.

a Engineers shall not knowingly associate with nor permit the use of their names nor firm names in business ventures by any person or firm which they know, or have reason to believe, are engaging in business or professional practices of a fraudulent or dishonest nature.

b Engineers shall not use association with non-engineers, corporations, nor partnerships as 'cloaks' for unethical acts.

7 Engineers shall continue their professional development throughout their careers, and shall provide opportunities for the professional development of those engineers under their supervision.

a Engineers shall encourage their engineering employees to further their education.

b Engineers should encourage their engineering employees to become registered at the earliest possible date.

c Engineers should encourage engineering employees to attend and present papers at professional and technical society meetings.

d Engineers should support the professional and technical societies of their disciplines.

e Engineers shall give proper credit for engineering work to those to whom credit is due, and recognize the proprietary interests of others. Whenever possible, they shall name the person or persons who may be responsible for designs, inventions, writings or other accomplishments.

f Engineers shall endeavor to extend the public knowledge of engineering, and shall not participate in the dissemination of untrue, unfair or exaggerated statements regarding engineering.

g Engineers shall uphold the principle of appropriate and adequate compensation for those engaged in engineering work.

h Engineers should assign professional engineers duties of a nature which will utilize their full training and experience insofar as possible, and delegate lesser functions to subprofessionals or to technicians.

i Engineers shall provide prospective engineering employees with complete information on working conditions and their proposed status of employment, and after employment shall keep them informed of any changes.

Accredation Board for Engineering and Technology
345 East 47th Street
New York, NY 10017

Engineers' Council for Professional Development

FAITH OF THE ENGINEER

I AM AN ENGINEER. In my profession I take deep pride, but without vainglory; to it I owe solemn obligations that I am eager to fulfill.

As an Engineer, I will participate in none but honest enterprise. To him that has engaged my services, as employer or client, I will give the utmost of performance and fidelity.

When needed, my skill and knowledge shall be given without reservation for the public good. From special capacity springs the obligation to use it well in the service of humanity; and I accept the challenge that this implies.

Jealous of the high repute of my calling, I will strive to protect the interests and the good name of any engineer that I know to be deserving; but I will not shrink, should duty dictate, from disclosing the truth regarding anyone that, by unscrupulous act, has shown himself unworthy of the profession.

Since the Age of Stone, human progress has been conditioned by the genius of my professional forbears. By them have been rendered usable to mankind Nature's vast resources of material and energy. By them have been vitalized and turned to practical account the principles of science and the revelations of technology. Except for this heritage of accumulated experience, my efforts would be feeble. I dedicate myself to the dissemination of engineering knowledge, and, especially to the instruction of younger members of my profession in all its arts and traditions.

To my fellows I pledge, in the same full measure I ask of them, integrity and fair dealing, tolerance and respect, and devotion to the standards and the dignity of our profession; with the consciousness, always, that our special expertness carries with it the obligation to serve humanity with complete sincerity.

Prepared by the Ethics Committee

* Now the Accrediting Board for Engineering and Technology (ABET).

MODEL GUIDE FOR PROFESSIONAL CONDUCT
AMERICAN ASSOCIATION OF ENGINEERING SOCIETIES

Preamble

Engineers recognize that the practice of engineering has a direct and vital influence on the quality of life for all people. Therefore, engineers should exhibit high standards of competency, honesty and impartiality; be fair and equitable; and accept a personal responsibility for adherence to applicable laws, the protection of the public health, and maintenance of safety in their professional actions and behavior. These principles govern professional conduct in serving the interests of the public, clients, employers, colleagues and the profession.

The Fundamental Principle

The engineer as a professional is dedicated to improving competence, service, fairness and the exercise of well-founded judgment in the practice of engineering for the public, employers and clients with fundamental concern for the public health and safety in the pursuit of this practice.

Canons of Professional Conduct

Engineers offer services in the areas of their competence and experience, affording full disclosure of their qualifications.

Engineers consider the consequences of their work and societal issues pertinent to it and seek to extend public understanding of those relationships.

Engineers are honest, truthful and fair in presenting information and in making public statements reflecting on professional matters and their professional role.

Engineers engage in professional relationships without bias because of race, religion, sex, age, national origin or handicap.

Engineers act in professional matters for each employer or client as faithful agents or trustees, disclosing nothing of a proprietary nature concerning the business affairs or technical processes of any present or former client or employer without specific consent.

Engineers disclose to affected parties known or potential conflicts of interest or other circumstances which might influence—or appear to influence—judgment or impair the fairness or quality of their performance.

Engineers are responsible for enhancing their professional competence throughout their careers and for encouraging similar actions by their colleagues.

Engineers accept responsibility for their actions; seek and acknowledge criticism of their work; offer honest criticism of the work of others; properly credit the contributions of others; and do not accept credit for work not theirs.

Engineers perceiving a consequence of their professional duties to adversely affect the present or future public health and safety shall formally advise their employers or clients and, if warranted, consider further disclosure.

Engineers act in accordance with all applicable laws and the ______________________[1] rules of conduct, and lend support to others who strive to do likewise.

[1] AAES Member Societies are urged to make reference here to the appropriate code of conduct to which their members will be bound.
Approved by AAES Board of Governors 12/13/84

CODE OF ETHICS
For Engineers

PREAMBLE

Engineering is an important and learned profession. The members of the profession recognize that their work has a direct and vital impact on the quality of life for all people. Accordingly, the services provided by engineers require honesty, impartiality, fairness and equity, and must be dedicated to the protection of the public health, safety and welfare. In the practice of their profession, engineers must perform under a standard of professional behavior which requires adherence to the highest principles of ethical conduct on behalf of the public, clients, employers and the profession.

I FUNDAMENTAL CANONS

Engineers, in the fulfillment of their professional duties, shall:

1 Hold paramount the safety, health and welfare of the public in the performance of their professional duties.
2 Perform services only in areas of their competence.
3 Issue public statements only in an objective and truthful manner.
4 Act in professional matters for each employer or client as faithful agents or trustees.
5 Avoid deceptive acts in the solicitation of professional employment.

II RULES OF PRACTICE

1 Engineers shall hold paramount the safety, health and welfare of the public in the performance of their professional duties.

a Engineers shall at all times recognize that their primary obligation is to protect the safety, health, property and welfare of the public. If their professional judgment is overruled under circumstances where the safety, health, property or welfare of the public are endangered, they shall notify their employer or client and such other authority as may be appropriate.

b Engineers shall approve only those engineering documents which are safe for public health, property and welfare in conformity with accepted standards.

c Engineers shall not reveal facts, data or information obtained in a professional capacity without the prior consent of the client or employer except as authorized or required by law or this Code.

d Engineers shall not permit the use of their name or firm name nor associate in business ventures with any person or firm which they have reason to believe is engaging in fraudulent or dishonest business or professional practices.

e Engineers having knowledge of any alleged violation of this Code shall cooperate with the proper authorities in furnishing such information or assistance as may be required.

2 Engineers shall perform services only in the areas of their competence.

a Engineers shall undertake assignments only when qualified by education or experience in the specific technical fields involved.

b Engineers shall not affix their signatures to any plans or documents dealing with subject matter in which they lack competence, nor to any plan or document not prepared under their direction or control.

c Engineers may accept assignments and assume responsibility for coordination of an entire project and sign and seal the engineering documents for the entire project, provided that each technical segment is signed and sealed only by the qualified engineers who prepared the segment.

3 Engineers shall issue public statements only in an objective and truthful manner.

a Engineers shall be objective and truthful in professional reports, statements or testimony. They shall include all relevant and pertinent information in such reports, statements or testimony.

b Engineers may express publicly a professional opinion on technical subjects only when that opinion is founded upon adequate knowledge of the facts and competence in the subject matter.

c Engineers shall issue no statements, criticisms or arguments on technical matters which are inspired or paid for by interested parties, unless they have prefaced their comments by explicitly identifying the interested parties on whose behalf they are speaking, and by revealing the existence of any interest the engineers may have in the matters.

4 Engineers shall act in professional matters for each employer or client as faithful agents or trustees.

a Engineers shall disclose all known or potential conflicts of interest to their employers or clients by promptly informing them of any business association, interest, or other circumstances which could influence or appear to influence their judgment or the quality of their services.

b Engineers shall not accept compensation, financial or otherwise, from more than one party for services on the same project, or for services pertaining to the same project, unless the circumstances are fully disclosed to, and agreed to by, all interested parties.

c Engineers shall not solicit or accept financial or other valuable consideration, directly or indirectly, from contractors, their agents, or other parties in connection with work for employers or clients for which they are responsible.

- **d** Engineers in public service as members, advisors or employees of a governmental body or department shall not participate in decisions with respect to professional services solicited or provided by them or their organizations in private or public engineering practice.
- **e** Engineers shall not solicit or accept a professional contract from a governmental body on which a principal or officer of their organization serves as a member.

5 Engineers shall avoid deceptive acts in the solicitation of professional employment.

- **a** Engineers shall not falsify or permit misrepresentation of their, or their associates', academic or professional qualifications. They shall not misrepresent or exaggerate their degree of responsibility in or for the subject matter of prior assignments. Brochures or other presentations incident to the solicitation of employment shall not misrepresent pertinent facts concerning employers, employees, associates, joint venturers or past accomplishments with the intent and purpose of enhancing their qualifications and their work.
- **b** Engineers shall not offer, give, solicit or receive, either directly or indirectly, any political contribution in an amount intended to influence the award of a contract by public authority, or which may be reasonably construed by the public of having the effect or intent to influence the award of a contract. They shall not offer any gift, or other valuable consideration in order to secure work. They shall not pay a commission, percentage or brokerage fee in order to secure work except to a bona fide employee or bona fide established commercial or marketing agencies retained by them.

III PROFESSIONAL OBLIGATIONS

1 Engineers shall be guided in all their professional relations by the highest standards of integrity.

- **a** Engineers shall admit and accept their own errors when proven wrong and refrain from distorting or altering the facts in an attempt to justify their decisions.
- **b** Engineers shall advise their clients or employers when they believe a project will not be successful.
- **c** Engineers shall not accept outside employment to the detriment of their regular work or interest. Before accepting any outside employment, they will notify their employers.
- **d** Engineers shall not attempt to attract an engineer from another employer by false or misleading pretenses.
- **e** Engineers shall not actively participate in strikes, picket lines, or other collective coercive action.
- **f** Engineers shall avoid any act tending to promote their own interest at the expense of the dignity and integrity of the profession.

2 Engineers shall at all times strive to serve the public interest.
  a Engineers shall seek opportunities to be of constructive service in civic affairs and work for the advancement of the safety, health and well-being of their community.
  b Engineers shall not complete, sign or seal plans and/or specifications that are not of a design safe to the public health and welfare and in conformity with accepted engineering standards. If the client or employer insists on such unprofessional conduct, they shall notify the proper authorities and withdraw from further service on the project.
  c Engineers shall endeavor to extend public knowledge and appreciation of engineering and its achievements and to protect the engineering profession from misrepresentation and misunderstanding.

3 Engineers shall avoid all conduct or practice which is likely to discredit the profession or deceive the public.
  a Engineers shall avoid the use of statements containing a material misrepresentation of fact or omitting a material fact necessary to keep statements from being misleading or intended or likely to create an unjustified expectation; statements containing prediction of future success; statements containing an opinion as to the quality of the Engineers' services; or statements intended or likely to attract clients by the use of showmanship, puffery, or self-laudation, including the use of slogans, jingles, or sensational language or format.
  b Consistent with the foregoing, Engineers may advertise for recruitment of personnel.
  c Consistent with the foregoing, Engineers may prepare articles for the lay or technical press, but such articles shall not imply credit to the author for work performed by others.

4 Engineers shall not disclose confidential information concerning the business affairs or technical processes of any present or former client or employer without his consent.
  a Engineers in the employ of others shall not without the consent of all interested parties enter promotional efforts or negotiations for work or make arrangements for other employment as a principal or to practice in connection with a specific project for which the Engineer has gained particular and specialized knowledge.
  b Engineers shall not, without the consent of all interested parties, participate in or represent an adversary interest in connection with a specific project or proceeding in which the Engineer has gained particular specialized knowledge on behalf of a former client or employer.

5 Engineers shall not be influenced in their professional duties by conflicting interests.
  a Engineers shall not accept financial or other considerations, including free engineering designs, from material or equipment suppliers for specifying their product.
  b Engineers shall not accept commissions or allowances, directly or indi-

rectly, from contractors or other parties dealing with clients or employers of the Engineer in connection with work for which the Engineer is responsible.

6 Engineers shall uphold the principle of appropriate and adequate compensation for those engaged in engineering work.

- **a** Engineers shall not accept remuneration from either an employee or employment agency for giving employment.
- **b** Engineers, when employing other engineers, shall offer a salary according to professional qualifications.

7 Engineers shall not attempt to obtain employment or advancement or professional engagements by untruthfully criticizing other engineers, or by other improper or questionable methods.

- **a** Engineers shall not request, propose, or accept a professional commission on a contingent basis under circumstances in which their professional judgment may be compromised.
- **b** Engineers in salaried positions shall accept part-time engineering work only to the extent consistent with policies of the employer and in accordance with ethical consideration.
- **c** Engineers shall not use equipment, supplies, laboratory, or office facilities of an employer to carry on outside private practice without consent.

8 Engineers shall not attempt to injure, maliciously or falsely, directly or indirectly, the professional reputation, prospects, practice or employment of other engineers, nor untruthfully criticize other engineers' work. Engineers who believe others are guilty of unethical or illegal practice shall present such information to the proper authority for action.

- **a** Engineers in private practice shall not review the work of another engineer for the same client, except with the knowledge of such engineer, or unless the connection of such engineer with the work has been terminated.
- **b** Engineers in governmental, industrial or educational employ are entitled to review and evaluate the work of other engineers when so required by their employment duties.
- **c** Engineers in sales or industrial employ are entitled to make engineering comparisons of represented products with products of other suppliers.

9 Engineers shall accept responsibility for their professional activities; provided, however, that Engineers may seek indemnification for professional services arising out of their practice for other than gross negligence, where the Engineer's interests cannot otherwise be protected.

- **a** Engineers shall conform with state registration laws in the practice of engineering.
- **b** Engineers shall not use association with a nonengineer, a corporation, or partnership, as a "cloak" for unethical acts, but must accept personal responsibility for all professional acts.

10 Engineers shall give credit for engineering work to those to whom credit is due, and will recognize the proprietary interests of others.

a Engineers shall, whenever possible, name the person or persons who may be individually responsible for designs, inventions, writings, or other accomplishments.

b Engineers using designs supplied by a client recognize that the designs remain the property of the client and may not be duplicated by the Engineer for others without express permission.

c Engineers, before undertaking work for others in connection with which the Engineer may make improvements, plans, designs, inventions, or other records which may justify copyrights or patents, should enter into a positive agreement regarding ownership.

d Engineers' designs, data, records, and notes referring exclusively to an employer's work are the employer's property.

11 Engineers shall cooperate in extending the effectiveness of the profession by interchanging information and experience with other engineers and students, and will endeavor to provide opportunity for the professional development and advancement of engineers under their supervision.

a Engineers shall encourage engineering employees' efforts to improve their education.

b Engineers shall encourage engineering employees to attend and present papers at professional and technical society meetings.

c Engineers shall urge engineering employees to become registered at the earliest possible date.

d Engineers shall assign a professional engineer duties of a nature to utilize full training and experience, insofar as possible, and delegate lesser functions to subprofessionals or to technicians.

e Engineers shall provide a prospective engineering employee with complete information on working conditions and proposed status of employment, and after employment will keep employees informed of any changes.

"By order of the United States District Court for the District of Columbia, former Section 11(c) of the NSPE Code of Ethics prohibiting competitive bidding, and all policy statements, opinions, rulings or other guidelines interpreting its scope, have been rescinded as unlawfully interfering with the legal right of engineers, protected under the antitrust laws, to provide price information to prospective clients; accordingly, nothing contained in the NSPE Code of Ethics, policy statements, opinions, rulings or other guidelines prohibits the submission of price quotations or competitive bids for engineering services at any time or in any amount."

Statement by NSPE Executive Committee In order to correct misunderstandings which have been indicated in some instances since the issuance of the Supreme Court decision and the entry of the Final Judgment, it is noted that in its decision of April 25, 1978, the Supreme Court of the United States declared: "The Sherman Act does not require competitive bidding."

It is further noted that as made clear in the Supreme Court decision:

1 Engineers and firms may individually refuse to bid for engineering services.
2 Clients are not required to seek bids for engineering services.
3 Federal, state, and local laws governing procedures to procure engineering services are not affected, and remain in full force and effect.
4 State societies and local chapters are free to actively and aggressively seek legislation for professional selection and negotiation procedures by public agencies.
5 State registration board rules of professional conduct, including rules prohibiting competitive bidding for engineering services, are not affected and remain in full force and effect. State registration boards with authority to adopt rules of professional conduct may adopt rules governing procedures to obtain engineering services.
6 As noted by the Supreme Court, "nothing in the judgment prevents NSPE and its members from attempting to influence governmental action. . . ."

Note: In regard to the question of application of the Code to corporations vis-à-vis real persons, business form or type should not negate nor influence conformance of individuals to the Code. The Code deals with professional services, which services must be performed by real persons. Real persons in turn establish and implement policies within business structures. The Code is clearly written to apply to the Engineer and it is incumbent on a member of NSPE to endeavor to live up to its provisions. This applies to all pertinent sections of the Code.

NSPE Publication No. 1102 as revised January 1987

INSTITUTE OF ELECTRICAL AND ELECTRONICS ENGINEERS

CODE OF ETHICS

PREAMBLE

Engineers, scientists and technologists affect the quality of life for all people in our complex technological society. In the pursuit of their profession, therefore, it is vital that IEEE members conduct their work in an ethical manner so that they merit the confidence of colleagues, employers, clients and the public. This IEEE Code of Ethics represents such a standard of professional conduct for IEEE members in the discharge of their responsibilities to employers, to clients, to the community and to their colleagues in this Institute and other professional societies.

ARTICLE I

Members shall maintain high standards of diligence, creativity and productivity, and shall:

1 Accept responsibility for their actions;

2 Be honest and realistic in stating claims or estimates from available data;

3 Undertake technological tasks and accept responsibility only if qualified by training or experience, or after full disclosure to their employers or clients of pertinent qualifications;

4 Maintain their professional skills at the level of the state of the art, and recognize the importance of current events in their work;

5 Advance the integrity and prestige of the profession by practicing in a dignified manner and for adequate compensation.

ARTICLE II

Members shall, in their work:

1 Treat fairly all colleagues and co-workers, regardless of race, religion, sex, age or national origin;
2 Report, publish and disseminate freely information to others, subject to legal and proprietary restraints;
3 Encourage colleagues and co-workers to act in accord with this Code and support them when they do so;
4 Seek, accept and offer honest criticism of work, and properly credit the contributions of others;
5 Support and participate in the activities of their professional societies;
6 Assist colleagues and co-workers in their professional development.

ARTICLE III

Members shall, in their relations with employers and clients;

1 Act as faithful agents or trustees for their employers or clients in professional and business matters, provided such actions conform with other parts of this Code;
2 Keep information on the business affairs or technical processes of an employer or client in confidence while employed, and later, until such information is properly released, provided such actions conform with other parts of this Code;
3 Inform their employers, clients, professional societies or public agencies or private agencies of which they are members or to which they may make presentations, of any circumstance that could lead to a conflict of interest;
4 Neither give nor accept, directly or indirectly, any gift, payment or service of more than nominal value to or from those having business relationships with their employers or clients;
5 Assist and advise their employers or clients in anticipating the possible consequences, direct and indirect, immediate or remote, of the projects, work or plans of which they have knowledge.

ARTICLE IV

Members shall, in fulfilling their responsibilities to the community:

1 Protect the safety, health and welfare of the public and speak out against abuses in these areas affecting the public interest;

2 Contribute professional advice, as appropriate, to civic, charitable or other nonprofit organizations;

3 Seek to extend public knowledge and appreciation of the profession and its achievements.

Approved February 18, 1979, by the Board of Directors of the Institute of Electrical and Electronics Engineers, Inc.

* Some revisions have been made to this code. Please contact the IEEE headquarters for this information.

BIBLIOGRAPHY

Adam, John A.: "The Sergeant York Gun: A Massive Misfire," *IEEE Spectrum,* February 1987, pp. 28–35.

Adam, John A.: "Dispute over X-Ray Laser Made Public, Scientists' Criticisms Shunted," *The Institute,* IEEE, February 1988, p. 1.

Adams, Donald D., and Walter P. Page (eds.): *Acid Deposition,* Plenum, New York, 1985.

Ahearne, John F.: "Nuclear Power After Chernobyl," *Science,* vol. 236, 8 May 1987, pp. 673–679; discussion by E. G. Silver and J. F. Ahearne, vol. 238, 9 Oct. 1987, pp. 144–145.

Alcorn, Paul A.: *Social Issues in Technology,* Prentice-Hall, Englewood Cliffs, New Jersey, 1986.

Alderman, Frank E., and Robert A. Schultz: *Ethical Problems in Consulting Engineering* (unpublished manuscript). Quotations in text used with permission of the authors.

Alger, Philip L., N. A. Christensen, and Sterling P. Olmsted: *Ethical Problems in Engineering,* Wiley, New York, 1965. Quotations used with permission of the publisher.

Alpern, Kenneth D.: "Moral Responsibility for Engineers," *Business and Professional Ethics Journal,* vol. 2, no. 2, 1983, pp. 39–48.

Anderson, Robert M., Robert Perrucci, Dan E. Schendel, and Leon E. Trachtman: *Divided Loyalties,* Purdue Univ. Press, West Lafayette, Indiana, 1980.

Aristotle: *The Nicomachean Ethics,* in Richard McKeon (ed.), *The Basic Works of Aristotle,* Random House, New York, 1941.

Arnould, Richard J., and Henry Grabowski: "Auto Safety Regulation: An Analysis of Market Failure," in *The Bell Journal of Economics,* vol. 12, no. 1, Spring 1981, pp. 27–48.

Asbrand, Deborah: "Engineer Unions Meet with Resistance but Continue to Push for Representation," *Electronic Design News* (EDN), vol. 80, April 18, 1985, pp. 411–414.

Backhouse, Constance: *Sexual Harassment on the Job*, Prentice-Hall, Englewood Cliffs, New Jersey, 1981.

Baier, Kurt, and Nicholas Rescher (eds.): *Values and the Future: The Impact of Technological Change on American Values*, Free Press, New York, 1969.

Bailey, Martin J.: *Reducing Risks to Life: Measurement of the Benefits*, American Enterprise Institute for Public Policy Research, Washington, D.C., 1980.

Bailyn, Lotte: *Living with Technology: Issues at Mid-Career*, MIT Press, Cambridge, Massachusetts, 1980. Quotation in text used with permission of the publisher.

Bane, Charles A.: *The Electrical Equipment Conspiracy*, Federal Legal Publications, New York, 1973.

Baram, Michael S.: "Regulation of Environmental Carcinogens: Why Cost-Benefit Analysis May Be Harmful to Your Health," *Technology Review*, vol. 78, July–August 1976, pp. 40–43.

Baram, Michael S.: "Trade Secrets: What Price Loyalty?" *Harvard Business Review*, November–December, 1968; reprinted in Barry, (see below), pp. 207–216.

Barnard, Chester I.: *The Functions of the Executive*, Harvard Univ. Press, Cambridge, Massachusetts, 1968.

Barnett, Chris: "Blowing the Whistle on Nuclear Power: Three Leave GE," *New Engineer*, May 1976, pp. 34–42.

Baron, Marcia: *The Moral Status of Loyalty*, Kendall/Hunt, Dubuque, Iowa, 1984.

Barry, Vincent: *Moral Issues in Business*, 2d ed., Wadsworth, Belmont, California, 1986.

Bartlett, Robert V.: *The Reserve Mining Controversy*, Indiana Univ. Press, Bloomington, 1980.

Barus, Carl: "On Costs, Benefits and Malefits in Technology Assessment," *IEEE Technology and Society Magazine*, March 1982, pp. 3–9.

Barzun, Jacques: "The Professions Under Siege," *Harpers*, vol. 257, no. 1541, October 1978, pp. 61–67.

Baum, Robert J.: "Engineering Services," *Business and Professional Ethics Journal*, vol. 4, no. 3–4, 1985, pp. 117–135.

Baum, Robert J.: *Ethics and Engineering Curricula*, The Hastings Center, New York, 1980.

Baum, Robert J. (ed.): *Ethical Problems in Engineering*, 2d ed., vol. 2: *Cases*, Center for the Study of the Human Dimensions of Science and Technology, Rensselaer Polytechnic Institutes, Troy, New York, 1980.

Baum, Robert J.: "The Limits of Professional Responsibility," in D. L. Babcock and C. A. Smith (eds.), *Values and the Public Works Professional*, 1980, available from the American Public Works Association, 1313 E. 60th St., Chicago, Illinois 60637; reprinted in Albert Flores (ed.), *Ethical Problems in Engineering* (see below).

Bayles, Michael D.: *Professional Ethics*, Wadsworth, Belmont, California, 1981.

Bazelon, David L.: "Risk and Responsibility," *Science*, vol. 205, July 20, 1979, pp. 277–280. Quotation used with permission of the author.

Beauchamp, Tom L., and Norman E. Bowie (eds.): *Ethical Theory and Business*, 2d ed., Prentice-Hall, Englewood Cliffs, New Jersey, 1983.

Beauchamp, Tom L.: "The Justification of Reverse Discrimination in Hiring," in ibid., pp. 625–635.

Beck, Melinda, et al.: "Could It Happen in America?" *Newsweek*, 17 Dec. 1984, pp. 38–44.

Bell, Trudy E.: "Wind Shear Cited as Likely Factor in Shuttle Disaster," *The Institute* (IEEE), vol. 11, May 1987, p. 1. Bell, Trudy E., and Karl Esch: "The Fatal Flaw in Flight 51-L," *IEEE Spectrum*, February 1987, pp. 36–51.

Benn, Stanley I.: "Privacy, Freedom, and Respect for Persons," in J. Roland Pennock and John W. Chapman (eds.), *Nomos XIII*, Atherton, New York, 1971, pp. 1–26.

Bignell, Victor, and Joyce Fortune: *Understanding Systems Failures*, Manchester Univ. Press, Dover, New Hampshire, 1984.

Binswanger, Hans Christoph: *Geld und Magie*, Edition Weitbrecht, Stuttgart, 1985.

Blackstone, William T.: "Ethics and Ecology," in William T. Blackstone (ed.), *Philosophy and the Environmental Crisis*, Univ. of Georgia Press, Athens, 1974.

Blackstone, William T.: "On Rights and Responsibilities Pertaining to Toxic Substances and Trade Secrecy," *The Southern Journal of Philosophy*, vol. 16, 1978, pp. 589–603.

Blades, Lawrence E.: "Employment at Will vs. Individual Freedom: On Limiting the Abusive Exercise of Employer Power," *Columbia Law Review*, vol. 67, 1967, pp. 1404–1435.

Blumberg, Phillip I.: "Corporate Responsibility and the Employee's Duty of Loyalty and Obedience: A Preliminary Inquiry," *Oklahoma Law Review*, vol. 24, August 1971, pp. 279–318. Condensed version printed in Beauchamp and Bowie, op. cit., pp. 132–138.

Boffey, Philip M.: "Teton Dam Verdict: Foul-up by the Engineers," *Science*, vol. 195, 21 Jan. 1977, pp. 270–272.

Boisjoly, Roger M.: Speech on shuttle disaster delivered to MIT students, 7 Jan. 1987. Printed in *Books and Religion*, Duke University, vol. 15, March–April, 1987, p. 3.

Bok, Sissela: "Whistleblowing and Professional Responsibilities," in Daniel Callahan and Sissela Bok (eds.), *Ethics Teaching in Higher Education*, Plenum, New York, 1980, pp. 277–295.

Bordewich, Fergus M.: "The Lessons of Bhopal," *Atlantic Monthly*, March 1987, pp. 30–33.

Borgmann, Albert: *Technology and the Character of Contemporary Life: A Philosophical Inquiry*, Univ. of Chicago Press, Chicago, 1984.

Borrelli, Peter, Mahlon Easterling, Burton H. Klein, Lester Lees, Guy Pauker, and Robert Poppe: *People, Power and Pollution: Environmental and Public Interest Aspects of Electric Power Plant Siting*, Environmental Quality Lab, California Institute of Technology, Pasadena, 1971.

Boulle, Pierre: *The Bridge Over the River Kwai*, trans. Xan Fielding, Vanguard, New York, 1954.

Bowie, Norman: *Business Ethics*, Prentice-Hall, Englewood Cliffs, New Jersey, 1982.

Brandt, Richard B.: *A Theory of the Good and the Right*, Clarendon, Oxford, England, 1979.

Brodeur, Paul: *Outrageous Misconduct: The Asbestos Industry on Trial*, Pantheon, New York, 1985.

Brummer, James: "Love Canal and the Ethics of Environmental Health," *Business and Professional Ethics Journal*, vol. 2, no. 4, 1983, pp. 1–22.

Buehler, John P.: "Allocation of Risks and Resolution of Disputes Under the Contract and Beyond the Contract," in *Exploratory Study on Responsibility, Liability, and Accountability for Risks in Construction*, National Research Council, National Academy of Sciences, Washington, D.C., 1978, pp. 91–100.

Burdick, Eugene, and Harvey Wheeler: *Fail-Safe*, Dell, New York, 1962.

Burke, John G.: "Bursting Boilers and the Federal Power," in M. Kranzberg and W. H. Davenport (eds.), *Technology and Culture*, Schocken, New York, pp. 93–118.

Burton, John F.: "Public Sector Strikes," in Richard T. De George and Joseph A. Pichler (eds.), *Ethics, Free Enterprise and Public Policy*, Oxford, New York, 1978, pp. 127–154.

Callaghan, Dennis W., and Arthur Elkins (eds.): *A Managerial Odyssey: Problems in Business and Its Environment*, 2d ed., Addison-Wesley, Reading, Massachusetts, 1978.

Camenisch, Paul F.: *Grounding Professional Ethics in a Pluralistic Society*, Haven, New York, 1983.

Camps, Frank: "Warning an Auto Company about an Unsafe Design," in Alan F. Westin, *Whistle-Blowing!* (see below), pp. 119–129.

Carter, Charles M.: "Trade Secrets and the Technical Man," *IEEE Spectrum*, vol. 6, no. 2, February 1969, pp. 51–55.

Cavell, Stanley: *The Claim of Reason*, Oxford Univ. Press, New York, 1979, p. 269.

Chalk, Rosemary, Mark S. Frankel, and Sallie B. Chafer (eds.): *AAAS Professional Ethics Project: Professional Ethics Activities in the Scientific and Engineering Societies*, American Association for the Advancement of Science, Washington, D.C., 1980.

Cherrington, David J.: *The Work Ethic*, AMACOM, New York, 1980.

Cogan, Morris L.: "The Problem of Defining a Profession," *Annals of the American Academy of Political Science*, vol. 297, 1955, pp. 105–111.

Cohen, Carl: "Why Racial Preference is Illegal and Immoral," in Burton M. Leiser (ed.), *Values in Conflict*, Macmillan, New York, 1981, pp. 373–395.

Cohen, Richard M., and Jules Witcover: *A Heartbeat Away: The Investigation and Resignation of Vice President Spiro T. Agnew*, Viking, New York, 1974.

Collins, Randall: *The Credential Society*, Academic, New York, 1979.

Comments from Professional Engineers: National Registration, *The American Engineer*, July 1964, pp. 25–30.

Comparato, Frank E.: *Age of Great Guns*, Stackpole, Harrisburg, Pennsylvania, 1965.

Conference on Engineering Ethics, American Society of Civil Engineers, New York, 1975.

Congressional Record: "Nuclear Regulatory Commission's Safety and Licensing Procedures," U.S. Senate Committee on Government Operations, 13 Dec. 1976. Printed in Baum, *Ethical Problems in Engineering*, op. cit., pp. 92–102.

Cooper, B. S., and D. P. Rice: "The Economic Cost of Illness Revisited," *Social Security Bulletin*, vol. 39, February 1976, pp. 21–36.

Council for Science and Society: *The Acceptability of Risks*, Barry Rose, Ringwood, Hants, England, 1977.

Cousins, Norman: *The Pathology of Power*, W. W. Norton, New York, 1987.

Cullen, F. T., W. J. Maakestad, and G. Cavender: *Corporate Crime Under Attack: The Ford Pinto Case and Beyond*, Criminal Justice Studies, Anderson, Cincinnati, Ohio, 1987.

Culver, Charles M. and Bernard Gerd: "Valid Consent," in Charles M. Culver and Bernard Gerd (eds.), *Conceptual and Ethical Problems in Medicine and Psychiatry*, Oxford University Press, New York, 1982.

Curd, Martin, and Larry May: *Professional Responsibility for Harmful Actions*, Kendall/Hunt, Dubuque, Iowa, 1984.

Davie, Michael: *The Titanic*, The Bodley Head, London, 1986.
Davies, Sandra L.: "Does Sex Make a Difference in Engineering Careers?" *Professional Engineer*, vol. 51, March 1981, pp. 30–32.
Da Vinci, Leonardo: *The Notebooks of Leonardo Da Vinci*, vol. I, Edward MacCurdy (ed.), George Braziller, New York, 1939.
Davis, Ruth M.: "Preventative Technology: A Cure for Specific Ills," *Science*, vol. 188, 18 Apr. 1975. Quotation in text used with permission of author.
De Camp, L. Sprague: *The Ancient Engineers*, Doubleday, Garden City, New York, 1963.
De George, Richard T.: "Ethical Responsibilities of Engineers in Large Organizations: The Pinto Case," *Business and Professional Ethics Journal*, vol. 1, no. 1, Fall 1981, pp. 1–14.
De George, Richard T.: *Business Ethics*, 2d ed., Macmillan, New York, 1986.
De Havilland, Sir G., and P. B. Walker, "The Comet Failure," adapted by R. R. Whyte for *Engineering Through Trouble*, The Institution of Mechanical Engineers, London, 1975. Quotations in text used with permission of the Council of the Institution of Mechanical Engineers.
DeVine, John C.: "A Progress Report: Cleaning Up TMI," *IEEE Spectrum*, vol. 18, no. 3, March 1981, pp. 44–49.
Dilley, Dean M.: "The Anxieties of Vulnerability: Liability Protection for Engineers in the Public Sector," *Professional Engineer*, vol. 49, no. 5, May 1979, pp. 12–14.
Donaldson, Thomas: *Case Studies in Business Ethics*, Prentice-Hall, Englewood Cliffs, New Jersey, 1984.
Donaldson, Thomas: *Corporations and Morality*, Prentice-Hall, Englewood Cliffs, New Jersey, 1982.
Donaldson, Thomas, and Patricia H. Werhane (eds.): *Ethical Issues in Business*, 2d ed., Prentice-Hall, Englewood Cliffs, New Jersey, 1983.
Drucker, Peter F.: *Management*, Harper and Row, New York, 1973.
Ducasse, C. J.: *A Philosophical Scrutiny of Religion*, Ronald, New York, 1953. Quotation in text used with permission of the publisher.
Dumas, Lloyd J.: *The Overburdened Economy*, Univ. of California Press, Berkeley, California, 1986.
Durbin, Paul T. (ed.): *A Guide to the Culture of Science, Technology, and Medicine*, Free Press, New York, 1980.
Dworkin, Ronald: *Taking Rights Seriously*, Harvard Univ. Press, Cambridge, Massachusetts, 1977.
Dynes, R. R.: *Organized Behavior in Disaster*, Heath, Lexington, Massachusetts, 1970.

ENR: *see* Engineering News Record
Eddy, Paul, Elaine Potter, and Bruce Page: *Destination Disaster: From the Tri-Motor to the DC-10, The Risk of Flying*, Quadrangle, New York, 1976.
"Edgerton Case," Reports of the IEEE-CSIT Working Group on Ethics & Employment and the IEEE Member Conduct Committee in the matter of Virginia Edgerton's dismissal as information scientist of New York City. Reproduced in *Technology and Soci-*

ety, no. 22, June 1978, pp. 3–10. See also *The Institute,* news supplement to *IEEE Spectrum,* June 1979, p. 6, for articles on her IEEE Award for Outstanding Public Service.

Edwards, Mike: "Chernobyl—One Year After," *National Geographic,* vol. 171, May 1987, pp. 632–653.

Elliston, Frederick A.: "Anonymous Whistleblowing," *Business and Professional Ethics Journal,* vol. 1, no. 2, 1982, pp. 39–58.

Elliston, Frederick A.: "Civil Disobedience and Whistleblowing," *Journal of Business Ethics,* vol, 1, no. 1, 1982, pp. 23–28.

Elliston, Frederick, John Keenan, Paula Lockhart, and Jane van Schaick: *Whistleblowing Research: Methodological and Moral Issues,*Praeger, New York, 1985.

Elliston, Frederick, John Keenan, Paula Lockhart, and Jane van Schaick: *Whistleblowing: Managing Dissent at the Workplace,* Praeger, New York, 1985.

Englebrecht, H. C., and F. C. Hanighen: *Merchants of Death: A Study of the International Armament Industry,* Dodd, Mead, New York, 1934.

ENR (Engineering News Record): "Suit Claims Faulty Bridge Steel," March 12, 1981, p. 14; see also 3.26 1981, p. 20; 4.23 1981, pp. 15–16; 11.19 1981, p. 28.

Esch, Karl: "How NASA Prepared to Cope with Disaster," *IEEE Spectrum,* March 1986, pp. 32–36.

Everest, Larry: *Behind the Poison Cloud: Union Carbide's Bhopal Massacre,* Banner, Chicago, 1985.

Ewing, David W.: *Freedom Inside the Organization,* McGraw-Hill, New York, 1977. Quotation in text used with permission of author and publisher.

Ewing, David W.: "IBM's Guidelines to Employee Privacy," *Harvard Business Review,* vol. 54, 1976, pp. 82–90.

Ewing, David W.: "What Business Thinks about Employee Rights," *Harvard Business Review,* vol. 55, 1977; reprinted in Westin and Salisbury (see below), pp. 21–42. Quotation in text used with permission of author.

Ezorsky, Gertrude (ed.): *Moral Rights in the Workplace,* State Univ. of New York Press, New York, 1987.

Fairweather, Virginia: "$80,000 in Payoffs: An Engineer Tells His Story," *Civil Engineering,* vol. 48, no.1, January 1978, pp. 54–55.

Faulkner, Peter: "Exposing Risks of Nuclear Disaster," in Westin, *Whistle-Blowing!* (see below), pp. 39–54.

Ferré, Frederick: *Philosophy of Technology,* Prentice-Hall, Englewood Cliffs, New Jersey, 1988.

Firmage, D. Allan: *Modern Engineering Practice,* Garland STPM, New York, 1980.

Fisher, John W.: *Fatigue and Fracture in Steel Bridges,* John Wiley & Sons, New York, 1984.

Fitzgerald, A. Ernest: *The High Priests of Waste,* W. W. Norton, New York, 1972.

Fitzgerald, Donald: "The Life and Times of Lawrence Tate," in Baum, *Ethical Problems in Engineering,* vol. 2, pp. 197–198.

Flores, Albert, and Deborah G. Johnson: "Collective Responsibility and Professional Roles," *Ethics,* vol. 93, April 1983, pp. 537–545.

Flores, Albert (ed.): *Designing for Safety: Engineering Ethics in Organizational Contexts,* Workshop on Engineering Ethics: Designing for Safety, Rensselaer Polytechnic Institute, Troy, New York, 1982.

Flores, Albert: "Engineering Ethics in Organizational Contexts: A Case Study—National Aeronautics and Space Administration," in ibid., pp. 41–82.
Flores, Albert: *Ethics and Risk Management in Engineering*, Westview Press, Boulder Colorado, 1988.
Flores, Albert (ed.): *Ethical Problems in Engineering*, 2d ed., vol. 1: *Readings*, Center for the Study of the Human Dimensions of Science and Technology, Rensselaer Polytechnic Institute, Troy, New York, 1980.
Flores, Albert: "Engineers' Professional Rights," in ibid., pp. 171–176.
Flores, Albert (ed.): *Professional Ideals*, Wadsworth, Belmont, California, 1988.
Florman, Samuel C.: *Blaming Technology: The Irrational Search for Scapegoats*, St. Martin's, New York, 1981.
Florman, Samuel C.: *The Existential Pleasures of Engineering*, St. Martin's, New York, 1976.
Florman, Samuel C.: "Moral Blueprints," *Harpers*, vol. 257, no. 1541, October 1978, pp. 30–33.
Florman, Samuel C.: *The Civilized Engineer*, St. Martin's, New York, 1987.
Ford, Daniel F.: *Three Mile Island: Thirty Minutes to Meltdown*, Viking, New York, 1982. Portions of this book appeared in *The New Yorker*, 6 Apr. and 13 Apr. 1981.
Frank, Nancy: "Murder in the Workplace," in Stuart L. Hills (ed.), *Corporate Violence*, Rowman and Littlefield, Totowa, New Jersey, 1987, pp. 103–107.
Frankena, William K.: *Ethics*, 2d ed., Prentice-Hall, Englewood Cliffs, New Jersey, 1973.
Frankena, William K.: "The Philosophy of Vocation," *Thought*, vol. 51, December 1976, pp. 393–408.
Frantz, Douglas: "B of A Abandons Costly Computer for Trust Clients," Los Angeles Times, 26 January 1988, pt. IV, pp. 1,8; see also, 7 February 1988, pt. *r*, pp. 6, 26–27.
Freedman, Benjamin: "A Meta-Ethics for Professional Morality," *Ethics*, vol. 89, 1978, pp. 1–19.
Freedman, Benjamin: "What Really Makes Professional Morality Different: Response to Martin," *Ethics*, vol. 91, 1981, pp. 626–630.
French, Peter A.: *Collective and Corporate Responsibility*, Columbia Univ. Press, New York, 1984.
French, Peter: "What is Hamlet to McDonnell-Douglas or McDonnell-Douglas to Hamlet: DC-10," *Business and Professional Ethics Journal*, vol. 1, no. 2, 1982, pp. 1–13.
Fried, Charles: *An Anatomy of Values*, Harvard Univ. Press, Cambridge, Massachusetts, 1970.
Friedlander, Gordon: "Bigger Bugs in BART?" *IEEE Spectrum*, vol. 10, no. 3, March 1973, pp. 32–37.
Friedlander, Gordon: "A Prescription for BART," *IEEE Spectrum*, vol. 10, no. 4, April 1973, pp. 40–44.
Friedlander, Gordon: "Nuclear Power Plant Safety," *IEEE Spectrum*, vol. 13, no. 5, May 1976, pp. 70–75.
Friedman, Milton: *Capitalism and Freedom*, Univ. of Chicago Press, Chicago, 1962.
Friedman, Milton: "The Social Responsibility of Business Is to Increase Its Profits," *New York Times Magazine*, 13 Sept. 1970. Reprinted in Donaldson and Werhane, op. cit., pp. 239–244.
Fruchtbaum, Harold: "Engineers and the Commonweal: Notes Toward a Reformation," in Flores, *Ethical Problems in Engineering*, vol. 1, pp. 257–259.
Fuller, John Grant: *The Gentlemen Conspirators*, Grove, New York, 1962.

Galbraith, John Kenneth: *The New Industrial State,* revised 2d ed., New American Library, New York, 1971.
Gansler, J.: *The Defense Industry,* MIT Press, Cambridge, Massachusetts, 1980.
Garrett, Thomas M., et al.: *Cases in Business Ethics,* Appleton Century Crofts, New York, 1968.
Geis, Gilbert: "The Heavy Electrical Equipment Antitrust Cases of 1961," in Gilbert Geis and Robert F. Meier (eds.), *White-Collar Crime: Offenses in Business, Politics, and the Professions,* revised ed., Free Press, New York, 1977, pp. 117–132.
Gini, A. R.: "Diablo Canyon: Nuclear Energy and the Public Welfare," in Thomas Donaldson (ed.), *Case Studies in Business Ethics,* Prentice-Hall, Englewood Cliffs, New Jersey, 1983, pp. 51–59.
Godson, John: *The Rise and Fall of the DC-10,* David McKay, New York, 1975.
Golding, William: *The Spire,* Harcourt, Brace & World, New York, 1964.
Goldman, Alan H.: *Justice and Reverse Discrimination,* Princeton Univ. Press, Princeton, New Jersey, 1979.
Goldman, Alan H.: *The Moral Foundations of Professional Ethics,* Rowman and Littlefield, Totowa, New Jersey, 1980.
Graham, F., Jr.: *Since Silent Spring,* Houghton Mifflin, Boston, 1970.
Gray, Mike, and Ira Rosen: *The Warning: Accident at Three Mile Island,* W. W. Norton, New York, 1982.
Greene, Crawford, Jr.: "The Case Against Fee Competition," *Consulting Engineer,* vol. 52, January 1979.
Greene, Graham: *A Burnt-Out Case,* Penguin, New York, 1977.
Grossman, Karl: "Red Tape and Radioactivity," *Common Cause,* July–August, 1986, pp. 24–27.
Grundy, Richard, Hanno C. Weisbrod, and Samuel S. Epstein: "Toxic Substances," chap. 3 in *Consumer Health and Product Hazards—Chemicals, Electronic Products, Radiation,* vol. I of S. S. Epstein and R. D. Grundy (eds.), *The Legislation of Product Safety,* MIT Press, Cambridge, Massachusetts, 1974, pp. 102–169.
Gubaryev, Vladimir: *Sarcophagus, a Tragedy,* trans. Michael Glenny, Vintage, New York, 1987.
Gueron, Henri M.: "Nuclear Power: A Time for Common Sense," *IEEE Technology and Society Magazine,* vol. 3, March 1984, pp. 3–9, 15–18.
Gunn, Alastair, and P. Aarne Vesilind (eds.): *Environmental Ethics for Engineers,* Lewis, Chelsea, Michigan, 1986.

Halamka, John D.: *Espionage in the Silicon Valley,* Sybex, Berkeley, California, 1984.
Hammer, W.: *Product Safety Management and Engineering,* Prentice-Hall, Englewood Cliffs, New Jersey, 1980. Diagrams adapted with permission of the publisher.
Hammurabi: *The Code of Hammurabi,* trans. R. F. Harper, Univ. of Chicago Press, Chicago, 1904, pp. 80–83.
Hardin, Garrett: *Exploring New Ethics for Survival,* Viking, New York, 1968.
Harding, C. Francis, and Donald T. Canfield: *Legal and Ethical Phases of Engineering,* McGraw-Hill, New York, 1936.
Harris, Robert C., Christoph Hohenemser, and Robert W. Kates: "The Burden of Technological Hazards," in G. T. Goodman and W. D. Rowe (eds.), *Energy Risk Management,* Academic, New York, 1979.

Hart, H. L. A.: *Punishment and Responsibility,* Clarendon, Oxford, England, 1973.

Haugen, Edward B.: *Probabilistic Approaches to Design,* Wiley, New York, 1968.

Hawkes, Nigel, Geoffrey Lean, David Leigh, Robin McKie, Peter Pringle, and Andrew Wilson: *Chernobyl—The End of the Nuclear Dream,* Vintage, New York, 1986.

Haydon, Graham: "On Being Responsible," *The Philosophical Quarterly,* vol. 28, 1978, pp. 46–57.

Hayward, David: "Lone Engineer Awaits Verdict on Florida Condominium Collapse," *New Civil Engineer International,* May 1981, pp. 36–40.

Hehir, J. Bryan: "The Relationship of Moral and Strategic Arguments in the Defense Debate," in Paul T. Durbin (ed.), *Research in Philosophy and Technology,* vol. 3, JAI Press, Greenwich, Connecticut, 1980, pp. 367–383.

Heilbroner, Robert L., et al.: *In the Name of Profit,* Doubleday, Garden City, New York, 1972.

Heller, Peter B.: *Technology Transfer and Human Values,* University Press of America, New York, 1985.

Herling, John: *The Great Price Conspiracy,* Robert B. Luce, Washington, D.C., 1962.

Higham, Robert: *The British Rigid Airship,* G. T. Foulis, London, 1961.

Hills, Stuart L. (ed.): *Corporate Violence,* Rowman and Littlefield, Totowa, New Jersey, 1987.

Hiltzig, Michael A.: "Case Gives Rare Glimpse of Silicon Valley Intrigue," *Los Angeles Times,* part I, p. 1, 22 March 1982.

Hoover, Herbert: "The Profession of Engineering," *The Memoirs of Herbert Hoover,* vol. 1, Macmillan, New York, 1961, pp. 131–134. Quotation in text used with permission of the Hoover Foundation.

Houston, Carl W.: "Experiences of a Responsible Engineer," *Conference on Engineering Ethics,* American Society of Civil Engineers, New York, 1975, pp. 25–30.

Howard, Robert T.: "A Bill of Professional Rights for Employed Engineers?" *The American Engineer,* vol. 36, no. 10, October 1966, pp. 47–50.

Hughson, Roy V., and Philip M. Kohn: "Ethics," *Chemical Engineering,* vol. 87, no. 19, 22 Sept. 1980, pp. 132–147. Quotations in text used with permission of McGraw-Hill Book Co.

Ibsen, Henrik: *The Master Builder and Other Plays,* trans. Una Ellis-Fermor, Penguin, New York, 1973.

Ihara, Craig K.: "Collegiality as a Professional Virtue," in Albert Flores (ed.), *Professional Ideals,* Wadsworth, Belmont, California, 1988, pp. 56–65.

Iijima Nobuko: *Pollution Japan,* Pergamon, New York, 1979.

Inhaber, H.: "Risk with Energy from Conventional and Nonconventional Sources," *Science,* vol., 23 Feb. 1979, pp. 718–723.

Inhaber, H.: *Energy Risk Assessment,* Gordon and Breach, New York, 1982.

James, Gene G.: "In Defense of Whistle Blowing," in W. Michael Hoffman and Jennifer Mills Moore (eds.), *Business Ethics,* McGraw-Hill, New York, 1984, pp. 249–260.

James, Gene G.: "Whistle Blowing: Its Nature and Justification," *Philosophy in Context,* vol. 10, 1980, pp. 99–117.

Jansen, Robert B.: *Dams and Public Safety,* Water and Power Resources Service, U.S. Department of the Interior, Denver, 1980.

Jenkins, A. H.: *Adam Smith Today,* Kennikat, Port Washington, New York, 1948.

Johnson, Deborah G.: *Computer Ethics,* Prentice-Hall, Englewood Cliffs, New Jersey, 1985.

Johnson, Deborah G., and John W. Snapper (eds.): *Ethical Issues in the Use of Computers,* Wadsworth, Belmont, California, 1985.

Jonas, Hans: *The Imperative of Responsibility: In Search of an Ethics for the Technological Age,* Univ. of Chicago Press, Chicago, 1984.

Jones, Russel C., et al. (eds.): *Ethics, Professionalism and Maintaining Competence,* American Society of Civil Engineers, New York, 1977.

Jung, C. G.: *Psychologie und Alchemie,* Zurich, 1946.

Kahn, Herman, William Brown, and Leon Martel: *The Next 200 Years,* Morrow, New York, 1976.

Kahn, Shulamit: "Economic Estimates of the Value of Life," *IEEE Technology and Society Magazine,* June 1986, pp. 24–29; reprinted in Flores, op. cit. (1988).

Kant, Immanuel: *Foundations of the Metaphysics of Morals, with Critical Essays,* Robert Paul Wolff (ed.), Bobbs-Merrill, Indianapolis, Indiana, 1969.

Kaplan, Gadi, and Ronald K. Jurgen: "Nuclear Power Plant Safety—1," *IEEE Spectrum,* vol. 13, no. 5, May 1976, pp. 52–69.

Kardos, Geza: *Heron Road Bridge,* Parts A-B-C-D, Engineering Case Library, ECL 133, Stanford Univ., Stanford, California, 1969.

Kates, Robert W. (ed.): *Managing Technological Hazards: Research Needs and Opportunities,* Institute of Behavioral Science, Univ. of Colorado, Boulder, 1977.

Kavaler, Lucy: *Freezing Point, Cold as a Matter of Life and Death,* John Day, New York, 1970.

Kavka, Gregory S.: "Nuclear Deterrence: Some Moral Perplexities," in *The Security Gamble,* Douglas Maclean (ed.), 1984; reprinted in Sterba (see below).

Keisling, Bill: *Three Mile Island: Turning Point,* Veritas, Seattle, Washington, 1980.

Kelly, Arthur L.: "Italian Tax Mores," in Donaldson and Werhane, op. cit., pp. 37–39.

Kemeny Commission Report: *Report of the President's Commission on the Accident at Three Mile Island,* Pergamon Press, New York, 1979.

Kemper, John Dustin: *Engineers and their Profession,* 3d ed., Holt, Reinhart and Winston, New York, 1982.

Kennan, George F.: "A New Way to Go," Albert Einstein Peace Prize Address, 1981; reprinted in *Fellowship,* vol. 47, September 1981, pp. 6–8.

Kettler, G. J.: "Against the Industry Exemption," in James H. Schaub and Karl Pavlovic (eds.), *Engineering Professionalist and Ethics,* Wiley, New York, 1983, pp. 531–534.

Kidder, Tracy: *The Soul of a New Machine,* Avon, New York, 1981.

Kipnis, Kenneth: "Engineers Who Kill: Professional Ethics and the Paramountcy of Public Safety," *Business and Professional Ethics Journal,* vol. 1, no. 1, 1981, pp. 77–91.

Klein, Heywood, and Hal Lancaster: "Major Flaws Persist in Big Buildings, Often Due to Pressure to Cut Costs," *Wall Street Journal,* 12 Feb. 1982.

Kleingartner, Archie: "Professionalism and Engineering Unionism," *Industrial Relations,* vol. 8, May 1969, pp. 224–235.

Kling, Rob: "Computer Abuse and Computer Crime as Organizational Activities," *Computer/Law Journal*, vol. II, Spring 1980, pp. 403–427.
Kling, Rob: *Social Issues and Impacts of Computing*, Univ. of California Press, Irvine, 1979.
Kohlberg, Lawrence: *The Philosophy of Moral Development*, vol. 1, Harper and Row, New York, 1971.
Kohn, Philip M., and Roy V. Hughson: "Perplexing Problems in Engineering Ethics," *Chemical Engineering*, vol. 87, no. 9, 5 May 1980, pp. 100–107. Quotations in text used with permission of McGraw-Hill Book Co.
Kotchian, A. Carl: "The Payoff: Lockheed's 70-day Mission to Tokyo," *Saturday Review*, 9 July 1977; reprinted in Donaldson and Werhane, op. cit., pp. 25–33.

Lachs, John: "'I Only Work Here': Mediation and Irresponsibility," in Richard T. De George and Joseph A. Pichler (eds.), *Ethics, Free Enterprise, and Public Policy*, Oxford, New York, 1978, pp. 201–213.
Ladd, John: "Loyalty," in Paul Edwards (ed.), *The Encyclopedia of Philosophy*, vol. 5, Macmillan, New York, 1967, pp. 97–98.
Ladd, John: "The Quest for a Code of Professional Ethics," in Chalk, Frankel, and Chafer, op. cit., pp. 154–159.
Ladenson, Robert F., J. Choromokos, E. d'Anjou, M. Pimsler, and H. Rosen: *A Selected Annotated Bibliography of Professional Ethics and Social Responsibility in Engineering*, Center for the Study of Ethics in the Professions, Illinois Institute of Technology, Chicago, 1980.
Ladenson, Robert F.: "Freedom of Expression in the Corporate Workplace: A Philosophical Inquiry," in Wade L. Robison, Michael S. Pritchard, and Joseph Ellin (eds.), *Profits and Professions*, Humana Clifton, New Jersey, 1983, pp. 275–286.
Ladenson, Robert F.: "The Social Responsibility of Engineers and Scientists: A Philosophical Approach," in D. L. Babcock and C. A. Smith (eds.), *Values and the Public Works Professional*, Univ. of Missouri-Rolla, 1980, available from the American Public Works Association, 1313 E. 60 St., Chicago, Illinois 60637.
Larmer, Brook: "Evacuation Plans Stymie A-Plant Builders," *Christian Science Monitor*, 2 Oct. 1986, p. 3.
Latta, Geoffrey W.: "Union Organization Among Engineers: A Current Assessment," *Industrial and Labor Relations Review*, vol. 35, no. 11, October 1981, pp. 29–42.
Laurendeau, Normand M.: "Engineering Professionalism: The Case for Corporate Ombudsmen," *Business and Professional Ethics Journal*, vol. 2, no. 1, 1982, pp. 35–45.
Lawless, Edward W.: *Technology and Social Shock*, Rutgers Univ. Press, New Brunswick, New Jersey, 1977.
Layton, Edwin T.: "Engineering Ethics and the Public Interest: A Historical View," in Flores, *Ethical Problems in Engineering*, vol. 1, pp. 26–29.
Layton, Edwin T.: "Engineering Needs a Loyal Opposition," *Business and Professional Ethics Journal*, vol. 2, no. 3, 1983, pp. 51–59.
Layton, Edwin T.: *The Revolt of the Engineers*, Case Western Reserve Univ. Press, Cleveland, Ohio, 1971.
Leiser, Burton M.: "Truth in the Marketplace: Advertisers, Salesmen, and Swindlers," in Burton M. Leiser (ed.), *Liberty, Justice, and Morals*, 2d ed., Macmillan, New York, 1979, pp. 262–297.

Leopold, Aldo: *A Sand County Almanac*, Oxford, New York, 1966.

Lide, D. R.: "Critical Data for Critical Needs," *Science*, vol. 212, 19 June 1969, pp. 1343–1349. Figure in text reproduced with permission of author and *Science*. The original appeared in C. Y. Ho, R. W. Powell, and P. E. Liley, *J. Phys. Chem. Ref. Data 3* (Suppl. 1), 1974.

Lockhart, T. W.: "Safety Engineering and the Value of Life," *Technology and Society* (IEEE), vol. 9, March 1981, pp. 3–5. Quotation in text used with permission of the author.

Logsdon, John M.: "The Space Shuttle Program: A Policy Failure?" *Science*, vol. 232, 30 May 1986, pp. 1099–1105.

Logsdon, Tom: *Computers and Social Controversy*, Computer Science Press, Potomac, 1980.

Lombardo, Thomas G.: "TMI: An Insider's Viewpoint," *IEEE Spectrum*, vol. 17, no. 5, May 1980, pp. 52–55.

Lord, Walter: *A Night to Remember*, illustrated edition, Holt, New York, 1976.

Lowrance, William W.: *Of Acceptable Risk*, William Kaufmann, Los Altos, California 1976.

Luegenbiehl, Heinz C.: "Codes of Ethics and the Moral Education of Engineers," *Business and Professional Ethics Journal*, vol. 2, no. 4, 1983, pp. 41–61.

Maccoby, Michael: *The Gamesman*, Bantam, New York, 1978.

MacDonald, John D.: *Condominium*, Fawcett, New York, 1977.

Machol, Robert E.: "Principles of Operations Research, 10: The Titanic Coincidence," in *Interfaces* (TIMS/ORSA), vol. 5, no. 3, May 1975, pp. 53–54.

MacIntyre, Alasdair: *After Virtue*, 2d ed., Univ. of Notre Dame Press, Notre Dame, Indiana, 1984.

MacIntyre, Alasdair: "Regulation: A Substitute for Morality," *Hastings Center Report*, February 1980, pp. 31–41.

MacKenzie, James J.: "Nuclear Power: A Skeptic's View," *IEEE Technology and Society Magazine*, vol. 3, March 1984, pp. 9–15, 18–21.

MacKinnon, Catherine A.: *Sexual Harassment of Working Women*, Yale Univ. Press, New Haven, Connecticut, 1978.

MacLean, Douglas (ed.): *Values at Risk*, Rowman & Allanheld, 1986.

Manchester, William: *The Arms of Krupp, 1587–1968*, Bantam, New York, 1970. Quotations in text used with permission of the publisher.

Maner, Walter: "The Management of Information in Political Campaigns," *Computers and Society*, vol. 11, 1980, pp. 2–9.

Manley, T. Roger, and Charles W. McNichols: "Scientists, Engineers, and Unions Revisited,"*Monthly Labor Review*, vol. 100, no. 11, November 1979, pp. 32–33.

Mantell, Murray I.: *Ethics and Professionalism in Engineering*, Macmillan, New York, 1964.

Margolis, Joseph: "Conflict of Interest and Conflicting Interests," in Beauchamp and Bowie (eds.), *Ethical Theory and Business*, 1st ed., Prentice-Hall, Englewood Cliffs, New Jersey, 1979, pp. 361–372.

Marples, David R.: *Chernobyl and Nuclear Power in the USSR*, MacMillan Press, London, 1986.

Marshall, Eliot: "Deadlock Over Explosive Dust," *Science*, vol. 222, 4 Nov. 1983, pp. 485–487; discussion p. 1183.

Marshall, Eliot: "Feynman Issues His Own Shuttle Report, Attacking NASA Risk Estimates," *Science,* vol. 232, 27 June 1986, p. 1596.

Marshall, Eliot: "Lightning Strikes Twice at NASA," *Science,* vol. 236, 22 May 1987, p. 903.

Marshall, Eliot: "The Scourge of Computer Viruses," *Science* vol. 240, 8 April 1988, pp. 133–134.

Martin, Daniel: *Three Mile Island: Prologue or Epilogue?,* Ballinger, Cambridge, Massachusetts, 1980.

Martin, Mike W.: "Professional Autonomy and Employers' Authority," in Flores, *Ethical Problems in Engineering,* vol. 1, pp. 177–181.

Martin, Mike W.: "Rights and the Meta-Ethics of Professional Morality" and "Professional and Ordinary Morality: A Reply to Freedman," *Ethics,* vol. 91, July 1981, pp. 619–625 and 631–633.

Martin, Mike W.: "Rights of Conscience Inside the Technological Corporation," in Otto Neumaier (ed.), *Wissen und Gewissen, Conceptus-Studien 4,* VWGO Wien, 1986, pp. 179–193.

Martin, Mike W.: *Self-Deception and Morality,* University Press of Kansas, Lawrence, Kansas, 1986.

Martin, Mike W.: "Why Should Engineering Ethics Be Taught?" *Engineering Education,* vol. 71, no. 4, January 1981, pp. 275–278. Some material from this essay is adapted in Chap. 1 of this text and used with permission of the American Society of Engineering Education.

Marx, Karl: *Economic and Philosophical Manuscripts,* trans. T. B. Bottomore, in Erich Fromm, *Marx's Concept of Man,* Frederick Ungar, New York, 1966.

Marx, Wesley: *Acts of God, Acts of Man,* Coward, McCann & Geoghegan, New York, 1977. Quotation in text used with permission of the author.

Mason, John F.: "The Technical Blow-By-Blow: An Account of the Three Mile Island Accident," *IEEE Spectrum,* vol. 16, no. 11, November 1979, pp. 33–42.

Matley, Jay, Richard Greene, and Celeste McCauley: "Health, Safety and Environment," *Chemical Engineering,* 28 Sept. 1987, pp. 108–120.

Matousak, Miroslav: *Outcome of a Survey of 800 Construction Failures,* Swiss Federal Institute of Technology, Zurich, 1977.

Mayer, Charles: "Appeals Court Bars Sex Bias by U.S. Firms to Please Foreign Customers," *Los Angeles Times,* Part IV, p. 2, 20 Aug. 1981.

Mayers, Teena K.: *Understanding Nuclear Weapons and Arms Control, a Guide to The Issues,* Pergamon-Brassey's, Washington, D.C., 1986.

Mazuzan, George T.: "'Very Risky Business': A Power Reactor for New York City," *Technology and Culture,* vol. 27, April 1986, pp. 262–284.

McConnell, Malcolm: *Challenger, a Major Malfunction,* Doubleday, Garden City, New York, 1987.

McGregor, Douglas: *The Human Side of Enterprise,* McGraw-Hill, New York, 1960.

McIntyre, Louis V., and Marion Bayard McIntyre: *Scientists and Engineers: The Professionals Who Are Not,* Arcola Communications, Lafayette, 1971.

McKaig, Thomas K.: *Building Failures: Case Studies in Construction and Design,* McGraw-Hill, New York, 1962.

McQuade, Walter: "Why All Those Buildings Are Collapsing," *Fortune,* 19 Nov. 1979, pp. 58–66.

Meese, George, P.E.: "The Sealed Beam Case: Engineering in the Public and Private Interest," *Business and Professional Ethics Journal,* vol. 1, no. 3, 1982, pp. 1–20.

Meisler, Stanley: "Glory in Uselessness. The Eiffel Tower: Joke's on Its Critics," *Los Angeles Times,* 28 Apr. 1987.

Melden, A. I., *Ethical Theories: A Book of Readings,* 2d ed., Prentice-Hall, Englewood Cliffs, New Jersey, 1967.

Melden, A. I.: *Rights and Persons,* Univ. of California Press, Berkeley, 1977.

Melman, Seymour: "A Note on: Safety Improvements as a Zero Defect Problem," in Flores, *Designing for Safety: Engineering Ethics in Organizational Contexts,* pp. 173–176.

Melman, Seymour: *Pentagon Capitalism,* McGraw-Hill, New York, 1970.

Meyer, Henry Cord: "Politics, Personality, and Technology: Airships in the Manipulations of Dr. Hugo Eckener and Lord Thomson, 1919–1930," *Aerospace Historian,* September 1981, pp. 165–172.

Milgram, Stanley: *Obedience to Authority,* Harper and Row, New York, 1974.

Mill, John Stuart: *Utilitarianism, with Critical Essays,* Samuel Gorovitz (ed.), Bobbs-Merrill, Indianapolis, Indiana, 1971.

Mironi, Mordechai: "The Confidentiality of Personnel Records," *Labor Law Journal,* vol. 25, May 1974, pp. 270–292.

Mitcham, Carl, and Alois Huning: *Philosophy and Technology II,* D. Reidel, Norwell, Massachusetts, 1986.

Moeller, Calvin E.: "Challenger Catastrophe," *Los Angeles Times,* Letters to the Editor, 11 March 1986.

Mogavero, Louis N., and Robert S. Shane: *What Every Engineer Should Know about Technology Transfer and Innovation,* Marcel Dekker, New York, 1982.

Moll, Richard A.: "Product Liability: A Look at the Law," *Engineering Education,* vol. 66, no. 4, January 1976, pp. 326–331.

Monsma, Stephen V. (ed.): *Responsible Technology, A Christian Perspective,* William B. Eerdmans, Grand Rapids, Michigan, 1986.

Morgan, Arthur E.: *Dams and Other Disasters,* Porter Sargent, Boston, 1971.

Morris, Joe Alex, Jr.: "Computer Age Has Yet to Dawn in Bonn," *Los Angeles Times,* Part I, p. 23, 15 June 1971.

Morrison, Carson, and Philip Hughes: *Professional Engineering Practice, Ethical Aspects,* 2d ed. McGraw-Hill Ryerson, Toronto, Canada, 1988.

Morrison, Robert, and Richard M. Vosburgh: *Career Development for Engineers and Scientists: Organizational Programs and Individual Choices,*Van Nostrand Reinhold, New York, 1987.

Moss, Thomas H., and David L. Sills, eds., *The Three Mile Island Nuclear Accident: Lessons and Implications,* Annals of the New York Academy of Sciences, vol. 365, New York, 1981.

Mostert, Noel: *Supership,* Alfred A. Knopf, New York, 1974.

Muir, John: *To Yosemite and Beyond,* R. Engberg and Donald Wesling (eds.), Univ. of Wisconsin Press, Madison, 1980.

Mullan, Fitzhugh: "Their Lives on the Line," a review of *Who Goes First? The Story of Self Experimentation in Medicine* by L. K. Altman (Random House, 1987), in the *New York Times Book Review,* 28 June 1987, p. 9.

Murdoch, William W. (ed.): *Environment,* Sinauer Associates, Sunderland, Massachusetts, 2d ed., 1975.

Nader, Ralph: "Responsibility and the Professional Society," *Professional Engineer,* vol. 41, May 1971, pp. 14–17.

Nader, Ralph, Peter J. Petkas, and Kate Blackwell: *Whistle Blowing*, Grossman, New York, 1972.

National Research Council: *Acid Deposition, Long-Term Effects*, National Academy of Sciences, Washington, D.C., 1986. Diagram in text used with permission of National Academy Press.

National Research Council: *Exploratory Study on Responsibility, Liability, and Accountability for Risks in Construction*, National Academy of Sciences, Washington, D.C., 1978.

Nelson, Carl, and Susan Peterson: "Are Cost-Benefit Analyses Immoral?" *ASEE/IEEE Frontiers in Education 1981 Conference Proceedings*, American Society for Engineering Education, Washington, D.C., and the Institute of Electrical and Electronics Engineers, New York, 1981.

Newhouse, John: *The Sporty Game*, Alfred A. Knopf, New York, 1982. An earlier version appeared in *The New Yorker* magazines of June 14, 21, and 28, and July 5, 1982, as "The Aircraft Industry."

Nielsen, Kai: "Alienation and Work," in Gertrude Exorsky (ed.), *Moral Rights in the Workplace*, State Univ. of New York Press, Albany, New York, 1987, pp. 28–34.

Nixon, F., N. E. Frost, and K. J. March: "Choosing a Factor of Safety," adapted by R. R. Whyte (ed.) for *Engineering Progress Through Trouble* (see below), pp. 136–139. Quotations in text used with permission of the Council of the Institution of Mechanical Engineers.

Noble, David: "Command Performance: A Perspective on Military Enterprise and Technological Change," Ch. 8 in Military Enterprise and Technological Change, ed. Merritt R. Smith, The MIT Press, Cambridge, Mass., 1985.

Nozick, Robert: *Anarchy, State, and Utopia*, Basic Books, New York, 1974.

NSPE Opinions of the Board of Ethical Review, National Society of Professional Engineers, Washington, D.C. Cases are published in the *Professional Engineer* and periodically republished in bound volumes. Quotations in the text are used with permission of NSPE.

Oldenquist, Andrew G.: *Moral Philosophy, Text and Readings*, 2d ed., Houghton Mifflin, Boston, 1978.

Oldenquist, Andrew G., and Edward E. Slowter: "Proposed: A Single Code of Ethics for All Engineers," *Professional Engineer*, vol. 49, May 1979, pp. 8–11.

O'Neill, Brian, and A. B. Kelley: "Costs, Benefits, Effectiveness, and Safety: Setting the Record Straight," *Professional Safety*, August 1975, pp. 28–34.

Otten, James: "Organizational Disobedience," in Flores, *Ethical Problems in Engineering*, vol. 1 pp. 182–186.

Papanek, Victor: *Design for the Real World*, 2d ed., Van Nostrand Reinhold, New York, 1984.

Parker, Donn B.: *Ethical Conflicts in Computer Science and Technology*, AFIPS Press, Arlington, Virginia, 1979. Case studies adapted in the text with permission of author and publisher.

Parnas, David L.: "Ex-SDI Software Expert Clarifies His Views," letter to *The Institute*, IEEE, November 1986, p. 2.

Perrow, Charles: *Normal Accidents: Living With High-Risk Technologies*, Basic Books, New York, 1984.

Perrucci, Robert, and Joel E. Gerstl: *The Engineers and The Social System,* Wiley, New York, 1969.

Perrucci, Robert, and Joel E. Gerstl: *Profession Without Community: Engineers in American Society,* Random House, New York, 1969.

Perry, Tekla S.: "Five Ethical Dilemmas," *IEEE Spectrum,* vol. 18, no. 6, June 1981, pp. 53–60. Quotations in text used with permission of the author and the Institute of Electrical and Electronics Engineers.

Peters, Charles, and Taylor Branch: *Blowing the Whistle,* Praeger, New York, 1972.

Peters, Tom: *Thriving on Chaos,* Alfred A. Knopf, New York, 1987.

Petersen, James C., and Dan Farrell: *Whistleblowing,* Kendall/Hunt, Dubuque, Iowa, 1986.

Petroski, Henry: *To Engineer Is Human: The Role of Failure in Successful Design,* St. Martin's, New York, 1985.

Pichler, Joseph A.: "Power, Influence and Authority," in Joseph W. McGuire (ed.), *Contemporary Management,* Prentice-Hall, Englewood Cliffs, New Jersey, 1974, pp. 400–434.

Plato: *Euthyphro,* trans. Lane Cooper, in Edith Hamilton and Huntington Cairns (eds.), *The Collected Dialogues of Plato,* Princeton Univ. Press, Princeton, New Jersey, 1971, pp. 169–185.

Popper, Norman N.: "Trade Secrets: How They Affect Your Job Mobility," *Chemical Engineering,* 7 Apr. 1980, pp. 101–104.

Press, Robert M.: "Southern Florida Alarmed by Drought, Studies Handling of Water Resources," *Los Angeles Times,* Part I-C, pp. 1 and 11, December 11, 1981.

Rabow, Gerald: "The Value of Human Lifetime—And Its Applications to Environmental and Energy Policy," *Technology and Society* (IEEE), vol. 9, March 1981, pp. 5–7.

Rachels, James: *The Elements of Moral Philosophy,* Random House, New York, 1986.

Ramo, Simon: *The Future Role of Engineering,* TRW Inc., Corporate Public Relations, Cleveland, 1976.

Rand, Ayn: *The Virtue of Selfishness,* New American Library, New York, 1964.

Randall, Adrian J.: "The Philosophy of Luddism: The Case of the West of England Woolen Workers, ca. 1790–1809," *Technology and Culture,* vol. 27, January 1986, pp. 1–17.

Ransom, W. H.: *Building Failures: Diagnosis and Avoidance,* E. & F. N. Spon, London, 1981.

Rasmussen, Norman C.: *Reactor Safety Study,* U.S. Atomic Energy Commission, WASH 1400, Draft of August 1974 (known as the "Rasmussen Report").

Raushenbakh, Boris V.: "Computer War," pp. 45–52 in *Breakthrough; Emerging New Thinking,* Anatoly Gromyko and Martin Hellman (eds.) for Beyond War, Walker, New York, 1988.

Raven-Hansen, Peter: "Dos and Don'ts for Whistleblowers: Planning for Trouble," *Technology Review,* vol. 82, May 1980, pp. 34–44.

Rawls, John: *A Theory of Justice,* Harvard Univ. Press, Cambridge, Massachusetts, 1971.

Reed, George L.; "Moonlighting and Professional Responsibility," *Journal of Professional Activities, Proceedings of the American Society of Civil Engineers,* vol. 96, September 1970, pp. 19–23.

Regan, Tom: *The Case for Animal Rights,* Univ. of California Press, Berkeley, 1983.

Regan, Tom (ed.): *Earthbound: New Introductory Essays in Environmental Ethics,* Random House, New York, 1984.

Reich, Charles: *The Greening of America,* Random House, New York, 1970.

Reiman, Jeffrey H.: "Privacy, Intimacy, and Personhood," in Richard A. Wasserstrom (ed.), *Today's Moral Problems,* Macmillan, New York, 1979, pp. 377–391.

Rescher, Nicholas: *Unpopular Essays on Technological Progress,* Univ. of Pittsburgh Press, Pittsburgh, Pennsylvania, 1980.

Riegel, J. W.: *Collective Bargaining as Viewed by Unorganized Engineers and Scientists,* Univ. of Michigan Press, Ann Arbor, 1959.

Rivlin, Alice M.: *Systematic Thinking for Social Action,* The Brookings Institution, Washington, D.C., 1971.

Roberts, Leslie: "Radiation Accident Grips Goiania," *Science,* vol. 238, 20 November 1987, pp. 1028–1031.

Roberts, Verne L.: "Defensive Design," *Mechanical Engineering,* September 1984, pp. 86–93.

Robinson, Douglas H.: *Giants in the Sky,* Univ. of Washington Press, Seattle, 1973.

Robison, Wade L., Michael S. Pritchard, and Joseph Ellin (eds.): *Profits and Professions: Essays in Business and Professional Ethics,* Humana, Clifton, New Jersey, 1983.

Roche, James M.: "The Competitive System, To Work, To Preserve, and To Protect," *Vital Speeches of the Day,* a periodical, vol. 37, May 1, 1971, p. 445.

Rogers Commission Report: *Report of the Presidential Commission on the Space Shuttle Challenger Accident,* U.S. Government Printing Office, Washington, D.C., 1986.

Rogovin, Mitchell, and George T. Frampton, Jr.: *Three Mile Island, A Report to the Commissioners and the Public,* vol. 1, Nuclear Regulatory Commission Special Inquiry Group, NUREG/CR-1250, Washington, D.C., January 1980. Diagram in text used with permission of Mitchell Rogovin.

Rosenbaum, Walter A.: *The Politics of Environmental Concern,* 2d ed., Praeger, New York, 1977.

Ross, W. D.: *The Right and the Good,* Oxford Univ. Press, Oxford, England, 1946.

Ross, Steven S.: "Technical Illiteracy," *New Engineer,* March 1978, p. 6. Quotation in text used with permission of author-editor.

Ross, Steven S.: *Construction Disasters,* McGraw-Hill, New York, 1984.

Rowe, William D.: *An Anatomy of Risk,* Wiley, New York, 1977.

Rowe, William D.: "What Is an Acceptable Risk and How Can It Be Determined?" in G. T. Goodman and W. D. Rowe (eds.), *Energy Risk Management,* Academic, 1979, pp. 327–344.

Ruckelshaus, William D.: "Risk, Science, and Democracy," *Issues in Science and Technology,* Spring 1985, pp. 19–38.

Rule, James, Douglas McAdam, Linda Stearns, and David Uglow: *The Politics of Privacy,* New American Library, New York, 1980.

Sagan, L. A.: "Human Cost of Nuclear Power," *Science,* vol. 177, 11 Aug. 1972, pp. 487–493.

Sagoff, Mark: *Risk-Benefit Analysis in Decisions Concerning Public Safety and Health,* Kendall/Hunt, Dubuque, Iowa, 1985. (Module Series in Applied Ethics, Center for the Study of Ethics in the Professions, Illinois Institute of Technology, Chicago)

Samuelson, Robert J.: "Industrial Espionage—Not to Worry," *Los Angeles Times,* Part II, p. 13, 15 July 1982.

Sayre, Kenneth (ed.): *Values in the Electric Power Industry,* Univ. of Notre Dame Press, Notre Dame, Indiana, 1977.

Schaub, James H., and Karl Pavlovic (eds.): *Engineering Professionalism and Ethics,* Wiley, New York, 1983.

Scherer, Donald, and Thomas Attig (eds.): *Ethics and the Environment,* Prentice-Hall, Englewood Cliffs, New Jersey, 1983.

Schinzinger, Roland: "The Engineer as an Agent of Change," (unpublished manuscript, 1973).

Schinzinger, Roland, and Mike W. Martin: "Engineering as Social Experimentation," in *1980 ASEE Annual Conference Proceedings,* vol. 2, American Society for Engineering Education, Washington, D.C., 1980, pp. 394–398.

Schinzinger, Roland: "The Experimental Nature of Engineering and Its Implications for Management," *Technology and Society* (IEEE), vol. 7, no. 27, September 1979, pp. 3–5.

Schinzinger, Roland, and Mike W. Martin: "The Experimental Nature of Engineering and Its Implications for the Style of Engineering Practice," in *1980 Frontiers in Education Conference Proceedings (Houston),* American Society for Engineering Education, Washington, D.C., and the Institute of Electrical and Electronics Engineers, New York, 1980, pp. 204–207.

Schinzinger, Roland, and Mike W. Martin: "Informed Consent in Engineering and Medicine," *Business and Professional Ethics Journal,* vol. 3 (Fall 1983), pp. 67–77.

Schinzinger, Roland: "Technological Hazards and the Engineer," *IEEE Technology and Society Magazine,* June 1986, pp. 12–16.

Schmandt, Jurgen, and Hilliard Roderick (eds.): *Acid Rain and Friendly Neighbors: The Policy Dispute Between Canada and the United States,* Duke Univ. Press, Durham, North Carolina, 1985.

Schumacher, E. F.: *Small Is Beautiful,* Harper and Row, New York, 1973.

Schwartz, Eugene S.: *Overskill,* Quadrangle, Chicago, 1971.

Schwarze, Sharon: "Intellectual Property and the Justification of Intellectual Property Rights" (unpublished manuscript).

Seiden, R. Matthiew: *Product Safety Engineering for Managers,* Prentice-Hall, Englewood Cliffs, New Jersey 1984.

Seidman, Joel: "Engineering Unionism," in Robert Perrucci and Joel E. Gerstl (eds.), *The Engineers and The Social System,* Wiley, New York, 1969, pp. 219–245.

Seldes, George: *Iron, Blood, and Profits,* Harper and Brothers, New York, 1934.

Senders, John W.: "Is There A Cure for Human Error?" *Psychology Today, vol. 33, April 1980, pp. 52–62.*

Sethi, S. Prakash: *Up Against the Corporate Wall,* 3d ed., Prentice-Hall, Englewood Cliffs, New Jersey, 1977.

Shapley, Deborah: "Unionization: Scientists, Engineers Mull over One Alternative," *Science,* vol. 176, 12 May 1972, pp. 618–621.

Shapo, Marshall S.: *A Nation of Guinea Pigs,* Free Press, New York, 1979. Quotation in text used with permission of Macmillan Publishing Company.

Shaw, Gaylord: "Bureau of Reclamation Harshly Criticized in New Report on Teton Dam Collapse," *Los Angeles Times,* 4 June 1977, Part I, p. 3.

Shedd, John A.: *Salt from My Attic,* Mosher, Portland, Maine, 1928, p. 20.

Shrader-Frechette, Kristin S.: *Science Policy, Ethics, and Economic Methodology: Some Problems of Technology Assessment and Environmental-Impact Analysis,* D. Reidel, Dordrecht, Netherlands, 1985.

Shrader-Frechette, Kristin S.: *Risk Analysis and Scientific Method: Methodological and Eth-*

ical Problems with Evaluating Societal Hazards, D. Reidel, Dordrecht, Netherlands, 1985.

Shrader-Frechette, Kristin S.: "The Conceptual Risks of Risk Assessment," *IEEE Technology and Society Magazine,* June 1986, pp. 4–11, reprinted in Flores, op. cit. (1988).

Shrader-Frechette, Kristin S.: *Nuclear Power and Public Policy: The Social and Ethical Problems of Fission Technology,* D. Reidel, Norwell, Massachusetts, 1982.

Shrivastava, Paul: *Bhopal, Anatomy of a Crisis,* Ballinger, Cambridge, Massachusetts, 1987.

Shue, Henry: "Exporting Hazards," *Ethics,* vol. 91, July 1981, pp. 579–606.

Shute, Nevil (pseudonym for Nevil Shute Norway): *No Highway,* Charter, New York, 1976.

Shute, Nevil: *Slide Rule,* William Morrow, New York, 1954.

Silverman, Milton, P. Lee, and M. Lydecker: *The Drugging of the Third World,* Institute for Health Policy Studies, Univ. of California Press, San Francisco, 1981.

Simon, Herbert A.: *Administrative Behavior,* 3d ed., Free Press, New York, 1976.

Simon, Herbert A.: "The Consequences of Computers for Centralization and Decentralization," in Michael L. Dertouzos and Joel Moses (eds.), *The Computer Age: A Twenty-Year View,* The MIT Press, Cambridge, Massachusetts, pp. 212–228.

Simrall, Harry C.: "The Civic Responsibility of the Professional Engineer," *The American Engineer,* May 1963, pp. 39–40.

Singer, Peter: *Animal Liberation,* Avon, New York, 1975.

Slade, Joseph W.: "The Man Behind the Killing Machine," *The American Heritage of Invention and Technology,* Fall 1986, pp. 18–25.

Slovic, Paul: "Perception of Risk," *Science,* vol. 236, 17 Apr. 1987, pp. 280–285.

Slovic, Paul, Baruch Fischhoff, and Sarah Lichtenstein: "Weighing the Risks: Which Risks Are Acceptable?" *Environment,* vol. 21, April 1979, pp. 14–20 and 36–39.

Slovic, Paul, Baruch Fischhoff, and Sarah Lichtenstein: "Risky Assumptions," *Psychology Today,* vol. 14, no. 1, June 1980, pp. 44–48. Quotations in text used with permission of Ziff Davis Publishing Co.

Slovic, Paul, Baruch Fischhoff, and Sarah Lichtenstein: "Weighing the Risks: Which Risks Are Acceptable?" *Environment,* vol. 21, May 1979, pp. 17–20 and 32–38.

Smith, Adam: *The Wealth of Nations,* Univ. of Chicago Press, Chicago, 1976.

Smith, Merritt Roe (ed): *Military Enterprise and Technological Change,* The MIT Press, Cambridge, Massachusetts, 1985.

Smith, Peter: "Designer Interest," in *Exploratory Study on Responsibility, Liability, and Accountability for Risks in Construction,* National Research Council, National Academy of Sciences, Washington, D.C., 1978, pp. 31–35.

Smith, R. Jeffrey: "Court Upholds Controversial Regulations," *Science,* vol. 213, 10 July 1981, pp. 185–188. Quotation in text used with permission of author.

Smith, R. Jeffrey: "Electroshock Experiment at Albany Violates Ethics Guidelines," *Science,* vol. 198, 28 October 1977, pp. 383–386.

Smith, R. Jeffrey: "Juarez: An Unprecedented Radiation Accident," *Science,* vol. 223, 16 March 1984, pp. 1152–1154.

Snow, C. P.: *The Two Cultures, A Second Look,* Cambridge Univ. Press, Cambridge, England, 1959.

Soderberg, C. Richard: "The American Engineer," in Kenneth S. Lynn (ed.), *The Professions in America,* Beacon, Boston, 1967, pp. 203–230.

Sowers, George B., and George F. Sowers: *Introductory Soil Mechanics and Foundations,*

3d ed., Macmillan, New York, 1970.
Squires, Arthur M.: *The Tender Ship: Governmental Management of Technological Change*, Birkhäuser, Boston, 1986.
Starna, William A.: "A Disaster's Toll," letter to the editor, *American Heritage of Invention and Technology*, Summer 1986, commenting on "A Disaster in the Making" in the Spring 1986 issue.
Starr, Chauncey, Richard Rundman, and C. Whipple: "Philosophical Basis for Risk Analysis," *Annual Review of Energy*, vol. 1, 1976, pp. 629–662. Graph in text used with permission of the *Annual Review of Energy*.
Starr, Chauncey: "Social Benefit Versus Technological Risk," *Science*, vol. 165, 19 Sept. 1969, pp. 1232–1238. Graph in text used with permission of the author.
Steinbrook, Robert: "Heart Valve Failures Prompt Concerns," *Los Angeles Times*, 1 Dec. 1985.
Stephens, Mark: *Three Mile Island*, Random House, New York, 1980.
Stevenson, Charles L.: "Persuasive Definitions," *Mind*, vol. 47, 1938, pp. 331–350.
Sterba, James P.: *The Ethics of War and Nuclear Deterrence*, Wadsworth, Belmont, California, 1985.
Stockholm International Peace Research Institute: *World Armaments and Disarmament*, SIPRI Yearbook 1980, Taylor & Francis, London, 1980; Crane, Russak, New York, 1980.
Stone, Christopher D.: *Where the Law Ends: The Social Control of Corporate Behavior*, Harper and Row, New York, 1975.
Storch, Lawrence: "Attracting Young Engineers to the Professional Society," *Professional Engineer*, vol. 41, May 1971, pp. 36–39.
Strobel, Lee P.: *Reckless Homicide? Ford's Pinto Trial*, And Books, South Bend, Indiana, 1980.
Sugarman, Robert: "Nuclear Power and the Public Risk," *IEEE Spectrum*, vol. 16, no. 11, November 1979, pp. 59–79.
Swenson, Gerald S.: "The Case for Fee Competition," *Consulting Engineer*, vol. 50, June 1978, pp. 90–96.

Taylor, Paul W.: *Principles of Ethics, An Introduction*, Dickenson, Encino, California, 1975.
Taylor, Paul W.: *Respect for Nature*, Princeton Univ. Press, Princeton, New Jersey, 1986.
Teich, Albert H. (ed.): *Technology and the Future*, 4th ed., St. Martin's, New York, 1986.
Thorpe, James F., and William H. Middendorf: *What Every Engineer Should Know About Product Liability*, Marcel Dekker, New York, 1979.
Thrall, Charles A., and Jerold M. Starr (eds.): *Technology, Power, and Social Change*, Lexington, Lexington, Massachusetts, 1972.
Trento, Joseph J., and Susan B. Trento: *Prescription for Disaster: From the Glory of Apollo to the Betrayal of the Shuttle*, Crown, New York, 1987.
Tversky, Amos, and Daniel Kahneman: "The Framing of Decisions and the Psychology of Choice," *Science*, vol. 211, 30 Jan. 1981, pp. 453–458.

Ui, Jun (ed.): *Polluted Japan*, Jishu-Koza Citizens's Movement, Tokyo, 1972.
Unger, Stephen H.: "The AAES Model Ethics Code," *IEEE Technology and Society Magazine*, June 1986, pp. 31–32.

Unger, Stephen H.: *Controlling Technology: Ethics and the Responsible Engineer*, Holt, Rinehart and Winston, New York, 1982.

Unger, Stephen H.: "How to be Ethical and Survive," *IEEE Spectrum*, vol. 16, no. 11, December 1979, pp. 56–57.

U.S. Catholic Bishops: "On the Use of Nuclear Weapons and Nuclear Deterrence," from *The Challenge of Peace, God's Promise and Our Response*, United States Catholic Conference, Washington, D.C., 1982, reprinted in Sterba op. cit., (1985).

Vandivier, K.: "Engineers, Ethics and Economics," in *Conference on Engineering Ethics*, American Society of Civil Engineers, New York, 1975, pp. 20–24.

Vaughn, Richard C.: *Legal Aspects of Engineering*, 3d ed., Kendall/Hunt, Dubuque, Iowa, 1977. Quotations in text used with permission of the publisher.

Veblen, Thorstein: *The Engineers and the Price System*, Viking, New York, 1965.

Velasquez, Manuel G.: *Business Ethics: Concepts and Cases*, 2d ed., Prentice-Hall, Englewood Cliffs, New Jersey, 1988.

Vetter, Betty M.: "Engineering: Science Outlook for Women," *Professional Engineer*, vol. 50, June 1980, pp. 29–31.

Vonnegut, Kurt, Jr.: *Player Piano*, Dell, New York, 1952.

Wade, Wynn C.: *The Titanic: End of a Dream*, Penguin, New York, 1980.

Walters, Kenneth: "Professionalism and Engineer/Management Relations," *Professional Engineer*, vol. 43, January 1973, pp. 41–42.

Walters, Kenneth: "Your Employees' Right to Blow the Whistle," *Harvard Business Review*, vol. 53, July 1975, pp. 26–34.

Walton, Richard E.: *The Impact of the Professional Engineering Union*, Harvard Univ. Press, Cambridge, Massachusetts, 1961.

Weber, Max: *The Protestant Ethic and The Spirit of Capitalism*, Charles Scribner's Sons, New York, 1958.

Weber, Max: *The Theory of Social and Economic Organization*, Talcott Parsons (ed.), Free Press, New York, 1947.

Weil, Vivian: *Action and Responsibility in the Engineering Profession*, C.S.E.P. Occasional Papers No. 2, Center for the Study of Ethics in the Professions, Illinois Institute of Technology, Chicago, 1979.

Weil, Vivian (ed.): *Beyond Whistleblowing: Defining Engineers' Responsibilities*, Center for the Study of Ethics in the Professions, Illinois Institute of Technology, Chicago, 1983.

Weil, Vivian: "The Browns Ferry Case," in Schaub and Pavlovic, op. cit. (1983), pp. 402–411.

Weinberg, Alvin M.: "The Many Dimensions of Scientific Responsibility," *Bulletin of the Atomic Scientist*, November 1976, pp. 21–25.

Weinstein, Deena: *Bureaucratic Opposition*, Pergamon, New York, 1979.

Wells, Paula, Hardy Jones, and Michael Davis: *Conflicts of Interest in Engineering*, Kendall/Hunt, Dubuque, Iowa, 1986.

Westin, Alan F., and Stephan Salisbury (eds.): *Individual Rights in the Corporation*, Random House, New York, 1980.

Westin, Alan F.: *Privacy and Freedom*, Atheneum, New York, 1967.

Westin, Alan F. (ed.): *Whistle-Blowing! Loyalty and Dissent in the Corporation*, McGraw-Hill, New York, 1981.

Whitbeck, Caroline: "Moral Responsibility and the Working Engineer," *Books and Religion,* Duke University, vol. 15, nos. 3 and 4, March–April 1987, p. 3.
Whitelaw, Robert L.: "The Professional Status of the American Engineer: A Bill of Rights," *Professional Engineer,* vol. 45, August 1975, pp. 37–41.
Whiteside, Thomas: *Computer Capers,* Crowell, New York, 1978.
Whyte, R. R. (ed.): *Engineering Progress Through Trouble,* The Institution of Mechanical Engineers, London, 1975. Quotations in text used with permission of the Council of the Institution of Mechanical Engineers.
Whyte, William H.: *The Organization Man,* Simon and Schuster, New York, 1956.
Wiener, Norbert: "A Scientist Rebels," *Atlantic Monthly,* vol. 179, January 1947, p. 46.
Weiner, Norbert: *God and Golem, Inc.,* The MIT Press, Cambridge, Massachusetts, 1964.
Wiesner, Jerome B., and Herbert F. York: "The Test Ban," *Scientific American,* vol. 211, 1964, pp. 27–35.
Wildavsky, Aaron: "No Risk Is the Highest Risk of All," in Albert Flores (ed.), *Ethical Problems in Engineering,* pp. 221–226.
Williams, Bernard, and J. J. C. Smart: *Utilitarianism: For and Against,* Cambridge Univ. Press, New York, 1973.
Winner, Langdon: *Autonomous Technology,* The MIT Press, Cambridge, Massachusetts 1977.
Wohl, Burton: *The China Syndrome,* Bantam, New York, 1979.
Wright, J. Patrick: *On a Clear Day You Can See General Motors,* Avon, New York, 1979.

"Yarrow Bridge," Editorial, *The Engineer,* vol. 210, 23 Oct. 1970, p. 415.
Zorpette, Glenn: "The Shoreham Saga," *IEEE Spectrum,* November 1987, pp. 24–37.

INDEXES

NAME INDEX

SUBJECT INDEX